Being Brains

forms of living

Stefanos Geroulanos and Todd Meyers, *series editors*

Being Brains

Making the Cerebral Subject

Fernando Vidal and Francisco Ortega

FORDHAM UNIVERSITY PRESS

NEW YORK 2017

Fordham University Press has no responsibility for the persistence or accuracy of URLs for external or third-party Internet
websites referred to in this publication and does not guarantee
that any content on such websites is, or will remain, accurate or
appropriate.

Fordham University Press also publishes its books in a variety of
electronic formats. Some content that appears in print may not
be available in electronic books.

Visit us online at www.fordhampress.com.

Library of Congress Cataloging-in-Publication Data available
online at http://catalog.loc.gov.

Printed in the United States of America

19 18 17   5 4 3 2 1

First edition

for Bethânia and Luisa
FO

for Nicole
FV

CONTENTS

How did the idea that humans are essentially their brains become think-able? Why should it be considered a "creed" and not the articulation or corollary of a proven "scientific fact"? How is it expressed in notions and practices across a wide variety of domains in many contemporary societies? Does it really fashion people's lives, and if yes, how and to what extent? These are the basic questions this book seeks to explore.

It is no news that since the "Decade of the Brain" of the 1990s, the brain has become a major focus of attention. Starting in the following decade, that focus became itself the object of considerable interest to scholars in human sciences such as anthropology, disability studies, history, and sociology, who have delved into aspects of what has been variously called the "neural turn," the "neuro-turn," and the "neuroscientific turn." While these labels some-times refer to academic developments in the human sciences themselves (for example, the appearance of neuroanthropology, neuroeducation, neurolaw, neurotheology, and others), they apply to a more widespread phenomenon.

Neuroscientific research has attracted lavish funding in North America and Europe; the World Health Organization considers neurological conditions one of the greatest threats to public health; and while it still speaks of "mental health," successive directors of the U.S. National Institute of Mental Health, the world's largest scientific organization in the area, have insisted that mental disorders must be understood and treated as brain disorders. Beyond research and the clinic, the brain and neuroscientific information form the core of a vast universe that ranges from crassly commercial enterprises to lofty metaphysical speculations. Within that universe, the undisciplined multiplication and the often comical abuse of the prefix *neuro-* are like the small signals that confirm the existence of a large-scale phenomenon.[1]

Over the years, in the course of presenting, individually or together, the topics that make up the present book, we have been complimented for being critical of the "neural turn" in matters of personhood, culture, and society—almost as much as we have been accused of being "against" the neurosciences or neuroimaging methods, of refusing fruitful interactions between the brain sciences and the human sciences, or of exaggerating the impact of the *neuro* in contemporary society. ("*Neuro*" is not the reification of anything but only a concise way of designating the universe discussed in this book.) Since we risk being again in the same position, let us, to begin with, try to be clear.

Beyond considerations about the intrinsic value of knowledge, how could anyone be "against" sciences and methods that demonstrate that a large number of persons diagnosed as being in a vegetative state may have minimal consciousness or against investigations that may contribute to the understanding and treatment of dementias or crucially strengthen what we know about the dramatic developmental effects of deprivation? We are, if anything, "for" them. But we could certainly be described as being "against" some of the most extravagant claims and some of the most injudicious applications of the *neuro*. We cannot be convinced that beauty is definable as a quality in objects that correlates with activity in the medial orbitofrontal cortex—or that such a definition can serve as the basis for assessing individual artworks or explaining aesthetic experience. Nor do we think that neurobiological approaches, whether inspired by conviction or by opportunism, are always appropriate or even relevant to account for complex human phenomena. Basically, we do not take the *neuro* as something that "comes naturally," and we question most interpretations of the commonplace according to which *The mind is what the brain does.*

Such a stance grew firmer in the course of our research, and it informs our account. Since the early 2000s there has been a considerable amount of work done by historians and social scientists from various disciplines on the rise, forms, and functioning of the *neuro* in contemporary culture. We have profited from it, but we do not deal with the history and sociology of the neurosciences, unravel the dynamics of "biopolitical governmentality" as such, or disentangle the many ways in which the *neuro* may represent technocratic ideals or accord "with neoliberal precepts of choice, flexibility, self-care and personal responsibility" (Cromby and Williams 2011, 217; see also Cooter 2014, Maasen and Sutter 2007, Pitts-Taylor 2010). Rather, we examine the roots of the notion that, as persons, humans are in principle reducible to their brains as well as some of that notion's main conceptual and practical forms, developments, and implications (related terms, such as *neurocultures* and *cerebral subject*, will appear and be explained later). Although we do not explicitly explore the potential biopolitical repercussions of the neurosciences or the "life-science boom" in political theory (Meloni 2012), *Being Brains* is "political" in the general sense that it deals, selectively but consistently, with processes that touch on people's lives, the constitution of subjectivities, and the distribution of power within societies. Its overarching political dimension concerns ways in which the *neuro* becomes involved in the "government of the living," informs interventions in human lives, and participates in processes of subjectivation (Rose and Abi-Rached 2014). This dimension cuts across all the contexts we examine here and encompasses other local and global processes involving interactions of knowledge, interests, and values, such as the rise of the global mental health movement or the impact of pharmaceutical companies in shaping healthcare.

We are here concerned with the history and forms of a modern creed. Yet how "neutrally" can they be explored? To reconstruct the genealogy of the late twentieth-century neural turns and to delineate the topography of the *neuro* not only opens the way to assessing them critically: it also makes criticism an integral part of the project. In using "genealogy" we do not intend to be systematic about it but merely to signal our affinity to Michel Foucault's enterprise and in particular to his idea of genealogy as a "history of the present"—as a mode of analysis and narrative that begins with a diagnosis and questions about a current situation and assumes that "writing a history *of* the present means writing a history *in* the present; self-consciously writing in a field of power relations and political struggle" (Roth 1981, 43; see also Garland 2014).

The medical historian Roger Cooter (2014, 147) rightly notes that "the neuro-turn stymies the will to, and possibility for, its own critique"; some authors explore the varieties of "neuroskepticism" from an epistemological point of view (Forest 2014), polemicize against the "neuro-enthusiasm" of contemporary "neuro-prophets" (Hasler 2013) or combat the "dictatorship of the brain" and the oppression of "cerebral fundamentalism" (Strasser 2014, 49). We have benefited from their work, trying to deal with the challenge of neutrality not in the distraught mode of the detractors of *neuromythology, neurotrash, neuromania,* or *neuromadness* but by uncovering implicit assumptions and the internal logic of the *neuro* and its applications.[2] In some cases, we show that the emperor has no clothes. Yet that is only one aspect, and probably not the most significant one, of our project. As Jan de Vos and Ed Pluth (2016, 2) aptly remark, just rejecting the claim that "we are our brains" overlooks "the crucial fact that we have already become brain-people, tremendously conscious that we are our brains." Rather than negating it, we take the ideology of the *neuro* as a social, cultural, and psychological reality.

We study different forms of brain-based subjectivation critically, but, contrary to the reproach that "we exaggerate," we remain aware that the neurocentric view of the human, while powerful and pervasive, is neither hegemonic nor monolithic and serves various, sometimes incompatible interests and values. We argue that this view is *ultimately not* dependent on scientific knowledge about the brain. In spite of that, and even if the *neuro* lacks the radically transformative impact often attributed to it, by the early twenty-first century it had taken shape in an immense range of products and initiatives, from amateurish self-help booklets to the one-billion-euro Human Brain Project and the three-billion-dollar BRAIN Initiative, both launched in 2013.[3]

A central feature of such range is that, instead of composing a hierarchy where the "scientific" merely inspires the "popular," it configures a sort of endless ribbon where forms of knowledge and practice circulate in all directions, sometimes colliding but generally driving one another and feeding into one another.[4] In the early twenty-first century, those processes belong in a framework where, as the projects just mentioned illustrate, neuroscience has turned into "big science" and increasingly into a science of "big data" (see, for example, Cunningham and Yu 2014) that has even opened itself to crowdsourcing and citizen participation.[5]

Moreover, as a clinical and laboratory discipline, neuroscience has become largely international.[6] Dominated by the United States between World War II and the 1990s, it expanded significantly in Europe, Australia, Canada, Brazil, and some Asian countries (Japan, and then China at the turn of the twenty-first century). This growth, which can be quantified via publications in international peer-reviewed journals (Abi-Rached, Rose, and Mogoutov 2010), can also be observed, though it has not been formally substantiated, at the level of neuroscientific studies of "human science" topics, science popularization, and straightforwardly commercial applications (such as the self-help industry), all of which can be found not only in the major players but also in countries as distant and different as Argentina and Taiwan, with many others in between.

This very range—conceptual, practical, methodological, geographical—demands that we specify, in a jargon-free manner, what we mean when we talk of "neurocultures" or refer somewhat globally to the *neuro*. Obviously, single-cell recordings in animals are not the same thing as neuroimaging studies of cultural difference or religious experience, and a "neurobics" institute does not represent the same kind of entity as an international conglomerate of major neuroscientific labs. They nevertheless share (at least in many of their practices, as well as in their self-presentations and promotional campaigns) the conviction that we are essentially our brains. This boosts the extreme porosity of their boundaries. The *neuro* encompasses scientific and nonscientific cultures permeated but not exclusively ruled by such a conviction, and this persuasion can in turn be expressed in many modes, from the proselytizing to the tentative. In the continent of neurocultures, local areas may be scrutinized, but it is not possible to demarcate in general the lands of the "good guys" who do the basic science and the territories of the "bad" ones who unduly market hype and hope. They all stand together within one system and are jointly liable for the consequences of their claims. There is in this respect no difference between the "brain gym" quack and the adviser to the BRAIN Initiative who claims that when humanity fully understands its brain, it will understand itself from within, and that such development will revolutionize culture and constitute "a new humanism."[7]

In 2011, we sketched a "neurocultural spectrum" and referred to the world of neurocultures as an "expanding universe" (Vidal and Ortega 2011). The expansion has kept going on, and that means that, although we have ranged widely, we have not spent the same amount of time in each place and have been necessarily selective in our travelogue. We focus on the view of the human as cerebral subject as it developed and became embodied in attempts to guide human behavior and explain its most complex expressions. We examine its different forms, fully aware that no one form is voiced or held by every individual involved, and that, as Nikolas Rose and Joelle Abi-Rached (2013, 223) point out, the *neuro* has not displaced our self-understanding as people with interior mental worlds that are causally related to our behaviors.

Again in connection with the ostensibly global nature of the *neuro*, it must be acknowledged that, as a set of concretely enacted concepts and beliefs, it does not exist everywhere, even if it has continuously expanded and its worldwide penetration can be expected to go on. But where it does exist, it displays, in contrast to the *psy*, remarkable homogeneity. Psychoanalysis provides a patent example. As it became international, it diversified and developed markedly idiosyncratic forms within different national contexts. Beyond a few common fundamental concepts, which can also take on some local color, psychoanalysis is not the same in Paris, New York, or Buenos Aires, and not even inside those cities (Damousi and Plotkin 2009). The *neuro* is. Not only does it globally share, as other sciences, a set of assumptions, concepts, and methods, but it everywhere derives epistemic and social value from its allegedly validating and making more real or objectively known phenomena that are well documented within the human sciences. Yet it seems clear that neuroimaging is not necessary for us to realize that meditation may have beneficial effects (we will come back to this), to learn that magical events in *Harry Potter* may surprise and give pleasure (Hsu et al. 2015), or to explain the "Pepsi paradox."[8] That is why numerous articles and lectures claiming to provide neuroscientific explanations for psychological phenomena do no more than present behavioral data juxtaposed with scant neuroscientific information—or merely accompany the description of *psy* processes with the assertion that they involve the brain. Clever marketing, but deceptive packaging.

Territories Traveled

Being Brains prospects distinct territories in which the *neuro* and "cerebralization" processes can be surveyed, selected so as to ensure diversity and breadth. The first chapter provides a historical, long-term view. It proposes to trace the distant roots of the cerebral subject (a notion we discuss in some detail) to the late seventeenth century and particularly to debates about the seat of the soul, the corpuscularian theory of matter, and John Locke's philosophy of personal identity. In the wake of Locke, who defined personal identity as a continuity of consciousness and memory, eighteenth-century authors began to assert that the brain is the only part of the body we need in order to be ourselves. In the nineteenth century, this form of deterministic essentialism contributed to motivate research into brain structure and function, and it in turn confirmed the brain-personhood nexus. Since then, from phrenology to functional neuroimaging, neuroscientific knowledge and representations have constituted a powerful support for prescriptive outlooks on the individual and society. A fundamental ideological continuity underlies technical, conceptual, and empirical advances in neuroscientific research and practices.

"Neuroascesis," as we call the business that sells programs of cerebral self-discipline, is a case in point. On the one hand, it appeals to the brain and neuroscience as bases for its self-help recipes to enhance memory and reasoning; fight depression, anxiety, and compulsions; improve sexual performance; achieve happiness; and even establish a direct contact with God. On the other hand, underneath the *neuro* surface lie beliefs and even concrete instructions that can be traced to nineteenth-century hygiene manuals. The vocabulary of fitness is transposed from the body to the brain, and traditional self-help themes and recommendations are given a neuroscientific luster.

The second chapter considers the emergence, since the 1990s, of fields whose names often combine the suffix *neuro* with the name of one of the human and social sciences, from anthropology and art history to education, law, and theology. These "disciplines of the *neuro*" (minor but vocal subspecialties within their parent fields) reframe the human sciences and their corresponding subjects on the basis of knowledge about the brain. Driven by the availability of imaging technologies, they look for neural correlates of behaviors and mental processes. Originally concerned mainly with the study of sensory and motor functions, brain imaging studies since the early 1990s have increasingly dealt with topics of potential ethical, legal, and social implications, such as attitudes, cooperation and competition, violence, political preference, or religious experience. Commercial enterprises like neuromarketing have developed concomitantly. The media, both popular and specialized, has given much room to these new fields, thus underlining how rapidly neuroscientific knowledge spreads beyond the confines of brain research proper into different areas of life and culture as a whole. We provide an overview of these fields as well as a more focused examination of neuroaesthetics and the "neurodisciplines" of culture. Though recurrently presented as a way of solving centuries-old riddles and offering solutions to supposed crises in the humanities, these new fields apply methods that are intrinsically inadequate to the objects and phenomena they claim to address.

The third chapter explores one area of major social impact: the cerebralization of psychological distress. The psychopharmacological revolution took place in the 1950s. Later on, the nosological biologization of mental disorders received a crucial impetus when *DSM III* (1980), the third edition of the American Psychiatric Association's influential *Diagnostic and Statistical Manual of Mental Disorders*, opened the way to redescribing in neurological terms disorders such as schizophrenia, autism, and depression. Behaviors previously considered merely awkward, such as shyness, or seen as having a major social component, like alcoholism or obesity, have become predominantly neurological conditions. We provide an overview of such a situation as well as a more detailed examination of the cerebralization of depression, which is a particularly complex cultural and biopolitical phenomenon. We shall also explore the consequences of the cerebralizing trend for the constitution of "forms of living." While biological psychiatry has been criticized as dehumanizing, it has also helped free patients and families from blame and stigma. Insofar as a problem resides within the brain, the individual bears no guilt; though organic, the disorder is externalized relatively to the person's identity. This absolution inspired and sustained a "neurodiversity" movement led by high-functioning autistics who believe that their condition is not a disease to be treated and, if possible, cured but rather a human specificity to be respected like other forms of difference (sexual, racial, and so forth). There is, however, no consensus on the cerebralization of psychological distress, which is fought out in a field characterized by strong tensions and conflicting interests.

The final chapter moves to forms of the *neuro* in popular culture. Film and literature have in many ways rehearsed the connection between personal identity, having a body and being a brain, and they have been major sites for elaborating and questioning the human as cerebral subject. Numerous works can be identified as "brain movies" and "brain novels": most Frankenstein films since the 1940s; B-series productions from the late 1950s to the early 1970s, in which brains themselves are protagonists; science fiction novels of the same period, which stage and exploit brain transplants or brains in vats. While we shall give room to these particular literary and filmic subgenres, our focus will be on later novels and films. We shall privilege works that explore existential, interpersonal, psychological, ethical, and scientific aspects of the relations between having a brain and being a person less through the basic structure of their plots or the direct display of physical brains than through stylistic and formal features. In both areas we demonstrate that even the productions that start out treating humans as cerebral subjects end up contesting brain reductionism and that such constitutive ambivalence is emblematic of the status of the cerebral subject in the modern and contemporary world.

A Threefold Argument

A threefold argument is refracted throughout this book and aims at providing an integrated perspective. Diverse disciplines and discourses presuppose that the mind or self are to different degrees reducible to brain states; they assume that the mind, together with the products of human action, is what the brain does and that we are essentially (though not exclusively) our brains. We first argue that such identification neither resulted from neuroscientific breakthroughs nor depends on knowledge about the brain but was made possible by early modern scientific and philosophical developments that affected notions of personhood and personal identity. We then observe that the corollary of the first historical and historiographical point is that while later neuroscientific research bolstered the "cerebralization" of personhood, it did not, despite many claims, substantiate it either conceptually or empirically. Rather, such cerebralization is an underlying presupposition that dictates the way research is done and its results are interpreted, generally well beyond anything legitimately allowed by the experimental settings and the data generated. Finally, despite its powerful rhetoric, the cerebralization of personhood has nothing necessary or inevitable about it, yet it acts as a connective tissue supporting and linking diverse materials, just as the same bedrock lies at variable depths and with different densities beneath disparate landscapes.

A trajectory through such landscapes poses many challenges at different levels. Two of the most uncomfortable ones concern the danger of being outdated and the possibility of maintaining distance. On the one hand, the corpus of relevant materials, not only from experimental laboratories and neuroimaging facilities but also from the media, the Internet, and the most varied businesses and human science departments, increases constantly and at an unmatchable rate. The timing of publication is such that any book or article analyzing as topical a subject as the *neuro* risks being partly outdated by the time it becomes publicly available. This is not merely a trivial practical problem: it potentially questions one's analyses and conclusions. Won't the rapid "advancement of science" quickly invalidate them? Surely this is what many *neuro* advocates would claim, but in doing so, they would be mainly reiterating the usual strategy of justifying the present by appealing to imagined, hoped-for futures. On the other hand, while even the most bizarre ideologies of the past are interesting objects to be examined and contextualized, the present touches us differently and prompts us to get involved. When understood on its own terms, nothing (or almost nothing) in the past seems absurd, vacuous, or nonsensical. But anachronism is not a risk for judgments about the present, and many neurocultural matters call for them. Such is the predicament of working on contemporary phenomena that have real effects on people's lives, and (as indicated above when we mentioned Foucault) we have considered it more productive to embrace it than to deny it.

In short, *Being Brains* approaches the *neuro* as a "cultural resource."[9] This it does in two ways. On the one hand, the *neuro* functions as an entry point into the history, conditions, and forms of modernity and thus as a resource for us as scholars. The price to be paid as a consequence is that those who critically work on the universe we here characterize as that of neurocultures inevitably contribute to its workings and perhaps even to its perpetuation. On the other hand, and more importantly, the *neuro* is a cultural resource in the same sense that evolutionism might be. As the details and specifics of neuroscientific results are blurred and transformed when they diffuse outside the labs, they provide vividly descriptive and interpretive pictures of the world as well as road maps and calls to act in the present and the future, including for those involved in neuroscientific research. The *neuro* thus ends up serving a multiplicity of interests in contexts ruled more by economic or political considerations (in the broadest sense of those terms) than by the ideals of logic, verifiability, and objectivity that—at least in an abstract view of science—govern the production of knowledge. But of course the two realms are not totally distinct: economics and politics permeate science as much as a certain sense of logic and objectivity may be pursued outside science. The vast realm of "the personal" matters crucially in the constitution, authority, and status of late modern technoscientific knowledge (Shapin 2008).

We have all too often heard that the brain is the most complex object in the universe and that the most important thing we have learned concerning that organ is how little we know about it. Beyond serving professional interests well, the combination of the Delphic "Know thyself" and the Socratic "I only know that I know nothing" has convinced many that the world is not totally disenchanted and compelled them to join the chorus. Ultimately, and beyond the narrower issues we deal with here, the ideology that tells us that we are essentially our brains claims to provide answers to a number of perennial questions about human nature and human destiny. We may object to those answers but don't need normatively to decide whether they are right or wrong, for the main thing about them is that, as William James said of God in the conclusion to *The Varieties of Religious Experience*, they are real since they produce real effects.

Genealogy of the Cerebral Subject

What "Is" the Cerebral Subject?

It may well be that nobody believes they literally are their brain. But when influential people proclaim it, we must take them at their word. Together with the brain in a vat, brain transplantation is one of the favorite thought experiments of philosophers of personal identity (Ferret 1993).[1] It is usual to observe that if the brain of A were transplanted into the body of B, then A would gain a new body, rather than B a new brain. Commenting on that commonplace, Michael Gazzaniga (2005, 31), a leading neuroscientist, serenely asserted: "This simple fact makes it clear that you are your brain." Yet what we have here is neither a fact nor anything simple; it is a profession of faith. The neurophilosopher Paul Churchland "carries in his wallet a colour picture of his wife. Nothing surprising in it," remarks the sociologist Bruno Latour, "except it is the colour scan of his wife's brain! Not only that," he continues, "but Paul insists adamantly that in a few years we will all be recognizing the inner shapes of the brain structure with a more loving gaze than noses, skins and eyes!" (Latour 2004, 224). Gazzaniga, Churchland, and many others who make similar claims express a widespread belief.[2] So widespread indeed, that saying, as the *New York Times* cultural commentator David Brooks did in June 2013, that "the brain is not the mind" immediately generates a flutter of suspicion about a religious and antiquated—even reactionary—dualistic antineuroscience backlash as well as self-confident reassertions of the assumption that "the mind is what the brain does" (Brooks 2013, Marcus 2013, Waldman 2013). The examples could be multiplied.

What is at stake here? Neither science nor ascertainable facts but an idea of the human being, the anthropological figure of the *cerebral subject*—an "ideology" in the plain sense of a set of notions, beliefs, values, interests, and ideals. Like any ideology, this one offers varieties and internal debates and inspires practices that are not necessarily compatible. Yet there is unity in diversity, so that the cerebral subject allows for a fairly unequivocal characterization, and even for a sort of formula: "Person P is identical with person P* if and only if P and P* have one and the same functional brain" (Ferret 1993, 79).[3] To have the same brain is to be the same person, and the brain is the only part of the body we need in order to be ourselves. As the philosopher Roland Puccetti (1969, 70) memorably put it: "Where goes a brain, there goes a person." Puccetti was not saying that a person *is* his or her brain but that insofar as the brain is the physical basis of personhood, one cannot be separated from the other. The brain is the somatic limit of the self, so that, as regards the body they need to be persons, humans are specified by the property of "brainhood" (Vidal 2009a), that is, the property or quality of *being*, rather than simply *having*, a brain.

Now we must go beyond definitions and ask, first, if there are any real, concrete cerebral subjects and, second, which magnitude (from hegemonic to inconsequential) the brainhood ideology may actually be said to have. In a first approximation, there is one answer to both questions, and it is: It depends. Yes, real people can see themselves as cerebral subjects and behave accordingly—but not necessarily all the time. The weight of the ideology depends on contexts and criteria.

The reason for thinking in terms of a "subject" is that views about what humans essentially are go hand in hand with concrete decisions about how to study them and how to treat them, and these decisions implicate processes of "subjectivation" (Foucault 1983; "subjectification" is sometimes also used). These are processes involved in the production of ways of being, in forms of reflexivity and "technologies of the self" (Foucault 1988); they make individuals what they are and contribute to shape their behavior and experience. In our case, then, they are processes whereby people think of themselves and others as primarily determined by their brains—and act, feel, and believe accordingly.[4] Individuation and subjectivation are rooted in sociohistorical contexts and, as we shall see, do not exclude the coexistence of different anthropological figures: cerebral selves, psychological selves, chemical selves, and others.

At the individual level, *cerebral subject* is not a label that can be permanently affixed to anyone but is rather a way of denoting notions and practices that may be operative in people's lives some of the time. In practice, no one conception of the human is monolithic or hegemonic in a given culture, and persons are not one kind of subject alone. For example, the developmental biologist Scott F. Gilbert (1995) contrasted four biological views of the body/self—the neural, immunological, genetic, and phenotypic—and put them in correspondence with different models of the body politic and different views of science. He thus highlighted how political debates mirror disputes over which body, and consequently which self, are the true body and self. "Immune selfhood" has a very rich history of its own (Tauber 2012), but writing in the mid-1990s, Gilbert noted that the genetic self had been recently winning over the other selves. These may be theoretical constructs, but they have real consequences. Thus, as Gilbert points out, in controversies over abortion, the self may be defined genetically (by the fusion of nuclei at conception), neurally (by the onset of the electroencephalographic pattern or some other neurodevelopmental criterion), or immunologically (by the separation of mother and child at birth). In each case, when affected by concrete medical decisions, individuals accomplish the "self" whose definitional criteria were used to reach the decisions.

Thus, it makes sense to refer to a "genetic self" when people's life and self-concept are largely defined by genetic conditions or by genetic testing, screening, and treatment (e.g., Peters, Djurdjinovic, and Baker 1999). Individuals are unlikely to reduce themselves and others to their genetic makeup. However, scientific authorities may suggest such a reduction in statements epitomizing beliefs that permeate a research field, inspire its quest, legitimize its promises, nourish expectations, and orient policy. This was the case when James D. Watson, the codiscoverer of the structure of DNA, uttered for *Time* an assertion that has been quoted hundreds of times: "We used to think our fate is in our stars. Today we know, in large measure, our fate is in our genes" (Jaroff 1989). The oracular claim was supposed to be universally valid, independently of particular individuals' sense of self. By the time the Human Genome Project was completed in 2004, the gene had long been a cultural icon; the HGP itself participated in the hype that the sociologists of science Dorothy Nelkin and M. Susan Lindee (1995) called the "DNA Mystique"—one that involved a basic posture of genetic essentialism and offered an overly optimistic picture of the future clinical applications of genetic research (Hubbard and Wald 1993).

In spite of the increasing convergence of neuroscience and genomics, by the late 1990s the brain had largely supplanted the genome as the source of foundational explanations for human features and behaviors as well as the source of scientific hype. Such a shift may appear justified. Since the brain and the nervous system seems more directly relevant than genetics to many of the philosophical and ethical questions raised by the Western philosophical tradition, including issues of personal identity, they are more likely to be felt as constitutive of one's self. Some occasions may prompt or sustain such a special relation. Thus, while people with genetic afflictions have been observed to "hiss and boo at pictures of genes or enzymes that cause these afflictions," sufferers of mental illnesses react to brain images of patients diagnosed with depression of schizophrenia with "care and concern," as if the image represented both the affliction and "the suffering of the afflicted" (Dumit 2003, 44–45).

As we shall see, such differences in attitude, as well as the precedence of brain over genes as far as human individuality is concerned, have deep roots in the history of notions of personal identity. Yet, again, this does not mean that brainhood is hegemonic. For example, on the basis of ethnographic research in a neuro-oncology clinic, the sociologist of science Sky Gross (2011) shows that while most brain tumor patients admit that the brain is the seat of "who they are," they tend to consider it as just another diseased organ. We must insist on this point, to which we return below, because there has been concern about the empirical accuracy and the interpretive traction of "totalising accounts of the neurological as determining subjectivity, as if the brain is the epicentre of personhood" (Pickersgill, Cunningham-Burley, and Martin 2011, 362).

Notions such as *cerebral subject, brainhood,* or *neurochemical self* are not meant to suggest that a neurobiological perspective dictates views of subjectivity always and absolutely but that, in some times and contexts, it effectively does, occasionally at a very large scale. The sociologist Nikolas Rose's example for neurochemical selves is the well-documented fact that millions of people around the world have come to think about sadness "as a condition called 'depression' caused by a chemical imbalance in the brain and amenable to treatment by drugs that would 'rebalance' these chemicals" (Rose 2003, 46; see here Chapter 3). However, as with "genetic self," it should be obvious that, in real life, everyday ontologies (in the loose sense of mainly implicit "theories about being") coexist, both inside a society and within a single individual. We shift registers in our ways of acting, experiencing, and interacting as well as thinking and speaking about ourselves and others, and this is why psychotherapies and antidepressants can live happily together, if perhaps not "ever after."

The coexistence of such ontologies and their related practices corresponds to what happens in the diachronic and historical dimension. When a phenomenon or area of knowledge is neurologized, it does not ipso facto cease to be what it previously or otherwise was. For example, in the neurobics industry examined below, "brain jogging" simply translates into training the mind, and the exercises proposed are basically the same as those long peddled to improve mental capacities. Nevertheless, when these exercises are relabeled *neurobics*, they realize the ideology of the cerebral subject. It may be a superficial instantiation of that ideology, where the *neuro* is no more than a marketing gimmick. That, however, does not abolish the fact that what is sold and bought belongs to a *neuro* business based on people believing (or at least being told) that they are essentially their brains.

In a medical context, individuals may share a condition but not its interpretation. For example, in her study of bipolar disorder patients, the anthropologist Emily Martin (2009) describes the clash between a dominant reductionist model and the individuals who challenged the idea that neurobiology sufficed to explain their experience. Grassroots diversity thus coexists with a more homogenous official discourse. As is well known, much of psychiatry, including scientists at the head of major national mental health agencies, assert that there are no mental diseases, only brain diseases. Different consequences could follow—one being an emphasis on pharmacological medication and a restriction of access to psychotherapies, with a huge impact on people's lives. A development such as the neurodiversity movement (Chapter 3 here) can only happen in a world where "mental disorders" have been redefined as "brain disorders that primarily affect emotion, higher cognition and executive function" (Hyman 2007, 725). In such a context, psychiatric patients are approached mainly as cerebral subjects, and this may contribute to modulate their self-understanding and how they live their lives.

However, the neuroscientific consensus does not automatically translate into public consent, and research confirms commonsense intuitions about the variety and coexistence of views and practices of the self. Emily Martin (2010, 367) noted that the uptake of brain-based explanations outside the neurosciences and in the wider public is "uneven" and that there is no full takeover by "a newly dominant paradigm." Such heterogeneity exists side by side with the development of brain-centered interventions in medicine, in the workplace, and in schools—interventions that may take place independently of how particular individuals understand themselves.

The sociologist Martin Pickersgill and his colleagues (2011) investigated how people draw on neuroscience and *neuro* ideas to articulate self-understanding. Working with patients suffering from epilepsy, head injury, and dementia as well as with neuroscientists and other professional groups (teachers, counselors, clergy, and foster care workers), they showed that individuals turn their attention to (popular) neuroscience mainly after some kind of neurological event, for example, a brain hemorrhage. This contingent interest, however, does not imply attributing to neuroscience an absolute capacity to define or explain subjectivity. Overall, attitudes are governed by pragmatism and personal relevance; rather than altering notions and practices of the self, neuroscientific concepts "seemed to simply substantiate ideas already held by individuals." The brain thus emerges "as an object of mundane significance," which sometimes helps one understand oneself but is "often far from salient to subjective experience" (Pickersgill, Cunningham-Burley, and Martin 2011, 358, 361–362). Using online questionnaires with Dutch adults diagnosed with ADHD, the sociologists Christian Broer and Marjolijn Heerings (2013) also noticed that although those individuals were interested in neurobiological explanations, they did not reduce their condition to a brain phenomenon. In the framework of the Dutch tradition of public debate and dissent over mental health issues, neurobiology did not colonize subjectivity and was invoked in different ways: as explanation or excuse but also as opening the possibility of governing the self "in the name of the brain" (Rose and Abi-Rached 2013, 8). A study of adults diagnosed with ADHD documented parallel discourses of self-regulation that did not rely on "brain talk" (Broer and Heerings 2013, 61). In Canada, adults diagnosed with major depression or bipolar disorder were asked their ideas about the potential role of neuroimages in stigma mitigation, moral explanations of mental illnesses, and the legitimation of psychiatric symptoms. The resulting interviews show the complex and ambivalent ways in which individuals integrate brain-based notions of mental disorders into their self-understanding; some assumed neurobiological explanations of their disorder yet struggled against pharmaceutical treatments (Buchman et al. 2013).

Studies with other populations produce similar results. Adolescents' explanations of their own behaviors and mental health issues emphasize personal, familial, and social contexts, rarely incorporating the brain or biology (Choudhury, McKinney, and Merten 2012). This may be partly attributable to a lack of information. When informed, however, teens do not refuse to include biological factors in their understanding of adolescent behavior. Rather, confronted with an overwhelmingly negative view of the "teenage brain" as defined by the incapacity to exert control over high-risk pleasure-seeking behaviors or by a deficit in the synchronization of cognition and affect (e.g., Steinberg 2008), they call for neuroscience to contribute to a positive view of their age of life and, in any case, do not generally see behavior in purely biological terms. In turn, on the basis of conversations with three groups (undiagnosed, diagnosed with ADHD but medicated, or diagnosed but not medicated) Ilina Singh (2013) described how children, including those of the two latter groups, did not subordinate their *I* to brain-based explanations but tended to depict the role of the brain in their lives in ways that emphasized personal agency. She thus confirmed that encounters with neuroscientific discourses or technologies do not necessarily cerebralize subjectivity. Similarly, fieldwork in a laboratory conducting fMRI research with children diagnosed with ADHD, learning disabilities, autism, and Tourette syndrome documented how subjects "appropriate lab-based descriptions of neurological difference to their own purposes, claiming a positive identity for themselves," and how "the effects of laboratory research and the metaphors used to describe them may serve expansive purposes in the practices of those who see their subjectivity embedded in research findings" (Rapp 2011, 3, 22).

In a review published in 2013, Cliodhna O'Connor and Helene Joffe examined the empirical evidence for three frequent claims: that neuroscience fosters a conception of the self based in biology, that it promotes conceptions of individual fate as predetermined, and that it attenuates the stigma attached to particular social categories. They concluded that "claims that neuroscience will dramatically alter people's relations with their selves, others and the world are overstated. In many cases, neuroscientific ideas have been assimilated in ways that perpetuate rather than challenge existing modes of understanding" (O'Connor and Joffe 2013, 262). Such bricolage will not surprise historians, who are used to the intertwining of continuities and discontinuities. They are nonetheless valuable for deflating fantasies about the subjective impact of the *neuro* and thus for disrupting "over-theorised accounts of the impact of ideas about the brain on personhood" (Pickersgill, Cunningham-Burley, and Martin 2011, 362).

A lot of this sociological literature has referred to our ideas about brainhood and the cerebral subject. We are thankful for such references but must also point out some misconceptions. One of us (FV) has been described as "one of the most outspoken critics of a cultural hegemony of the 'neuro'" (Besser 2013, 48). However, arguing that the neuroscientific level of explanation is not always the most appropriate or questioning claims that the neurosciences will radically alter our view of the human is not the same thing as maintaining that the *neuro* is hegemonic. Another misinterpretation concerns the level at which the *neuro* exerts its power. Notions of the self and identity are not limited to self-conceptions, which is what the sociological research we just mentioned is about. When, to give just one example, the director of the U.S. National Institute of Mental Health (NIMH) proclaims that illnesses categorized as "mental" or "behavioral" actually are brain disorders, that diagnoses should be aligned with neural systems, and that psychiatry must become a neuroscientific discipline (e.g., Insel 2012, Insel and Quirion 2005), his statements reflect a position that, regardless of its explicit incorporation into people's self-concept, regulates public health policy and the allocation of resources. Whether individuals like it or not, NIMH considers them cerebral subjects, and that has a significant effect on their lives—and even more so since Thomas Insel, NIMH's director for over a decade, became in 2015 head of the new life sciences unit of Alphabet, the company better known as Google (Regalado 2015). If that were not the case, there would be no debate around these issues.

Finally, because our focus is indeed on recent and contemporary contexts, our temporal perspective has been misapprehended. We grasp the scope of the confusion when we read that "the contemporary salience of the brain does not mark the emergence of new conceptions of personhood as 'brainhood' (as suggested by Vidal 2009[a])" (Rose and Abi-Rached 2013, 22). For the argument was, precisely, that the cerebral subject as an anthropological figure is *not* attributable to the contemporary prominence of the brain, nor is it anything "natural," but exactly the other way around: the cerebral subject was enabled by an early modern reconceptualization of personal identity, independently of any naturalistic knowledge about the brain.[5]

The Cerebral Subject in the Longue Durée

As all historical phenomena, the emergence of the cerebral subject is multilayered and overdetermined, and it involves different time scales. In the perspective of the recent past, the appearance of the "neurochemical self" has been considered as "one element of a more widespread mutation in which we in the west, most especially in the United States, have come to understand our minds and selves in terms of our brains and bodies" (Rose 2004, 90). The conception of health and illness in terms of treatable bodily malfunctions is linked to a "more profound transformation in personhood," whereby our sense as psychological individuals is "supplemented or displaced" (109) by a tendency to redefine crucial aspects of the self in bodily terms. Such a turn toward "somatic individuality" constitutes "a shift in the presuppositions about human beings that are embedded in and underpin particular practices" in human genetics, molecular biology, and genetic medicine and biotechnology (Novas and Rose 2000, 485–486).

Parallel to the rise of somatic individuality, Nikolas Rose and Joelle Abi-Rached identify the emergence, in the 1960s, of a "neuromolecular gaze" resulting from the hybridization "of different styles of thought, practices and knowledges in the investigation of the brain, mind and behaviour and the introduction of a reductionist and predominantly molecular approach to the realm of the nervous system" (Abi-Rached and Rose 2010, 31). Such an "epistemological shift," they suggest, "was accompanied by a shift in the mode of governance; with the state, the industry and the scientific community gathering around the same object of interest ('the brain') albeit with different aims, drives, expectations, and motivations" (26).

While such narratives depict the ascendancy of the brain and the scientific, political, and institutional contexts that have sustained it since the mid–twentieth century, they leave open the question: Why the brain? We have already suggested an apparently obvious answer, namely that the brain seems most directly relevant to many of the issues that, in the Western philosophical, moral, and political traditions, have been central for questions of personal identity. This answer, however, implies that scientific discoveries about the brain have inspired views about personhood and attributes to the advancement of science the choice of the brain as the organ of the soul/self. This is a widespread view. To give just one example, a distinguished specialist of medical humanities explains that

> anatomical and physiological understandings of the structure and function of the brain have further established it as the "seat of the soul" because of an increased understanding of its cognitive powers. As an organ of reflection, meditation, and memory, the brain becomes synonymous with what defines the self through the existence of consciousness—the mind. (Dolan 2007, 2)

As historical narrative, this description is untenable. The brain did not become the seat of the soul because it was better known but because, at a certain point, the self was defined in terms of functions that were associated with processes located inside the head. Obviously we are not advancing a radical constructionist argument according to which the choice of the brain rather than the ankle as the organ of thought is purely "ideological." After all, head injuries have long been linked to alterations of personality, cognition, and emotion. The point is rather that by naturalizing historically contingent definitions of self and personhood, the received accounts turn the metaphysical claim that "we are our brains" into a factual statement.

A longer-term perspective helps undo such an illusion. At the same time, it suggests that the preeminence of the brain is more deeply rooted and results from a more protracted history than is usually suggested.

To begin with, brainhood is rooted in a Western context, albeit now almost universally disseminated through the circulation of originally European forms of knowledge and systems of values. Let us look briefly at a major instance of such globalization: the definition of death according to brain-based criteria, which has been increasingly accepted since the late 1960s and predominates everywhere in clinical practice (De Grazia 2011). National legislations and medical guidelines generally allow cardiopulmonary criteria to apply but tend to define death on the basis of irreversible loss of brainstem or (more commonly) whole-brain function. Present controversies concern the coherence of the brain death concept, the extent of necessary neuronal damage, and (especially within the medical profession) the tests required to prove irreversibility (Bernat 2009, 2013). A survey of eighty countries published in 2002 documented the existence of practice guidelines for brain death in seventy countries but considerable differences in diagnostic procedures (Wijdicks 2002); the variability persists and keeps prompting calls for an elusive worldwide consensus (Smith 2012).

Arguments from religious traditions modulate both attitudes and practices (Bernat 2005, Bülow et al. 2008). The brain death definition is officially accepted in the various Christian churches as well as in Judaism and Islam; some groups in all traditions oppose it. In Japan, where the 1997 Organ Transplant Law admits it, a significant proportion of people reject it and do not think that "the essence of humans lies in self-consciousness and rationality" (Morioka 2001, 44). Thus, as the medical anthropologist Margaret Lock (2002, 8) explains, in Japan "the cognitive status of the patient is of secondary importance to most people," and even if an individual suffers from irreversible brain damage and loss of consciousness, many people do not recognize him or her as dead.

In Western medical and philosophical contexts there is an ongoing debate concerning persons who are in VS (vegetative state, now also called "unresponsive wakefulness syndrome"; Laureys et al. 2010). These persons have irreversibly lost the capacity for consciousness but retain some autonomic functions, such as unaided breathing. In the eyes of relatives and physicians, their ontological status is unclear—they seem neither distinctly alive nor unequivocally dead (Holland, Kitzinger, and Kitzinger 2014). While many people also tend to see early death as better than being in VS, positions "may hinge upon our tendency to see minds and bodies as distinct. . . . Advocates of terminating life support may frame vegetative patients as bodies, while those who advocate continued life support may highlight their mental capacities" (Gray et al. 2011). At a more philosophical level, it has been argued that the default position of *not* withdrawing artificial nutrition and hydration should be reversed: Insofar as there is no evidence that VS patients have a "compelling interest in being kept alive," we "commit a worse violation of autonomy" by maintaining them alive than by not doing so (Constable 2012, 157, 160).

We mention these debates because they highlight two of the features that permeate the notion of personhood as brainhood: dualism (at least insofar as positions are framed according to a distinction of mind and body) and an emphasis on autonomy. But there are other instances, for example, the way in which the courts have treated conjoined twins as autonomous individuals competing for scarce resources (Barilan 2002, 2003). Discussions around brain death in the contexts in which it *is* accepted go in the same direction: the "higher-brain" death criterion has not been adopted as the legal standard anywhere, but the fact that it is theoretically envisaged underlines the kinds of features that are considered essential for personhood.

What counts for grasping the cultural significance of the cerebral subject is the fact that identifying the end of personhood with the loss of higher brain functions became *imaginable*. It implies that a state of the brain may define not only the end of a person's life but also the beginning (Sass 1989). If neuromaturation *could* provide biomedical indicators of personhood, then, as human persons distinct from merely living organisms, we would exist essentially from "brain life" to "brain death" (Jones 1989, 1998). As is well known, the redefinition of death as "brain death" was prompted by advances in life-sustaining technology and related challenges in obtaining organs for transplantation. These issues were placed center stage in the 1968 Harvard Medical School landmark report that proposed "to define irreversible coma as a new criterion of death" (Beecher et al. 1968, 337). What marked the end of life was irreversible brain damage, a state of the body in which the patient's heart continued to beat but he or she had suffered "permanent loss of intellect" (337).

In the Harvard report, *intellect* clearly stood for a complex of psychological features, such as memory, consciousness, and self-awareness, whose possession defines both our individual personal identity and human personhood in general. Despite the appellation "brain death," it is the permanent cessation of those functions, not a state of the brain per se, that signals the end of the human being as a living person. To the extent that these features reside in or are a product of the brain, we may indeed be seen as "being our brains." But the *historical* point is that personhood was *not* always reduced to those psychological features and that therefore, as long as personhood was not thus redescribed, it could *not* be conceived of in terms of brainhood. Anthropologists who study conceptions and practices related to the beginnings and ends of life make a similar point when they notice that "producing persons is an inherently social project" and that "personhood is not an innate or natural quality but a cultural attribute" (Kaufman and Morgan 2005, 320–321).

"Personhood as brainhood" was enabled by early modern systems of thought that conceptualized self and body in such a way that the body, while experientially significant, became ontologically derivative. Being an *I* or having a self was equated with memory, consciousness, and self-awareness. This is the "modern" self, and it is in the framework of its early development that the cerebral subject became the anthropological figure of modernity (Vidal 2009a).

Though a notoriously malleable concept, "modernity" is generally understood to include the rise, during the seventeenth century, of a new concept of selfhood—specifically, the notion of a "detached" and autonomous self, which has self-awareness as a constitutive property and is characterized by reflexivity, self-distancing, a sense of inwardness, a first-person standpoint, and disengagement from body and world (Taylor 1989). Related to this modern self is "possessive individualism," a conception of the individual "as essentially the proprietor of his own person or capacities, owing nothing to society for them" (Macpherson 1962, 3). The British philosopher John Locke (1632–1704) provided its founding formula when, in the *Second Treatise of Government* (1690, §27), he wrote that "every Man has a Property in his own Person."

Locke too, in a revolutionary move, reconceived "person" as a continuity of memory and consciousness. It followed that personhood could in principle be attached to any material substance. However, since memory and consciousness were associated with the contents of the head, the brain became the organ of the self or, more precisely, the only organ we need in order to be ourselves as persons. Such localization of personhood was independent from empirical knowledge of the brain and predated by over three centuries the emergence of the "somatic individuality" thought to supplement our sense of ourselves as psychological individuals. In short, as a view of the human being, the cerebral subject derives neither from neuroscientific progress nor from a late twentieth-century biopolitical mutation. Both are relevant, but, as far as their import for personhood is concerned, they are best understood in a long-term historical perspective. For only that perspective shows that, contrary to what neuroscientists often assert or imply, the conviction that "we are our brains" is neither a corollary of neuroscientific advances nor an empirical fact. Rather, it is a position, philosophical or metaphysical, even if some claim it is dictated by science, that depends on views about what it is to be a human person.

"From Nothing Else but the Brain Come Joys, Delights, and Sorrows"

Some timelines documenting awareness of the link between brain and self go back to the Edwin Smith surgical papyrus (dated ca. 1700 BCE but datable in part to 3000–2500 BCE), which includes reports about twenty-seven cases of head injuries.[6] It is more common to trace it to Hippocrates in the fifth century BCE and then to the second-century Greek physician Galen of Pergamon. Such chronologies smooth out context, and the continuity they give to *mind, soul, brain,* and *body* masks significant transformations in the meanings of these terms and in the concepts and practices associated with them.

In the Aristotelian frameworks that largely dominated Western scholarly thought from the thirteenth to the seventeenth century, the soul was a principle of life, that which animated potentially live matter. In Aristotle's analogy (*De Anima* 412a–413a), if the eye was an animal, then sight would be its soul: It would actualize the material eye's potential to see, rendering it capable of fulfilling its intrinsic function. Soul was therefore responsible for the basic functions, faculties, or powers of living beings, known as nutritive or vegetative, perceptive or sensible, appetitive or desiderative, motor or locomotive, and rational or intellective (Michael 2000). Sometimes these faculties were attributed to different souls, and their possession defined a hierarchy: Human beings had all of them, nonhuman animals lacked a rational soul, and plants had only a vegetative soul. Yet all were "animals," or ensouled bodies. That is why the word *psychology*, in use by 1590, originally designated the science of all living beings (Vidal 2011).

As the Aristotelian frameworks disintegrated in the seventeenth century, the soul ceased to be responsible for organic functions; most famously in the philosophy of René Descartes (1596–1650), it became equal to the mind. Even though this was a radical transformation of the concept of soul, the interaction of soul and body remained understood through the humoral theory derived from Galen (Temkin 1973). In Galenism, four bodily humors (blood, yellow bile, black bile, and phlegm) were made up of mixtures of the four elements (air, fire, earth, water) and shared in their basic qualities (warm and humid, warm and dry, cold and dry, cold and humid). The "temperaments," or proportions and mixtures of the humors, dictated individual temperaments, in the sense of "characters" (respectively the sanguine, choleric, melancholic, and phlegmatic). Physiology thus elucidated someone's personality and aptitudes as well as soul-body interactions in general.

According to Galen, as the blood passed through various organs, it was transformed into increasingly subtle and thin fluids, or "spirits." It first became a "natural spirit," responsible for nutrition and growth. After combining with air in the lungs, it passed into the heart, where a portion was transformed into the "vital spirit" on which motor and life-sustaining functions depended. The final refinement took place in the cerebral ventricles, with the formation of the "animal spirits," thus called because they sustained the sensitive and intellectual functions of the *anima* or soul. The qualities of these spirits, such as their temperature, humidity, or density, depended on those of the humors and determined in turn those of the mind. If a person's blood was too cold, the animal spirits would also be cold, and the mental acts that relied on them would lack "heat" and be correspondingly weak and slow. Thus, it was not the brain and nerves but the humors via the animal spirits that held body and soul firmly together.

There is a myriad of early modern examples of such a psychophysiological theory. In English, virtually every page of Robert Burton's famous *Anatomy of Melancholy*, first published in 1621, illustrates the claim that

> as the body works upon the mind by his bad humours, troubling the spirits, sending gross fumes into the brain, and so per consequens disturbing the soul, and all the faculties of it . . . so on the other side, the mind most effectually works upon the body, producing by his passions and perturbations miraculous alterations, as melancholy, despair, cruel diseases, and sometimes death itself.[7]

The animal spirits, together with the rest of the humors, determine a person's character and capacities. The same humoral determinism is the foundation of another late Renaissance bestseller, the Spanish physician Juan Huarte de San Juan's *Examen de ingenios para las ciencias*, or (as the title of the second English translation puts it) *The Tryal of Wits: Discovering the Great Difference of Wits Among Men, and What Sort of Learning Suits Best with Each Genius* (Huarte 1698).

First published in 1575, then censored and expurgated in subsequent editions, by the end of the seventeenth century its 1594 edition had been reprinted several times in Spain and variously translated into Latin, French, Italian, and English (followed by German in the eighteenth century). Huarte (1698, 92) reproduced Galen in explaining that, for the rational soul to perform its actions adequately, the brain needs a "good configuration" and "unity of parts," its substance should "be composed of very fine and delicate Parts," and neither should its heat exceed its coldness nor its moisture surpass its dryness. While attention was paid to the organ's morphology, individual dispositions were dictated by its degree of heat, moisture, and dryness, by way of correspondences between humidity and memory, dryness and the understanding, heat and the imagination. For example, "Old Men have a good Understanding, because they are very dry; and . . . they have no Memory, because they have no moisture" (146). Partly through their effect on brain substance, the bodily humors and their qualities were ultimately responsible for an individual's "wits" and psychological features. The title of Galen's treatise *Quod Animi Mores Corporis Temperatura Sequantur* (*That the Traits of the Soul Follow the Temperaments of the Body*) transparently expresses the doctrine.

The corresponding theory of mental functioning provides another instance of the predominance of fluids. The animal spirits were believed to reside in and move among the brain ventricles (cavities filled with cerebrospinal fluid), which were therefore considered the seat of mental faculties. From the front to the back of the head, these were the "common sense" where sensory information was collected, the imagination and fantasy, the judgment and intellect, and memory (Clarke and Dewhurst 1996, Harvey 1975, Kemp 1990). The brain was primarily a factory and storehouse of the animal spirits; Galen considered it the *hegemonikon* precisely because of the role of the ventricles in producing them (Rocca 2003). Yet, again, the ultimate causes of a person's character and personality were to be found in the qualities of the animal spirits and the humors.

That what we call "mental faculties" somehow depends on what is inside the skull has presumably been intuited since the first member of the genus *Homo* hurt his or her head. That, however, does not amount to "knowledge about the brain" nor makes the humoral theory a direct predecessor or (as in Arikha 2007) an approximate equivalent of modern theories of enzymes or neurotransmitters. It is not difficult to find apparent continuities. Hippocrates often appears as the father of the idea that the brain is the organ of the mind. Yet, as Stanley Finger (2000, chap. 3) notes in his history of neuroscience "pioneers," this tends to be done by taking out of context a few lines from Hippocratic treatises. In *On the Sacred Disease*, written around 400 BCE, we certainly find the oft-quoted sentence, "Men ought to know that from nothing else but the brain come joys, delights, laughter and sports, and sorrows, griefs, despondency, and lamentations" and the claim that "by the same organ we become mad and delirious." Nevertheless, Hippocrates goes on to say that we endure these things when the brain "is more hot, more cold, more moist, or more dry than natural." And he further explains that brain disorders arise from phlegm and bile (and therefore reflect the classic temperaments), such that "those who are mad from phlegm are quiet" and "those from bile are vociferous."[8]

In sum, while it is true that behavior and psychological functions have long been associated with processes taking place inside the head, the philosophical and psychomedical traditions that remained dominant in Western learned cultures until about the end of the seventeenth century defined the human being as a composite of body and soul and made humors circulating within the body, rather than particular brain structures, responsible for the features of the individual self.

A Huron's Soul and Montesquieu's Brain

As mentioned, the breakdown of the Aristotelian frameworks in the seventeenth century entailed the reduction of soul to mind and its consequent localization in the brain. The so-called seat of the soul was not a physical place where the soul materially resided but the organ where it interacted with the body. Descartes, in several letters as well as in his *Treatise of Man* (written before 1637) and *The Passions of the Soul* (1649), speculated that the soul exerted its functions "immediately" at or through the pineal gland. His model was hydrostatic. When the soul desires something, it makes the pineal gland move in such a way that it displaces the animal spirits to obtain the required effect. Memory, for example, was explained by the flow of animal spirits through pores in the brain substance: The flow widens the pores, and the widened pores then function as memory traces that are activated when the pineal gland pushes the animal spirits through them.

In contrast to Descartes, the English anatomist and physician Thomas Willis (1621–1675) proposed a distributed localization of the faculties. Celebrated as the founder of modern neuroanatomy and clinical neuroscience, he provided seminal descriptions of many structures, notably the vasculature at the base of the brain, known as the circle of Willis, as well as the cranial nerves; he also described morphological abnormalities in pathological cases, for example, congenital mental retardation and unilateral paralysis (Molnár 2004, Rengachary et al. 2008). Postmortem study of brain lesions provoked by a loss of blood supply as well as comparisons between the cortex of humans and other animals led him to conclude that the cerebrum was the seat of the rational soul in humans and of the sensitive soul in animals.

Nevertheless, in his Oxford lectures of the 1660s on the anatomy, physiology, and pathology of the nervous system, Willis explained functions such as the will or memory by the circulation of the animal spirits in the cerebral convolutions. In *The Anatomy of the Brain and Nerves*, first published in Latin in 1664, he accounted for the difference in cerebral convolutions between humans and animals by "the dispensation of the animal Spirits." His explanations combine a basically humoral physiology with a new emphasis on the "substance" of the brain:

For as the animal Spirits, for the various acts of Imagination and Memory, ought to be moved within certain and distinct limited or bounded places, and those motions to be often iterated or repeated through the same tracts or paths: for that reason, these manifold convolutions and infoldings of the brain are required for these divers manners of ordinations of the animal Spirits. . . . Hence these folds or rollings about are far more and greater in man than in any other living Creature, to wit, for the various and manifold actings of the superior Faculties. . . . Those Gyrations or Turnings about in four-footed beasts are fewer, and in some, as in a Cat, they are found to be in a certain figure and order: wherefore this Brute thinks on, or remembers scarce any thing but what the instincts and needs of Nature suggest. In the lesser four-footed beasts, also in Fowls and Fishes, the superficies of the brain being plain and even, wants all crankling and turning about: wherefore these sort of Animals comprehend or learn by imitation fewer things, and those almost only of one kind. (Willis 1681, chap. 9, 59–60)

The smoother the brain, the simpler the mind; amount and degree of convolution correlated with an organism's degree of "perfection." In a more convoluted cortical surface, the animal spirits circulated more freely and were less limited to one pathway; such a surface offered more spaces for the storage of mental representations and hence for learning. To the extent that they regulated the circulation of the spirits, structures and morphology gained precedence over the humors and their qualities and acquired more causal significance. Willis (1683, 209) thus remarked that "Stupidity is excited by the mere solitary fault of the Spirits" and that "the Brain it self is found to be first in fault." He justified his opinion by describing abnormal features of the cerebral substance.

In short, it was possible to retain a Galenic physiology and still locate the seat of the soul in structures that had more consistency and materiality than the hollow reservoirs of the humors. The redefinition of soul as mind and the turn to "solidism" stimulated empirical research and a lively localization debate that lasted until the late eighteenth century. Contrary to matter, the soul was defined as "simple" and indivisible. Many therefore believed that its seat must be a discrete point or area inside the brain where the nerves converged, and provided anatomical and clinical support for various localizations: the corpus callosum, the semioval center, the walls of the ventricles. Materialists, on the contrary, considered the quest for the seat of the soul as "one of the chimeras of ancient and modern philosophy" (D'Alembert 1767, 273).

This quest did not lead to any reliable anatomical conclusion. But neither did it weaken the connection between the self and the brain. In fact, it strengthened it, though not thanks to any empirical evidence. Despite considerable advances in brain and nerve anatomy during the seventeenth century, the first identifiable formulation of "brainhood" did not derive from neuroscientific discoveries but from a combination of Locke's theory of personal identity and the corpuscular theory of matter. Not neuroscience but a mutation in the concept of person opened the way for anchoring the self in the brain.

On the one hand, corpuscularianism, the theory of matter associated with the Scientific Revolution of the seventeenth century, explained natural phenomena by the size, local motion, shape, and contrivance of microscopic corpuscles of matter (Eaton 2005). Differences among physical bodies no longer originated in the essential nature of their substance but in the "mechanical affections" of its component particles. Consequently, body A at time T_1 did not have to be made of the same matter as body A at time T_2 in order to be the same. Material continuity thus lost its earlier significance as a constitutive element of the identity and sameness of material bodies. This, as Locke realized, applied also to persons and to the very definition of personhood (Thiel 2011).

In a radical philosophical innovation introduced in the second edition of his *Essay Concerning Human Understanding* (1694, book 2, chap. 27), Locke separated substance and personal identity, the "man" and the person. The identity of the man, he wrote, consists in "a participation of the same continued life, in succession vitally united to the same organized body" (§6). The person, in contrast, is "a thinking being, that has reason and reflection, and can consider itself as itself, the same thinking thing, in different times and places" (§9). Thus, if the soul of a prince, containing the consciousness of the prince's past life, is transferred into a cobbler's soulless body, then the being who resembles the cobbler would in fact be the prince (§15). In Locke's view, personal identity requires the capacity to recognize one's actions and accept responsibility for them. In turn, this capacity implies a continuity of memory and consciousness, which the philosopher identified to "the sameness of a rational being." It follows that "as far as this consciousness can be extended backwards to any past action or thought, so far reaches the identity of that person" (§9). In other words, personal identity depends exclusively on the "same consciousness that makes a man be himself to himself," regardless of the substances to which it might be "annexed" (§10).

We just mentioned the cobbler and the prince, an example of Locke's strategy of psychologizing personal identity with the help of thought experiments. Another such experiment concerns the little finger: If my consciousness is located in my little finger, and this finger is cut off my hand, then, Locke claimed, "it is evident the little finger would be the person, the same person; and self then would have nothing to do with the rest of the body" (§17). In short, bodies become things we have, not things we are; in turn, personal identity becomes purely psychological and distinct from bodily identity. Such a loss of body broke with the Christian tradition, which, founded on the doctrine of the Incarnation, insisted on the essential corporality of the self. Not surprisingly, some of the earliest objections to Locke's theory of personal identity were formulated by divines defending the notion that resurrected persons must regain the *same* body they had on earth if they were to be the same persons they had been while alive.

Despite the depth of the theoretical rupture, disincarnation in practice could not be complete. Locke speculated about a conscious little finger or a cobbler's body with a prince's soul but knew that the nerves conveyed sensory information "to their Audience in the Brain, the mind's Presence-room" (1694, 2.3.1). Some later authors were more explicit as to the brain's role and emphasized the union of soul and brain as requirement for personal identity. Thus, in his *Analytical Essay on the Faculties of the Soul* (1760, §771), the Genevan naturalist and philosopher Charles Bonnet (1720–1793) wrote, "If a Huron's soul could have inherited Montesquieu's brain, Montesquieu would still create." The native North American was an Enlightenment paradigm of the savage, yet if his soul were joined to Montesquieu's brain, then one of the era's greatest thinkers would, for intellectual purposes at least, be still alive. It did not matter that the soul and body were those of a "primitive," provided the brain was the philosopher's own.

In short, the conviction that the brain is the only organ indispensable for personal identity emerged independently or, at most, marginally connected to empirical neuroscientific advances. Bonnet's 1760 statement about

Montesquieu and the Huron declares exactly the same thing as Puccetti's aphorism of 1969, "Where goes a brain, there goes a person," or Gazzaniga's confident assertion of 2005, "you are your brain." A good number of twentieth- and twenty-first-century (neuro)scientists and (neuro)philosophers claim that their convictions about the self are based on neuroscientific data. That may be so for them personally. Historically, however, things happened the other way around: Brainhood predated reliable neuroscientific discoveries and has all the appearance of having been a motivating factor of brain research. As it advanced, this research legitimized and reinforced the brainhood ideology.

"Man Will Increasingly Become a Brain Animal"

Looking back to the early modern period reveals that the rise of the cerebral subject was not prompted by knowledge about the brain and that the neural turn of the late twentieth century is in fundamental respects neither a novelty nor the result of scientific progress. By the time Bonnet wrote his striking aphorism in 1760, "brainomania" (Rousseau 2007) had been developing for about a century. The early modern and Enlightenment "nervous wave" (170) housed the mind in the cerebrum; it placed the brain at the core of selfhood but never apart from soul and mind. Whether in a positive or negative, apologetic or offensive, Christian or atheistic vein, addressing the soul was a constitutive element of that early neural turn. When the soul later dropped out of the picture, it was not (as Francis Crick [1994], for one, suggested in *The Astonishing Hypothesis*) because brain research proved it did not exist.

On the contrary, in the eighteenth century, the psychological theories that gave most room to the brain and the nerves in explaining the mind were authored by convinced Christians, such as Bonnet and David Hartley (1705–1757), who proclaimed their belief in an immortal and immaterial soul. However, they insisted on not discussing its nature or union with the body (which they of course assumed) but only the observable results of its "commerce" or interaction with the body. This interaction, they explained, took place in the brain and by the intermediary of the nerves, and precisely that gave the nervous system its paramount significance (Vidal 2011). Rather than requiring a materialist stance, localizing mental contents or functions in the brain was compatible with the definition of the human as a composite of body and soul, matter and spirit, and had over materialism the advantage of accounting (though indeed mysteriously) for the unity of mental life (Kaitaro 2004).

Brain research was of course not foreign to such an intellectual configuration. To begin with, John Locke, himself a physician, attended Thomas Willis's Oxford lectures, and it is largely through his notes that the lectures have been preserved (Dewhurst 1980).[9] The immense neuroscientific progress that has taken place since then has variously strengthened the conviction that "we are our brains." Yet it has not crucially modified its initial form. Replace *soul* by the functional equivalent of your choice, and you readily update Bonnet's fantasy of 1760, that "if a Huron's soul could have inherited Montesquieu's brain, Montesquieu would still create."

As far as the social and cultural role of brain research, nothing sounds more like statements by late twentieth-century advocates of the *neuro* than the prophecies of their late nineteenth-century predecessors (Meloni 2011). The main difference is that the former consider that their prophecies have better bases and are closer to being fulfilled. For example, in 1907, the Swiss psychiatrist, neuroanatomist, and social reformer Auguste Forel (1848–1931) characterized neurobiology as "a science of the human in man" and as "the basis of the object of the highest human knowledge which can be reached in the future" and depicted its growth as the condition for social progress (quoted in Hagner 2001, 553). Similarly, in 1912, the German neurologist Oskar Vogt (1870–1959) announced that "man will increasingly become a brain animal [*Der Mensch wird immer mehr ein Hirntier werden*]" and anticipated that "in our further development, the brain will play an increasingly significant role" (553–554). It would be invidious to select here, for their similarity in content with these early proclamations, a few quotations from the neuroscientific literature since the mid-1990s. On the one hand, there are endless possibilities; on the other, numerous examples are to be found throughout this book and its bibliography. The point is that new neuroscientific data, theories, and techniques have allegedly substantiated but not crucially affected an ideology that in its modern form dates from the late seventeenth century. That is why the cultural history of the cerebral subject is largely independent from the history of brain science. This is particularly obvious in its early instances: it is clear that Bonnet's aphorism about Montesquieu's brain did not derive from neuroscientific investigation but from a conception of personhood.

Nineteenth-century brain scientists refined anatomical description and pursued functional localization as one of their main goals. The bond of brain to self and personhood was thereby confirmed but not reframed. Phrenology is a case in point (Clarke and Jacyna 1987, Renneville 2000, van Wyhe 2002). It also illustrates how psychological theory—in this case, one that emphasizes individual differences—orients discourses and research about the brain. Based on the theories of the Viennese physician Franz Joseph Gall (1758–1828), who called it "organology" and "doctrine of the skull" (*Schädellehre*), phrenology assumed that the brain is the organ of the mind; that the mind is composed of innate faculties; that each faculty, from amativeness and benevolence to secretiveness and wit, has its own brain "organ" (twenty-seven in Gall's original scheme); that the size of each organ is proportional to the strength of the corresponding faculty and that the brain is shaped by their differential growth; and, finally, that since the skull owes its form to the underlying brain, its "bumps" reveal psychological aptitudes and tendencies. Phrenology and the accompanying practices of cranioscopy and cranial palpation remained hugely popular into the 1840s, and phrenological publications appeared steadily until after World War I.

Gall (1835, 1:55) noted that "as the organs and their localities can be determined by observation only, it is also necessary that the form of the head or cranium should represent, in most cases, the form of the brain, and should suggest various means to ascertain the fundamental qualities and faculties, and seat of their organs." The deductive form of his claim points to the lack of empirical connection between organology and brain research. Yet Gall, together with his disciple Johann-Caspar Spurzheim (1776–1832), carried out significant neuroanatomical investigations, innovated in dissection methods, contributed to demonstrations that the nerves stem from gray matter, and described the origins of several cranial nerves (Rawlings and Rossitch 1994, Simpson 2005). All of this, however, had no empirical connection to their phrenological localizations. After Gall and Spurzheim presented their neuroanatomical researches at the Institut de France in 1808, a committee discussed them in a report. Though notoriously ungenerous to the authors, its members, including such celebrities as the alienist Philippe Pinel and the naturalist Georges Cuvier, were right to observe that, even if the connection between the brain and psychological functions was undeniable, neuro-anatomy had so far not contributed to elucidate it.

Spurzheim and Gall emphasized that physiology must be grounded on anatomy and that anatomy should lead to physiology. Several times, however, they declared that function is not directly observable or deducible from structure and that knowledge of the former precedes that of the latter—in the same way, they said, that we know the function of the eye before understanding its structure or learning anything about the optic nerve. "It is also without the assistance of anatomical dissection," they wrote, "that we made most of our physiological discoveries; and those discoveries could persist for centuries before their concordance with the material organization of the brain is known" (Gall and Spurzheim 1809, 246). They admitted that their anatomical findings were inspired by their "physiological and pathological views," including the fundamental assumption that moral and intellectual qualities are innate, and they added,

> it is precisely the perfect concordance of mental phenomena with the material conditions of their existence that will guarantee for ever the duration of our anatomical and physiological doctrine. . . . It is one thing to say that the discovery of the brain's functions was made independently from knowledge of its structure, another to claim that those functions do not have an immediate and necessary connection with its structure. (249–250)

As far as the distributed localization of mental faculties, inclinations, and personality features is concerned, their main general conclusion was that, since most brain structures are double, and since nerves neither originate in nor lead to the same point, "there is not, and there cannot be, a common center of all sensations, all thoughts and all desires." It followed, in their view, that "the unity of the self will forever remain a mystery" (168).

Discussions around that "mystery"—its interpretation, mechanisms, and relationship to phenomenal consciousness—have not subsided (see, e.g., Metzinger 2009 or, for a larger audience, Ananthaswamy 2015). Beyond apparent mysteries, the persistent question is whether the gap between psychological and neuroscientific analyses and explanations is inherent to the problem at hand or a temporary state of science that can be superseded. As of 2015, the best theories about the brain and about some aspect of mind "do not seem to share any properties" (Phillips et al. 2015, 367), and while it is clear that psychological theories largely contribute to inflect neuroscientific investigation, the extent to which brain research techniques such as neuroimaging can inform psychology remains debated (e.g., Coltheart 2013, Moran and Zaki 2013, Uttal 2015).[10]

Localization

Nineteenth-century experimental psychophysiology and pathological anatomy fueled the localization project and at the same time contributed to the demise of the phrenological enterprise. While phrenology correlated behavior or dispositions with cranial shape, the anatomo-clinical method searched for correlations between symptoms and brain lesions and was common to the partisans of discrete loci of mental faculties and those who insisted on the unity of intelligence and the integrated nature of brain action. The case of "Tan," an aphasic patient studied in the late 1850s by the French anatomist and physical anthropologist Paul Pierre Broca (1824–1880), is paradigmatic of the anatomo-pathological method and of mid-nineteenth-century localization debates.

"Tan, tan," accompanied by hand gestures, was Monsieur Leborgne's response to all and any questions. His clinical history and the postmortem study of his brain led Broca to conclude that the faculty of articulate language was possibly located in the second or third frontal convolution. It was clear to him that the higher "brain faculties," such as judgment, reflection, comparison, and abstraction, had their seat in the frontal lobes, whereas feelings, inclinations, and passions depended on the temporal, parietal, and occipital lobes. Broca (1861, 338) recognized "that the major areas of the mind correspond to major areas of the brain." At the same time, he found that differences in the localization of lesions inducing loss of articulate language were incompatible with the phrenological *système des bosses* yet consistent with the "system of localizations by convolutions."

Moreover, Broca's demonstration of the unilateral localization of language (in the left hemisphere) opened the way to the formulation of promising new dichotomies (Harrington 1987, 1991). The right brain ended up associated with "animality," femininity, and the emotions, the left with humanness, masculinity, and the "rational" faculties of will, intelligence, consciousness, and understanding. As we shall see below, hemispheric lateralization and dominance were to be assimilated into the discourses of "neuroascesis." They inspired a vast personal development and self-help literature for cultivating the supposedly neglected right brain and even neuropolitical considerations about the catastrophic future of a society tyrannized by left-hemisphere values (Harrington and Oepen 1989).

For nineteenth-century British and German brain scientists, the method of correlating clinical and pathological phenomena was suspiciously reminiscent of the craniological approach (Young 1990). Few, however, would have denied that the extraordinary positive or negative qualities of geniuses, criminals, and the mentally ill were somehow inscribed in their brain's fleshy substance. This brand of localizationism, with its galleries of exceptional individuals and its collections of preserved brains, matched the nineteenth-century development of anthropometry and the related elaboration of physiognomic, cranial, and bodily typologies; closely connected to craniometry, the measurement of differences in brain weight and size dates back to the early days of physical and racial anthropology and was a truly international fad (Hagner 2004, Podgorny 2005, Rafter 2008). By the late nineteenth century, cerebral localization, functional differentiation, and the correlation of site and effect, or structure and function, had become investigative principles.

Starting in the 1950s, cybernetics provided abstract models of brain neurophysiology; a decade later, artificial intelligence and cognitive science fostered the brain-as-computer paradigm (Pickering 2011). While circuit diagrams and flowcharts became tools for thinking about brain structure and function, the quest for localizationist explanation did not lose its appeal, even though it picked up only later. The saga of Albert Einstein's brain is extreme but emblematic. After the physicist's death in 1955, the pathologist Thomas Harvey cut his brain into 240 cube-shaped blocks from which microscopic slides were prepared; like relics of a medieval saint, some of these pieces and slides were sent over the years to devotees around the world. By the time of Einstein's death, the relic status of "elite brains" was nothing new. Investigations into the gross anatomy of genius's brains was underway by the mid–nineteenth century, and after Lenin's death in 1924, Oskar Vogt sliced his brain more finely than Harvey would slice Einstein's.

Three decades after Einstein's death, a contested but well-publicized histological analysis claimed that the left inferior parietal area of Einstein's brain contained more glial cells per neuron than the average (Diamond et al. 1985). A 1996 article described Einstein's cortex as thinner and more densely populated with neurons than control brains; a few years on, an equally disputed study stated that in the posterior end of the Sylvian fissure, Einstein's brain is 15 percent wider than controls (the parietal lobes were singled out for study because neuroimaging techniques had allegedly confirmed that these areas are responsible for mathematical reasoning as well as for visual and three-dimensional representation; Witelson et al. 1999). In the meantime (1994), the BBC produced Kevin Hull's hilarious documentary *Einstein's Brain*, about the Japanese Einstein worshipper Kenji Sugimoto's quest for a piece of the genius's brain.[11]

The saga continues: Newly discovered photographs of Einstein's entire brain prompted a revival of interest and led to detailed descriptions of the physicist's "extraordinary prefrontal cortex" (Falk, Lepore, and Noe 2012). On the same basis, a highly technical study of his corpus callosum found it thicker and displaying enhanced connectivity, so that, it concluded, "Einstein's intellectual gifts were not only related to specializations of cortical folding and cytoarchitecture in certain brain regions, but also involved coordinated communication between the cerebral hemispheres" (Men et al. 2013, e7). These findings have been widely covered in the media and dozens of online sites; the *Los Angeles Times* celebrated the "wonder of connectedness" (M. Healy 2013), and *New Scientist* announced that a "new look at Einstein's brain pictures show his genius" (Carver 2012). No recent case illustrates more eloquently the persistence of hopes for reading mind from brain than these technologically updated revivals of the nineteenth- and early twentieth-century morphological approaches.

From nineteenth-century phrenologists palpating head bumps, through electroencephalography starting in the 1930s, and up to today's brain scans, the hope of being able to read the mind and the self through brain recordings has not subsided (Borck 2005, Uttal 2003); the late twentieth-century comeback of the cerebral localization of mental aptitudes and inclinations "is due to a cohabitation of new visualization techniques with old psychological parameters" (Hagner and Borck 2001, 508). At the same time, these techniques confirm the anatomical, functional, and developmental evidence that the brain is neither a mosaic of minute sites nor a hard-wired collection of neuronal circuits but an array of interconnected and parallel networks, highly plastic and capable of developing and repairing itself.

Cognitive functions, in particular, turn out to be dispersed in various cortical areas, and the networks that represent them seem highly mobile, both functionally and anatomically. This does not invalidate complex forms of the localizationist approach (Zawidski and Bechtel 2005), which emphasize circuits and their "plasticity." Since the 1990s, studies of how diverse activities, from taxi driving to meditating, correlate with anatomical changes in the brain as well as discoveries about the brain's capacity for recovery, repair, and self-reprogramming after injury or amputation have turned *neuroplasticity* into a powerful motivator in rehabilitation and geriatric medicine and stimulated research on learning and cognition, aging and development, brain injury, addiction, and such brain-related disorders as Alzheimer's, Parkinson's, autism, and depression (e.g., Doidge 2015; Merzenich, Nahum, and van Vleet 2013; Merzenich, van Vleet, and Nahum 2014; Schwartz and Begley 2002; for discussions, see Choudhury and McKinney 2013; Droz 2011; Pickersgill, Martin, and Cunningham-Burley 2015; Rees 2010; Rose and Abi-Rached 2013).

Neuroplasticity has become a central neurocultural keyword not only inside but also outside the neurosciences. We shall see that it plays a role in "neuroarthistory"; in philosophy, it is one of the best allies of "neuropragmatism" (Solymosi and Shook 2014). In works for general audiences, such as *Brave New Brain: Conquering Mental Illness in the Era of the Genome* (2004) and *The Creating Brain: The Neuroscience of Genius* (2006), both by the neuroscientist Nancy Andreasen, neuroplasticity appears as the basis for creativity and therapy. According to the Canadian psychiatrist Norman Doidge (2007, xv) in his bestseller *The Brain That Changes Itself*, neuroplasticity is "one of the most extraordinary discoveries of the twentieth century." As "proof" that the mind indeed alters the brain, neuroplasticity substantiates convictions about the mind's power to bring about illness or cure (on whose history see Harrington 2008), which the same Doidge (2015) now markets as "neuroplastic healing."

In 2003, with wonderful irony, the conceptual artist Jonathon Keats copyrighted his brain as a sculpture created thought by thought (Singel 2003); the following year, a professional philosopher claimed, "Humans make their own brain, but they do not know that they make it" (Malabou 2008, 1) and repeatedly linked neuroplasticity to our "sculpting" our brains. As we show below, the "neurobics" industry, with its slogan "Change your brain, change your life," has effectively incorporated the idea into its strategies for marketing brain fitness. The point here is not to scorn scientific accomplishments or deride therapeutic hopes but rather to highlight how the ideology of brainhood feeds on the most diverse pieces of evidence and the most varied beliefs.

In short, the claim that "the success of the scientific method partially replaced older notions of the soul or mind-body dualism with the doctrine that mind . . . is the brain's exclusive output" (Lepore 2001) is as commonplace as it is false. The substitution in question is rooted in developments that have nothing to do with brain science (though brain science subsequently reinforced it), and the absorptive capacity of the brainhood ideology derives precisely from its *not* being the result of neuroscientific progress. In short, as Cathy Gere (2011, 236–237) has noted, the cerebral subject

> is not a historically contingent outcome of research into localization of brain function: it is the aim and object of the whole enterprise. Over the course of its one hundred and fifty year history, localization theory has consistently posited the cerebral subject as an a priori commitment: the question is not so much "can psychological phenomena be translated into the language of brain function?" but rather "where can we locate those functions that define human personhood in our neural topography?"

Neuroascesis: Health for the Cerebral Subject

Once these functions are localized—whether on solid or weak bases, in discrete spots or distributed across complex circuits—practical consequences rapidly follow. Genuine or spurious, knowledge about the brain has not only prompted further empirical, theoretical, and applied research but also given a new lease on life or new directions to more or less doubtful businesses. One such business is automated lie detection, which since the early 1900s has evolved from polygraphs measuring blood pressure, pulse, respiration, and skin conductivity to twenty-first-century "neurotechnologies of truth" such as brain fingerprinting and "No Lie MRI" (Pugliese 2010, chap. 5). Although calling these latter-day brain technologies respectively "neurognomics" and "digital phrenology" underlines the persistence of the belief that truth can be automatically read from outward bodily signs, the analogies to earlier techniques are no more than suggestive. Brain-based approaches embody an old goal but are in themselves a recent development—so much so, that in the early 1990s, the brain had not yet made it into the history of lie detection (Hanson 1992, Littlefield 2011).

At the level of practices asking from persons that they treat themselves as cerebral subjects, the self-help advice industry provides a much stronger instance of continuity, accompanied with renewal via *neuro* discourses. Manufacturers of self-help products have been appealing to the brain for a long time, but two periods stand out: the second half of the nineteenth century and the decades since 1990. The 1960s, for example, also witnessed the emergence of prescriptions upgrading the "mind-power" strain of self-help by way of rhetoric drawn from the cybernetic brain-as-computer model (McGee 2005, chap. 2). Yet it is mainly in the earlier and later periods that the brain itself was placed center stage. That is why we can speak of "neuroascesis." Insofar as *ascesis* refers to self-discipline, to the regulation of one's life for the sake of improvement, *neuroascesis* may designate the practices of the self aimed at the brain or pursued by way of behaviors purported to affect the brain directly. In neuroascesis, we benefit ourselves by acting on our brains. Obviously, everything we do has to do with them. But we are here talking about regimens and prescriptions that, even before the appearance in the 1990s of terms such as "neurobics" or "brain fitness," were advertised as having been specifically designed to enhance brain function.

Exercises for the Double Brain

A number of nineteenth-century authors considered that some mental pathologies were to be explained by the independent and disharmonic functioning of both "brains." Before Broca's discovery in the 1860s of the left-hemisphere location of language ability, it was indeed believed that the hemispheres were functionally identical and worked harmoniously together.

The notion of a double brain without lateralization of function inspired explanations of mental illness and neuroascetic proposals for attaining brain health.

The Brighton clinician Arthur Wigan (1785–1847) provides a prominent example. His *A New View of Insanity: The Duality of the Mind Proved by the Structure, Functions, and Diseases of the Brain and by the Phenomena of Mental Derangement, and Shewn to Be Essential to Moral Responsibility* (1844) illustrates the idea, not uncommon in the British medical context at the time, that madness was attributable to the uncoordinated, asymmetrical functioning of the two "brains" (Clarke 1987). Wigan saw each hemisphere as a distinct organ, complete in itself, and therefore capable of exerting independent volitions. While the organism remained healthy, one of the brains exerted control over the other; in pathological conditions, each brain followed its own way and could oppose the other. Curing mental illness required "presenting motives of encouragement to the sound brain to exercise and strengthen its control over the unsound brain" (Wigan 1844, 22). Brainpower, according to Wigan, could be indefinitely potentiated through "exercise and moral cultivation." By means of a "well-managed education," it was possible to "establish and confirm the power of concentrating the energies of both brains on the same subject at the same time; that is, to make both cerebra carry on the same train of thought together" (22, 23).

The Duality of the Mind proposed a system of cerebral ascesis that emphasized the importance of exercising and cultivating the brain for augmenting its power. The tasks and abilities involved, requiring exercise, self-control, and dedication, were moral as much as pedagogical. The brain must be constantly attentive, always watchful, and one of the hemispheres should permanently fulfill the role of "sentinel" (52, 298); "self-indulgence," "excess," or a "neglected education" would make such cerebral pedagogy fail (207–208). Training and perfecting the brain were according to Wigan the "great duty of man" (295). Programs of cerebral self-improvement should be incorporated into the treatment of the mentally ill as well as into the legal and educational systems. In the latter, for instance, arithmetical calculations could contribute to the "education of the cerebral fibres"; such training would bring about a "real physical change" in the exercised brain parts and produce "alterations in the external form of the skull" (343–344).

On the Continent, a major figure of double-brain neuroascesis was Charles-Édouard Brown-Séquard (1817–1894), Claude Bernard's successor at the Collège de France. While Wigan flourished before Broca's discovery of cerebral asymmetry, Brown-Séquard wrote at a time when language ability had already been located in the left hemisphere (Aminoff 1993, Clarke 1987, Harrington 1987). That, however, did not prevent Brown-Séquard from becoming Wigan's main advocate in the second half of the nineteenth century. He was especially interested in the possible application of Wigan's theory for "educating" the cerebral hemispheres (Brown-Séquard 1874a, 1874b, 1890). He recognized hemispheric functional differences, but instead of considering them as innate and structural, he believed they were attributable to educational failures. "We find," Brown-Séquard (1874b, 10) declared, "that it is owing to that defect in our education that one-half of our brain is developed for certain things, while the other half of the brain is developed for other things."

So the issue was clear-cut: "If we have two brains, why not educate both of them?" (1). Indeed, "if children were thus trained, we would have a sturdier race, both mentally and physically" (Brown-Séquard 1874a, 333). Training the brain would not only improve its efficacy but also increase its size, since "every organ which is put into use for a certain function becomes developed" (Brown-Séquard 1874b, 15–16). The exercises proposed, primarily motor, were meant to affect each hemisphere by means of activities of the contra-lateral side of the body:

> Try to make every child, as early as possible, exercise the two sides of the body equally—to make use of them alternately. One day or one week it would be one arm which would be employed for certain things, such as writing, cutting meat, or putting a fork or spoon into the mouth or in any of the other various duties in which both hands and the feet are employed. (Brown-Séquard 1874b, 20)

Brown-Séquard's neuroeducational program, like some contemporary counterparts, anticipated the "ambidexterity movement" that would become popular in the early twentieth century.

In his 1900 *New Methods in Education*, James Liberty Tadd (1854–1917), the headmaster of the Philadelphia Public School of Industrial Art, proposed a regimen based on an ambidextrous program that also valued hemispheric symmetry. He explained:

If I work with the right hand I use the left side of the brain. In truth, I exercise some special region or center of the brain and in every conscious movement I make and in every change of movement I bring into play some other center. If, by performing any such action with energy and precision, I aid in the development of the accordant center, I am improving the cerebral organism, building for myself a better and more symmetrical mental fabric. (Tadd 1900, 48)

Such a view of brain structure and function grounded an entire neuroascetic and neuroeducational perspective.

In 1903, John Jackson, a grammar school teacher in Belfast, founded the British Ambidextral Culture Society, whose goals he defined in *Ambidexterity, or, Two-Handedness and Two-Brainedness: An Argument for Natural Development and Rational Education* (1905). Here, Jackson blended Wigan, Brown-Séquard, and Tadd to elaborate a neuroeducational system that would improve the functioning of both hemispheres (Harrington 1987; Harris 1980, 1985). Future generations, he stated, quoting a member of the society, "must utilize to the utmost every cubical line of brain substance, and this can only be done by a system of education which enforces an equal preeminence to both sides of the brain in all intellectual operations" (Jackson 1905, 103–104). The implication was that we do not use all of our brains and that individual and social progress depends, at least in part, on no longer wasting our precious cerebral substance.

The mechanism was straightforward: while you exercise both hands, "the motor cells of the controlling side of the brain [will] be stimulated, strengthened, and developed" (84). As a result, brainpower will be duplicated, and the brain will be able to perform independent activities simultaneously. "If required, one hand shall be writing an original letter and the other shall be playing the piano; one hand shall be engaged in writing phonography, and the other into making a pen-and-ink sketch" (225). Jackson even imagined that training both cerebral hemispheres would not only increase brainpower but also lead to the growth of new language centers in the right hemisphere, thus preventing aphasias and hemiplegias. In the following decade, several authors in the United Kingdom and France claimed to supply evidence in favor of ambidexterity as a treatment for aphasia and several kinds of brain damage. By the 1920s, the ambidextral perspective as a source of neuroeducational goals and practices had been marginalized (Harris 1980, 1985), but there was no dearth of *neuro* beliefs to nurture the neuroascetic imagination. Both a general notion that the brain is plastic and the myth of the underutilized brain proved remarkably widespread and durable (Boyd 2008).

Phrenological Discipline

Phrenological self-help emerged in parallel to the double-brain approach. Phrenologists speculated that brain "organs" functioned like a muscular system, so that the action of disturbed organs could be compensated by the contrariwise exercise of the healthy ones. They therefore came up with a neuroeducational program based on training, redirecting, and strengthening specific brain organs. Contrary to what the *New York Times* suggested in 2006, "brain calisthenics" was nothing new (Belluck 2006). For phrenologists as for latter-day promoters of "brain fitness," mental health consisted of exercising all organs daily; both inactivity and excessive exercise were considered unhealthy. The difference between the mid–nineteenth century and the early twenty-first is that phrenologists asked schools to encourage sobriety, moderation, chastity, and personal amelioration. With the help of phrenological self-discipline, individuals could cultivate and enhance virtues favored by Victorian society while strengthening their capacity to inhibit vices and pernicious inclinations.

Phrenology's social and individual moral significance derived from its perfect fit with the Victorian ideals of self-knowledge, self-control, and self-improvement (Cooter 1984, De Giustino 1975). It provided guidelines for how to lead one's life and offered a panacea for mental and physical ills. The brain emerged as the clue to manifold queries, from personal talents to exercise, whom one should trust and whom not, how to raise one's children, how to go about sexual education and choose one's spouse or even appropriate servants. Every aspect of an individual's social and personal life could be phrenologically approached (Stern 1971).

Phrenology had a considerable impact on educational reform, particularly through the action of George Combe (1788–1858) and his brother Andrew (1797–1847). George was largely responsible for the transformation of phrenology into a scientifically respectable vehicle for ideas on social life and its organization (Cooter 1984, Van Wyhe 2004). For him, the cerebral organs had to be treated like muscles. The best way to increase their strength and energy was to train them regularly but judiciously, "according to the laws of their constitution"; as a result, "when the cerebral organs are agreeably affected, a benign and vivifying nervous influence pervades the frame, and all the functions of the body are performed with increased pleasure and success" (Combe 1828, 115, 117–118).

Exercise would also enlarge cerebral organs. Andrew Combe (1836–1837, 7) claimed that "even in mature age the size of the individual organs of the brain may be increased by adequate exercise of the corresponding faculties." James Deville (1841), a well-known practical phrenologist, offered many examples of increases of up to half an inch in the size and the diameter of particular cerebral organs as a result of training. Phrenology therefore looked like an efficient philosophy of education, one based on the idea that the organs of the brain need as much training as those of the body and can be affected in targeted ways by physical exercise. This is the very premise of twenty-first-century "brain gyms," whose pseudoneuroscientific bases have been debunked without apparent effect on their commercial success.[12]

Phrenology was credited with the power of contributing to general good health, and an avalanche of phreno-physiological literature sustained belief in such a power. This literature was at the same time moral. For example, by showing the noxious effects of alcohol and sexual depravity on the brain, it encouraged temperance and sexual moderation as rational prescriptions for a healthy life; the natural laws of health converged entirely with social norms, and the achievement of good health depended on following the organic laws that governed both body and brain (Cooter 1984, Van Wyhe 2004). These views inaugurated a recurrent neuroascetical motif and introduced a number of prescriptions that have close late twentieth-century analogs. Today's neuroascesis, like its phrenological ancestor, claims to pursue cerebral improvement. The practices it recommends include dieting, physical exercises, and a healthy life in the broadest sense—in short, they target the body as a whole and not the brain alone. Yet their proponents insist that it is the brain that undergoes training and is thereby enhanced. This offers other similarities with earlier proposals.

A major contribution toward the popularization of phrenological neuroascesis was made by Sylvester Graham (1794–1851), one of the founders of the natural food movement in the United States, who set city life and industrialization in opposition to the virtues of traditional agricultural (and vegetarian) life (Nissenbaum 1980, Sokolow 1983). In Graham's view, the improvement of individual health enhanced moral capacities and vice-versa. The self-discipline and self-control required to lead a healthy life were seen as acts of moral excellence (Gusfield 1992); the moral and medical spheres went hand in hand, and both conveyed traces of more ancient wisdoms. For instance, the most popular American phrenologists, Lorenzo and Orson Fowler, took up Graham's conviction that phrenology opened the way to health reform (Fuller 1989, Stern 1971). In their teaching they phrenologically reformulated the old belief that a carnivorous diet fosters a carnivorous temperament, turning it into the idea that meat's stimulating power circulates through the nerves, inflames the lower regions of the brain, and strengthens the organs of "Combativeness" and "Destructiveness" (cited in Whorton 1982, 125).

One of Graham's main followers was the physician and Seventh-Day Adventist John Harvey Kellogg (1853–1943), a prolific writer and inventor of corn flakes, who continued the Grahamites' crusade for natural food and sexual purity (Carson 1957). In the chapter on "How to Keep the Brain and the Nerves Healthy" of his *First Book in Physiology and Hygiene*, Kellogg outlined a neuroascetic program, again aimed at training the brain as if it were a muscle. "We should exercise the Brain," he wrote, and he explained:

> What do we do when we want to strengthen our muscles? We make them work hard every day, do we not? The exercise makes them grow large and strong.
> It is just the same with our brains. If we study hard and learn our lessons well, then our brains grow strong and study becomes easy. But if we only half study and do not learn our lessons perfectly, then the study does not do our brains very much good. (Kellogg 1887, 203)

Brain gymnastics were to be supplemented by physical exercise, a balanced diet, and a sufficient amount of sleep; toxins, alcohol, and drugs were of course to be avoided. Children should not "eat freely of meat," which "excites the brain" and irritates the nerves, and avoid spicy food, which tends to "injure brain and nerves" (204). Psychological and moral habits also had to be disciplined. Becoming angry does the "brain and nerves great harm," and every child must refrain from swearing or use slang phrases, for "the brain after a while will make him swear or use bad words before he thinks" (205).

These various facets of Kellogg's cerebral self-help program reappear literally in many of the neuroascetic manuals of the late twentieth century.

Cerebral Self-Help

The phreno-physiological wave was of paramount importance for the emergence of the self-help movement in the nineteenth century. The brothers Combe's emphasis on personal responsibility, both physical and cerebral, and on the role of education and self-control announced the movement's fundamental values (Van Wyhe 2004). Essential topics, such as rationalism, natural laws, education, health, hygiene, self-knowledge, and self-development, all contributed to the very concept of "self-help," as can be found, for example, in the 1859 bestseller of that title by the Scottish social reformer Samuel Smiles (1812–1904). As already mentioned, the emphasis on developing one's mental faculties through exercise belongs to the basic credo of phrenology as much as to Victorian morals.

By the end of the nineteenth century, the ethics of individual self-help and self-improvement had become more important to phrenologists than the dimension of social reform that had characterized the movement in earlier decades (Cooter 1984). Phrenology accompanied the growth of "self-healing" and other forms of alternative, heterodox popular medicine, and it sometimes combined forces with spiritualism and various forms of occultism. Especially in the United States, phrenology merged with interest in the paranormal. The "psychologization of esotericism" prepared some of the ground for the New Age movements of a century later (Hanegraaf 1998). Around 1890, the New Thought or Mind Cure crusade, which borrowed from Samuel Smiles's self-help outlook, generated dozens of books mingling metaphysical spirituality with self-help training programs (Braden 1963; Fuller, 1982, 1989, 2001).

The quest for health and spiritual integration embodied in the "Mind-cure movement," as William James called it in *The Varieties of Religious Experience*, readily incorporated elements of neuroascesis. A major instance is to be found in the works of Warren Felt Evans (1817–1889), an American Methodist minister turned Swedenborgian. The basic idea of his doctrine was that illness originates in the mind because of false beliefs and can be overcome by way of openness to God. He developed it in books with such titles as *Mental Cure* (1869), *Mental Medicine* (1871), *The Divine Law of Cure* (1881), *The Primitive Mind Cure* (1885), and *Esoteric Christianity and Mental Therapeutics* (1886). Evans believed it possible to tap the healing resources of divine energy by getting in contact with the unconscious mind, whose healing power, he thought, corresponded to the *kerygma*, the preaching of the early Christian church. The principles of mind cure combined an idealistic tradition that referred back to the Hindu Vedas, according to which the only reality is thought itself; a Swedenborgian transcendentalist mysticism; elements of pantheism; and occult, gnostic-like interpretations of Christianity (Fuller 1989, 2001; Teahan 1979). Following on the steps of earlier mental healing and self-help systems, Evans made individuals responsible for their own physical and mental condition. For him, the only reason why external circumstances seem to exert an influence upon us is that we believe they do. Thought, he claimed, can change and shape any situation in the real world. And thought depends on the brain.

Evans picked up several gimmicks from phrenology and phrenomagnetism. For example, he claimed that touching the skull could increase the action of the underlying cerebral organ: "Touch the organ that you wish to excite, or any part of the brain whose activity you may desire to augment," he wrote, "and silently will or suggest that they feel happy, or calm, or strong, or hopeful, as the case may require, and it will have its effect in inspiring the proper mental state" (Evans 1874, 74). His recommendation for those prone to despondency and despair was the following: "Let us fix the attention upon the part of the cerebrum which is the organ of hope and, if need be, place your finger upon it and a joyful sunshine will light your darkness" (75). The old healing touch magic combines here with the religious laying on of hands as the means to reach the patient's innermost being. Over one hundred years later, one of the basic brain gym exercises still consists of laying one's fingertips on the "positive points" above each eye, halfway between the hairline and the eyebrows, in order to "bring blood flow from the hypothalamus to the frontal lobes, where rational thought occurs" (Dennison, Dennison, and Teplitz 1994, 32). In the French-speaking world, the Coué method provides another instance of *neuro* varnish: According to the websites of some French coaching agencies, the self-improvement method by way of conscious autosuggestion promoted by the French pharmacist Emile

Coué (1857–1926) is in fact a *programmation positive du cerveau*, a positive programming of the brain.

Contemporary Neurobics

Brain gymnastics is one of the many nineteenth-century neuroascetical ingredients that reappear in updated garb in contemporary cerebral self-help. But the continuity at the level of practices must not mask the difference in contexts. Sociologically, neuroascesis always involves the development of "objective selves," a process of "objective self-fashioning" (Dumit 2004) whereby individuals and categories of people are transformed through the assimilation and application of expert knowledge. However, today that process is bolstered by factors that at the time of phrenology or the New Thought movements were absent, weaker, or qualitatively different—among others, the role of the media, brain imaging techniques, the pursuit of a "strong" neuroscientific program, and an extremely assertive global pharmacological industry (Ehrenberg 2004, Healy 2002, Rose 2003). These factors have sustained the emergence of contexts where seeing oneself as a cerebral subject functions as a biosocial criterion of personal identity (see here Chapter 3).

We have already noted the neurocultural significance of brain plasticity, which "neurobusiness" has been using for its own benefit (Wolbring 2007). The vague claims about the effect of mental and physical activity on the brain asserted in nineteenth-century regimens have been replaced by more precise information. For example, research has found that aerobic activities are beneficial beyond their well-known effects on the cardiovascular system and in cases of depression. The brains of rats that exercise have over twice as many new neurons and show more interconnections than the brains of sedentary rats (Brownlee 2006a, 2006b; Cotman and Berchtold 2002). It allegedly follows that physical exercise helps healthy brains function at an optimum level and may increase their performance and plasticity. Exercise, it has been claimed, may also delay the progression of Alzheimer's disease and the onset of Parkinson's. Similar effects were observed in connection with feeding habits involving low rates of saturated fatty acids and a high content of Omega-3 factor. The basic idea is simple: As Carol E. Greenwood, a

Toronto University specialist of nutrition and the aging process, puts it, "by taking care of your body, your brain also benefits" (Brownlee 2006b). This statement illustrates a recurrent trope in the history of neuroascesis, namely the ontological subordination of the body to the brain, as if brain and body were actually separable. Exercise trains the body as a whole, but its real target is the brain; hence, for example, the redescription of healthy food as "powerful brain medicine" (see the discussion in D. Johnson 2008).

The Posit Science Corporation is a good example of how neuroascetic firms take cerebral plasticity as a point of departure. Posit explains that its goal is "to help people flourish throughout their lives," and it adds: "We do this by providing effective, non-invasive tools that engage the brain's natural plasticity into improving brain health."[13] Its "brain fitness program" focuses on increasing the speed, precision, and intensity with which the brain receives, registers, and remembers information. This program is a first step that can be followed by a more complete "brain gym" to train the totality of the motor and cognitive systems.

Posit's advertisement resembles that of cosmetic products: neuroascesis promises to "rejuvenate" the brain's "natural" plasticity and postpone mental decline for as much as ten years. Not coincidentally, the elderly make up Posit Science Corporation's main target audience. The company does not advertise a fountain of youth but claims to supply "a part of the solution." More astutely, it predicts an increased "brainspan" or "cerebral longevity," something particularly valuable at a time of aging populations and growing life expectations (Anonymous 2006). Like bodily fitness, cerebral fitness involves a moral dimension: Exercises are said to demand a great deal of discipline, willpower, and self-motivation—and they are all said to be indispensable for neuroascesis to "reverse the brain's aging process" (Olney 2006). Neuroplasticity research thus legitimates a market for brain gymnastics and cerebral self-help; some of it even comes out directly of Posit Science Corporation's own "Brain Plasticity Institute" (Merzenich, Nahum, and van Vleet 2013).

The neuroascesis market offers a vast range of products. Some are books by neuroscientists, cognitive psychologists, and well-known psychiatrists who explain recent neuroscientific advances while offering programs to enhance brainpower, prevent mental decay, and improve perception, short- and long-term memory, and logical, verbal, visual, and spatial abilities (Chafetz

1992, Goldberg 2001, Mark and Mark 1991, Winter and Winter 1987). "Brain training" programs thus sustain a multimillion-dollar industry whose efficacy remains unproven. Results from a six-week online study involving 11,430 participants who trained several times each week on specific cognitive tasks aiming at improving reasoning, memory, planning, visuospatial skills, and attention provided "no evidence to support the widely held belief that the regular use of computerised brain trainers improves general cognitive functioning in healthy participants beyond those tasks that are actually being trained" (Owen et al. 2010, 777).

Always on seemingly neuroscientific bases, other products, by psychological self-help authors converted to neuroascesis, lead their buyers to expect more: to identify hidden meanings in people's conversation, absorb facts "like a sponge" and reproduce them intact years later, read and understand any book in half an hour, or easily memorize facts, images, and even complete works. Among the authors of these products, those closest to a New Age imaginary also employ an apparently scientific vocabulary but promise to reach any possibly desired result. After all, some claim, on the basis of a crude oversimplification of quantum mechanics, that since reality is no more than an illusion created by our brains, "the universe is the mind and the mind is the universe" (Spotts and Atkins 1999, 80). The exercises they propose presume to allow the individual brain to connect to the forces of the universe and a superior intelligence, a Cosmic or Divine Mind. It is all at once instructing, amusing, and alarming to see the extent to which this quack neuroascetic literature reproduces with an updated scientific-sounding vocabulary the main topics of older self-help literature.

Commonplaces in this framework are an emphasis on creativity as a means to engender reality, the idea of an "internal self" that can be cultivated by means of cerebral exercises, and the insistence upon autonomy, responsibility, and self-control not only of one's personal destiny but even of reality itself, all to be attained by means of brain practices. In cerebral self-help literature, the absolute irreducibility of the individual goes hand in hand with a belief in the reducibility of reality to the designs of thought. Finally, the essentially cerebral nature of the self renders other people as well as the social and cultural environment obsolete. The brain takes over, so that the old slogan "You are what your mind is" is replaced by the basic assumption of the neurocultural universe: "You are your brain."

As we have seen, the notion of a divided mind, embodied in a divided brain in conflict with itself, goes back to Wigan and others in the nineteenth century. After Broca, the left hemisphere came to be considered superior because it was seen as responsible for the intellectual, civilized activities predominant in white European males, while the right one was thought to dominate in women, criminals, Indians, blacks, madmen, and homosexuals (Harrington 1987). Cerebral self-help bestsellers reproduced and exploited the right-brain boom that emerged during the 1960s in the context of counterculture movements, but they also referred to the split-brain research that was emerging at the time, which could itself make room for Wigan (the neurophysiologist Joseph Bogen [1971, 1985] reprinted *The Duality of Mind* and described his own position as "neowiganism"). The self-help market is full of titles relating the right hemisphere to the most varied phenomena, from the classification of artists, musicians, politicians, and dictators according to their cerebral "orientation" to tantric sexuality, mediumistic capacities, and other paranormal activities supposedly enabled by the right brain (Capacchione 2001, Ehrenwald 1984, Spotts and Atkins 1999, Wells 1989).

Since the late 1960s, several authors in the area of education have insisted on the countless advantages of a school that would focus on the right brain and have criticized traditional pedagogy for its emphasis on left-hemisphere capabilities (Edwards 1979, Gainer and Gainer 1977, Hunter 1976). Such proposals for a "hemispheric balance in the curriculum" that would avoid the didactic failures of left-brain educational programs hark back to nineteenth-century pedagogical crusades and revive many of the assumptions of Brown-Séquard in France and the Ambidextral Culture Society in the United Kingdom. For all their success among teachers, more recent brain-based ideas about teaching and learning are no less scientifically specious and no more relevant or effectual for their stated purposes than their predecessors (Becker 2006, Bowers 2016). Yet no amount of failure dampens the hopes of bringing about an "integrative framework" through "constructive interdisciplinary dialogue" (Busso and Pollack 2015).

Important as it might be to differentiate science from quackery, the genealogy of cerebral self-help brings to light the porosity of the distinction and the extent to which twentieth- and twenty-first-century neuroascesis reproduces with an updated appearance the commonplaces of much earlier self-help discourse. But there are some major differences. At the end of the nineteenth century, the aim of Kellogg's workout for the brain was to resist a weakening of the social fabric; the disorders of the physical, social, and political bodies were to be countered by neuroascetic practices. The brain fitness movement at the time wished to salvage an individual and collective moral order seen as eroded by the rise of industrial society and by the concomitant loss of traditional sources of authority and legitimacy (Gusfield 1992). In contrast, contemporary neuroascesis is not aimed at restoring or saving an allegedly endangered social order; rather, it instantiates the values of an individualistic somatic culture. Yet the spirit of neuroascetical prescriptions and practices remains largely the same then and now. In light of Foucault's (1986, 1990) depiction of technologies of the self in the transition from paganism to Christianity, the fact that contrasting goals and frameworks sustain similar practices does not come as a surprise.

As highlighted by the very ideas of "brain fitness" or "neurobics" present in so many titles since the 1990s, the muscular-fitness model offers another element of continuity between the nineteenth century and recent decades (Cohen and Goldsmith 2002; Dennison, Dennison, and Teplitz 1994; Mark and Mark 1991; Winter and Winter 1987). The brain is a muscle: "Just as weight lifting repetitions in the gym or jogging strengthen certain muscle groups, mental exercises appear to strengthen and enhance cognitive functions over time" (Tannen n.d.). It is common to praise the "mental weight lifting" one can do in the "Brain Gym" (CBS 2006). Train your "cerebral muscles" (Goldberg 2001, 255), but in such a way that you avoid "brain cramps" (Chafetz 1992, 72). Do regularly the "brain stretches" that will help you "burn some synaptic calories" and prevent you from becoming a "mental couch potato" (Parlette 1997, 16); this is a challenging goal because mental muscles enjoy television, a true "bubble-gum for the brain" (152–153). Most neurobics authors establish distinctions among levels of brain accomplishment or mental prowess, since "you do not have to attain the brain equivalents of Steffi Graf's or Michael Jordan's level of physical fitness to be quicker in conversation, better at solving problems, have richer memories, and livelier associations" (Chafetz 1992, 23). For "those of you who wish to exercise your brain systematically as an athlete would exercise various muscle groups," manuals provide well-ordered cerebral training programs and recommend hiring a cerebral "marathon trainer" and keeping "brain workout diaries"

(213–214). The vocabulary of bodily fitness is thus extrapolated to the brain itself. Causally and rhetorically, bodily and cerebral fitness go hand in hand.

But none of this can be explained by invoking neuroscientific advances, not even those connected to cerebral plasticity, which have come to play such a central role in contemporary neuroascetical discourse. Rather, the genealogy of neuroascesis is best seen as an episode in the development of views about the human as well as of forms of sociality and subjectivation that involve notions and practices of the self and its relationships with one's own body and other people. In short, neuroascetical practices are tools whereby persons constitute themselves as cerebral subjects, and that is why to do their genealogy amounts to throwing critical light on that particular form of being human.

TWO

Disciplines of the *Neuro*

The previous chapter dealt in part with neuroascesis as a technology of the self and with neurobics as a package of self-care practices allegedly based on scientific knowledge and capable of acting directly on its user's brain. From mid–nineteenth century exercises for the double brain or the phrenological organs to the twenty-first-century brain gym, cerebral self-help has been a commercial industry. But even if the regimens it sells are widely shared, those who buy them pursue the exclusively personal goal of improving or maintaining themselves. Thus, insofar as neuroascesis and neurobics impel believers and consumers to adopt regimens for activities such as eating and exercising, they incarnate individual forms of "being brains." In Chapter 3 we shall explore another context for the making of cerebral subjects: the neurobiologization of psychic distress. At the global and institutional level, such a context is more consequential than that of neuroascesis. Yet, contrary to the adoption of neurobics, which requires the voluntary acquisition of commercial products, the cerebralization of psychological suffering involves, as we shall see, choices and decisions by individuals and collectives, but it also affects subjectivities independently of those choices and decisions. It thus stands between, as it were, the world of neuroascesis and the more purely academic universe we shall explore here, namely that of the "disciplines of the *neuro*."

The message that we are essentially our brains does not always require a deliberate commitment to regimens supposed to affect our brains directly or the unintentional and often unnoticed participation in mechanisms that shape us individually as cerebral subjects. Indeed, it may be independent from both and take the form of more or less professionalized and institutionalized programs of research and teaching in the human sciences. The actors of these programs do not necessarily implement the cerebralizing doctrine in their private lives. But what they advocate in their writings and investigative practices, though representing a minority viewpoint in their fields, has become a conspicuous element in the larger landscape of the *neuro* and the academic embodiment of the brainhood ideology. Moreover, by virtue of the "looping effects" the human sciences may have (Hacking 1995, 2006), it is conceivable that the disciplines of the *neuro* influence some aspects of subjectivities even beyond their actors' scientific persona, that is, beyond a "cultural identity that simultaneously shapes the individual in body and mind and creates a collective with a shared and recognizable physiognomy" (Daston and Sibum 2003, 2).[1]

The disciplines of the *neuro* are varied but have several features in common. The most immediately visible for many of them is the way they are designated. Since the 1990s, intellectual and institutional projects whose names combine the prefix *neuro-* with the name of one of the human or social sciences have multiplied. An incomplete alphabetical list of the "disciplines of the *neuro*" or "neurodisciplines" (a collective we shall also sometimes call "neuroX") may start with neuroanthropology, neuroarcheology, and neuroarthistory and finish with neurosociology and neurotheology; in between, we could place neuroeconomics, neuroeducation, neuroaesthetics, neuroethics, neurolaw, neuromarketing, neuropolitics, neuropsychoanalysis, and more. Such a list would be merely suggestive, since it would leave out older labels, such as neurophilosophy, as well as undertakings such as "affective" or "cultural neuroscience," which lack the prefix *neuro-* yet are forms of the same pursuit. This chapter will focus on two of those fields, made up of the *neuro* approaches to aesthetics and to culture (in the anthropological sense).

To understand the neuroX it is not enough to describe their premises and promises. We have to look at what they actually do. That has been to some extent undertaken from the points of view of the anthropology, ethnography, and history of science as well as from that of the sociology of the professions and the scientific field. Our approach has instead been to range very broadly across the *neuro* spectrum—that is what allows some generalizations—but also to examine a few areas in sufficient detail to understand and analyze their inner "logic."

We limited ourselves to areas that emerged during the Decade of the Brain and have therefore excluded those, such as neuropsychiatry and neurophilosophy, that were well established by the 1990s. The newer disciplines of the *neuro* share the basic premises and goals of the older ones. However, they constitute a major expansion of the range of application of neurobiological concepts and methods to problems that had been the traditional preserve of the humanities and the human sciences. It is the scale of the phenomenon sometimes characterized as a "neuro-turn" (e.g., Cooter 2014, Pedersen 2011), rather than particular instances, that makes it significant for the recent history of the cerebral subject. At the same time, concrete instances had to be chosen to show how that "turn" is carried through in scientific work and communication strategies.

Given the inherently political nature of the processes that shape persons' lives, neuropolitics might have been an appropriate choice. The term seems first to have been used in 1977 by the psychologist Timothy Leary, the legendary American advocate of psychedelic drugs, to capture the belief that political problems can be traced to psychological issues based in brain chemistry. He declared: "The solutions to our predicament are neurological. We must assume responsibility for our nervous systems" (Leary 1977, 57), and he imagined recreating on that basis the relationship between individuals and the social order: "Young minds exposed to neurological freedom and the free spray of electronic information suddenly blossom like flowers in the spring" (Leary 1980, 33, 6). A less romantic notion emerges from William Connolly's 2002 *Neuropolitics*, a cryptic Deleuzian manifesto that defines its subject as "the politics through which cultural life mixes into the composition of body/brain processes. And vice versa" (Connolly 2002, xiii). The predominant uses of "neuropolitics" since the 1990s correspond neither to Leary's nor Connolly's. Rather, in the spirit of the "neuro-turn," the term designates a neurobiological (and predominantly neuroimaging) approach to questions of applied political science, such as people's political attitudes or how they assess candidates and choose to vote. Similarly to neuroethics, which has been defined as the neuroscience of morality and the ethics of neuroscience, neuropolitics concerns both the neurologization of the political field and the study of the biopolitical implications of neuroscientific theories and practices. Considering such a double understanding, several ways in which the brain has been mobilized in politics can be identified (Meloni 2012 and Vander Valk 2012b offer taxonomies that partially overlap with ours).

One approach proposes to take "political affect" seriously in the framework of the "nonmechanistic materialism" it attributes to the contemporary neurosciences (Protevi 2009). Such "embodied" political theory tends to contradict the model of the "rational subject" said to prevail as the foundation for "neoliberal thought" (Lakoff 2008, 4). This brand of neuropolitics has been criticized for "honoring the neurosciences a good deal too much" and for "generously giving credit before actual research findings have emerged, and surely before alleged findings have settled into well-founded explanatory theories" (Slaby, Haueis, and Choudhury 2012, 60). A second approach emphasizes the role of empathy in political thinking and practice (Coles 2012; Olson 2008, 2013) and hopes that mirror neurons will provide an "Archimedean moral point from which to lever public discourse toward an appreciation of our true nature, which in turn might release powerful emancipatory forces" (Olson 2008).

A third approach focuses on the so-called political brain (Western 2008), mainly to anticipate voter behavior and preferences; it is an academic enterprise but also, as would be expected in an area teeming with consultants, a commercial one (Randall 2015). It too claims to show that political decisions do not rest on cost-benefit evaluations and argues (for example on the basis of studies on amygdala responses as detected by fMRI) that "the neural basis for political decisions extends across cultures" (Blank 2013, 269). A fourth approach, which largely overlaps with neuroethics, addresses the policy implications of neuroscientific research (see Blank 1999, 2013), for example in the context of the increasing social and economic costs of neurodegenerative disease in aging populations and the use of a range of interventions, from psychopharmacological enhancement to brain implants and neurogenetic procedures.

A fifth approach, which discusses the plastic brain in a political key, further illustrates the interpretive leeway the *neuro* provides. On the one hand, the widely covered notion of neuroplasticity is said to reflect the neoliberal emphasis on autonomy and self-responsibility; on the other, it is presented as designating a feature that allows us to "make" and "sculpt" our brains freely and "change our life" however we choose (Malabou 2008; see critiques by Pitts-Taylor 2010 and Rees 2011). Critical neuroscience can be seen as another neuropolitical approach. Assuming that "there is substantially more politics in the 'neuro' than there is 'neuro' in politics" (Slaby, Haueis, and Choudhury 2012, 64), it examines how "brain facts" are appropriated in different social domains, seeks to increase awareness of the factors that contribute to stabilize scientific worldviews and make them seem inevitable, and scrutinizes how people are "made up" according to brain categories. Finally, the neurodiversity movement is essentially a politics of the brain. Considering that what society usually classifies as psychiatric illness is a particular form of being derived from unusual brain "wiring," the movement advocates the recognition of diagnosed individuals as different yet fully entitled members of society.

In Chapter 3 we examine the neurodiversity movement as a form of social action rooted in cerebralized understandings of subjectivity. Here, we explore and assess the potentially transformative action of the *neuro* by focusing on disciplines chosen because they derive from classic fields that played historically foundational roles among the human sciences: on the one hand, aesthetics and art history (by way of neuroaesthetics); on the other hand, and in order to tip the balance in the direction of social science fieldwork and non-Western societies, the disciplines of culture (by way of neuroanthropology and cultural neuroscience).

The NeuroX: An Overview

"Neuroskeptics" have taken the neuroX as the premier instance of the *neurobabble, neuromythology, neurospeculation, neurotrash, neuromania,* or *neuro-*

madness of the late twentieth century.[2] However, if madness there is, it has logic in it, and it is precisely this logic we wish to examine. Indeed, in spite of their diversity, the disciplines of the *neuro* may be considered as a single constellation held together by a set of core common beliefs and foundational elements. We first outline those commonalities, derived inductively from countless articles and books across the neuroX spectrum.

First, the neurodisciplines postulate that "the mind is what the brain does." This, however, does not function as a working assumption but as a general and established truth. It follows (implicitly) that only cerebral mechanisms are truly constitutive of the studied phenomena, while the rest, including all aspects of culture and society, only "modulates" universal neurobiological processes. Second, the disciplines of the *neuro* have a common goal, namely to discover neurobiological "foundations" or "substrates," and assume that uncovering the processes that "underlie" the phenomena they study will unveil causes and thereby provide explanatory depth. The preferred tool to pursue such a goal is neuroimaging, particularly fMRI (functional magnetic resonance imaging); this is a third common element. Fourth, as a consequence of the predominant methodology, the results of neuroX research come mainly in the form of neural correlates of the investigated processes and behaviors; these correlates are said to reveal the brain structures "involved" in those processes and behaviors.

Nevertheless (fifth feature), given the correlational nature of the data and other features of fMRI, the neurodisciplines cannot explicate the meaning of the data and the role of the identified structures. This intrinsic limitation conflicts with a widespread desire for causality (a sixth common feature), which manifests itself in frequent interpretive slippages from correlations to causes. Hence (a seventh aspect) the great disparity between the apparent rigor of the methodologies on the one hand and on the other hand the range, generality, and speculative character of the programmatic statements and discussion of results.

More than the most widely shared utensil of the neurodisciplines, neuroimaging (in its well-documented overhyped mode) has been their condition of existence.[3] It has enabled their emergence and sustained their development but has also been a major source of their drawbacks. We arrive here at two further shared characteristics. One (the eighth of our list) is the impression of irrelevance the neurodisciplines give, insofar as they seem incapable of providing a nontrivial answer to the question *So what?* with regard to their empirical findings. The other (ninth feature), which emerges from and epitomizes the others, is that the neuroX are not merely, as others have argued, irrelevant or ideologically problematic: they are basically self-defeating.

By "self-defeating" we mean that the assumptions and methods of the neuroX are inadequate to study the topics they ostensibly want to understand. One may object that they redefine their objects in ways that make them amenable to neuroscientific inquiry. This is perhaps so in some cases, but even then the objects turn out to be so transformed that they end up having little to do with what they originally were said to be or to embody. Whatever they are about, fields such as neuroaesthetics or cultural neuroscience are definitely *not* about the things, such as art, beauty, or cultural difference, whose neurobiological foundations they claim to be looking for. Yet, ironically, the disciplines of the *neuro* convey the belief (which they do not formulate explicitly) that revealing the neurobiological "substrates" of a phenomenon not only means knowing it "objectively" but also confers upon it the ontological consistency it allegedly lacks when treated with the tools of the human and social sciences.

These simultaneously epistemological and ontological persuasions bolster another common characteristic of the neuroX, namely (despite allegations to the contrary) that they place the neurobiological approach at the top of a hierarchy of forms of knowledge. The brain sciences are presented as dealing with the fundamental level, with the mechanisms that ultimately account for phenomena the human and social sciences merely describe and interpret. This takes us back to the premise that humans are what they are and do what they do because of the way their brains function. That premise justifies the methodological and interpretive preeminence of the *neuro*— even if it should be obvious that it is one thing to admit that we cannot create culture without our brains and another to posit that culture is a "product" of the brain and that therefore the neurobiological is the most essential and ultimate level of analysis for cultural phenomena.

From the hierarchy of approaches and modes of explanation follows a lack of attention to the productions of the humanities and social sciences—a neglect, most flagrantly reflected in the dearth of human science and social science research from the bibliographies of neuroX publications. To engage in neurodisciplinary activities (but also to study and eventually to critique them), scholars from the humanities and human sciences must become conversant with relevant neuroscientific conceptual, technical, and empirical issues; with very rare exceptions, the opposite does not happen. Such an asymmetry is understandable: Why would you study, say, philosophical aesthetics or theories of culture if those approaches are precisely what you want brain science to supersede? And why bother with historical details if you are convinced that history is fundamentally a consequence of brain processes? Even though the disciplines of the *neuro* pursue intellectual agendas dictated by philosophy and the human sciences, at the level of practice they neglect and even disdain them.

Finally, the neurodisciplines are usually expected to make conceptually, empirically, and methodologically positive contributions to the corresponding human sciences, and they are sometimes given the redemptive function of driving them out of their supposed crises and dead ends. (The art historian Norman Bryson [2003, 14], for example, found in the neural turn distinct advantages "over the broad family of accounts of the real that are based on the primacy of the signifier"—accounts said to have disconnected the human sciences from material reality and pushed them to the edge of irrelevance.)

The Case of Neuroethics

Neuroethics occupies a special place in this landscape. Unheard of in the 1990s, it required a few years in the early 2000s to become a recognized, autonomous, and professional discipline, with its own networks, platforms, societies, journals, academic centers, and teaching and research programs (Conrad and De Vries 2011, Hoyer 2010). Although it has been described as covering two related domains, the neuroscience of morality and the ethics of neuroscience, *neuroethics* is almost exclusively reserved for the discipline that examines the implications of brain science for society as well as the actual and anticipated ethical, social, political, and legal consequences of neuroscientific knowledge and its applications, including the ethics of neuroscientific research. By 2015, neuroethics publications amounted to hundreds of articles as well as (to mention only the English language) single-authored books (Levy 2007, Racine 2010) and seven multiauthored volumes, including two massive handbooks, three anthologies, and two collections of original essays (Chatterjee and Farah 2013, Clausen and Levy 2015, Farah 2010a, Giordano and Gordijn 2010, Glannon 2007, Illes 2006, Illes and Sahakian 2011). To this abundance corresponds an extremely broad range of inquiry, from the detailed empirical examination of particular cases, contexts, and circumstances to far-reaching ethical and philosophical reflections. We shall not deal in detail with this thriving, expansive, and sophisticated field but only suggest its function among the disciplines of the *neuro*.

To a very large extent, the existence of neuroethics as a research program with normative aspirations has thriven on the expanding application range of functional neuroimaging and therefore depends on the assumption, considered valid, that "imaging the brain provides information about the mind" (Farah 2010b, 4). Neuroethics fully shares such an assumption, which it reinforces even as it critically analyzes some of its instantiations. It may welcome resistance to "explanatory reductionism," but at the same time it does not examine the premise that since "thought and behavior cannot be separated from their neurobiological underpinning," the "empirical correctness of how the brain enables behavior is necessary to make sense of how we think and act" and therefore that people must be understood as behaving the way they do "because of their brains" (Glannon 2011, 191–192).

The usually implicit way neuroethics interprets *separated, make sense,* and *because* (to take the reasoning we just quoted) buttresses the features we identified as being common to the disciplines of the *neuro* and (intentionally or not) turns it into a protective barrier for their funding interests, constitutive beliefs, and methodological options. Whoever recommends, as the neuroethics group set up within the U.S. BRAIN Initiative, "Avoid Hype, Overstatement, and Unfounded Conclusions," should also examine in great detail the claim that neuroscience "presents an unparalleled opportunity to gain a deeper understanding of the human brain and mind, including our cognition, behavior, memory, learning, mood, and social interactions"— which is instead taken for granted (Presidential Commission 2015, 9, 2). Such an "organic" function of neuroethics appears in its most subtle yet conspicuous aspect under the form of "proactive ethics." [4]

For example, a survey among health providers and patients diagnosed with major depressive disorder found "high receptivity to brain scan for treatment tailoring and choice, for improving understanding of and coping with disease, and for mitigating the effects of stigma and self-blame" (Illes et al. 2008). The authors recognize that, as regards depression, there is no "translation" of neuroimaging research to the clinic and observe that "fMRI is still a long way from being used in individuals." They nonetheless justify their work by the "rapid innovation" that, they claim, is taking place along the "trajectory of discovery to implementation." The authors wish to promote "the development of responsible social and public policies in response to new diagnostic and prognostic capabilities for the benefit of patients and their families" and explain that the early identification of future challenges related to the clinical use of imaging technologies "can maximize benefit and prevent false hope, mitigate hype, and curtail their premature use and even misuse in the private sector." However, by riding the "translation" hype (discussed below), unequivocally advertising the assumption that "fMRI promises significant benefit to the diagnostic process for major depression," and confidently announcing the future realization of that promise without hinting at the doubts it has inspired, the study accomplishes just the opposite.

Neuroethics thus further legitimates itself. The consolidation of the discipline involved not only a vigorous research program, a strong media presence, and an extraordinarily rapid institutionalization but also claims to a "neuroethical exceptionalism" that rationalized its being separate from bioethics (see in particular Illes and Racine 2005 and the responses by Buford and Allhoff 2005, Doucet 2005, Schick 2005, Wilfond and Ravitsky 2005). The debates around such claims highlight the extent to which the stakes of neuroethics' disciplinary autonomy are not merely professional or financial but, at least at the level of explicitly given reasons, primarily ontological. Indeed, the autonomous character of neuroethics has been justified by an appeal to neurotechnology's capacity to modify the interaction between personal identity, responsibility, and free will more radically than genetics, and thus to transform our ideas about personhood, as well as by the "hypothesis," which appears both "inevitable" and "omnipresent," according to which "the mind is the brain" (Illes and Racine 2005, 12). More examples of similar assertions could be given.

Neuroethics has succeeded in being recognized as distinct from bioethics by defending its claims to be exceptional on grounds that represent ontological beliefs offered as empirical fact. Indeed, the discipline's special status is said to derive from the "intimate connection" between brain and behavior, the "peculiar relationship between our brains and our selves," and "the intuition that our ever increasing understanding of the brain mechanisms underlying diverse behaviors has unique and potentially dramatic implications for our perspective on ethics and social justice" (Roskies 2002, 21). Neuroethicists claim that, by virtue of those same connections and intuitions, the neurosciences will end up redefining "our sense of selfhood and brain-body relations" (Wolpe 2002, 8), radically transforming ancient philosophical questions, and giving rise to unheard-of moral and legal challenges (Illes and Racine 2005, 6). In spite of considerable criticism (some of it emanating from neuroscientists and neuroethicists) of the hype surrounding the *neuro*, in 2015 the Oxford Centre for Neuroethics still presented itself as follows:

> Neuroscience has made enormous advances in recent years, challenging our traditional understanding of consciousness, responsibility, well-being and morality. Our newfound knowledge of the brain and the mind undermines previous beliefs about a number of areas of private and public life, including addiction and its treatment, criminal responsibility, the treatment of vegetative patients, medical decision making and the enhancement of normal human capacities. It also raises a new question: what are the moral limits of the use of such technology? Neuroethics is a new discipline, addressing these urgent issues.[5]

This paragraph illustrates how the self-promotion strategies that brought about and have successfully maintained neuroethics contribute to create the phenomena the discipline deals with, thus confirming the observation that neuroethics is part of the problem it supposedly seeks to address (Singh and Rose 2006, 100).

This observation, however, could be made differently: The unique ability of neuroethics has been to give itself as one of its fundamental tasks the analysis of the assumptions that legitimate its existence. For instance, an influential study of fMRI "in the public eye" (Racine, Bar-Ilan, and Illes 2005) identified in the media's handling of neuroimaging research three

main features: *neurorealism*, which refers to "how coverage of fMRI investigations can make a phenomenon uncritically real, objective or effective in
the eyes of the public" (160); *neuroessentialism*, or the belief that neuroimaging has direct access to the mind and thus to what we "really" think and
feel; and *neuropolicy*, a term to designate attempts to use fMRI to influence
public policy. The method is flawless and the analysis, insightful, but the
distancing from those positions is superficial.

In short, having instituted their discipline by arguing in favor of neuro-
exceptionalism, neuroethicists established themselves as competent for guid-
ing both the public (including policy makers) and the scientists. To manage
that identity, they must remain close to the sensibilities of both; they must
be farsighted enough to capture the former and prudent enough to be trusted
by the latter. They have succeeded on both counts, feeding public hopes and
fears while adhering to the basic tenets of the *neuro* and supporting its advance. Indeed, neuroethicists have proclaimed that brain imaging will renew
beliefs about human nature and will provide new and scientifically grounded
responses to questions that have been "traditionally" examined by philosophy and the human sciences (Illes, Racine, and Kirschen 2006). In so doing,
they have placed their field at the very heart of the *neuro* enterprise.

To recapitulate, in spite of the diversity of topics they deal with, the disciplines of the *neuro* share a complex of interconnected features:

1. A *postulate*: The mind is what the brain does.
2. A *goal*: To discover neurobiological "foundations" or "substrates."
3. A *tool*: Neuroimaging, especially fMRI.
4. A *product*: Neural correlates.
5. *Mystification* as to the meaning of the correlations and therefore the
   substrates they are said to unveil.
6. A *desire for causality* manifest in the sliding from correlations to
   causes.
7. *Disparity* between the methodology, the stated goals, and the
   interpretation of the results.
8. *Irrelevance* insofar as they cannot answer the question *So what?*
   addressed to their empirical results.

9. A *self-defeating logic* whereby the neurodisciplines miss or annihilate (conceptually) their purported objects of study.

10. *Claims about objectivity and reality,* according to which to demonstrate neurobiological substrates amounts to knowing a phenomenon objectively and makes it more real.

11. An *epistemic hierarchy* in which, since the neurosciences ultimately account for psychological, social, and cultural phenomena, the neurobiological approach ranks above other forms of inquiry and knowledge.

12. A *neglect* of the concepts and productions of the humanities and social sciences.

13. A *redemptory role,* since the neuroX may help save the humanities and social sciences from their theoretical and methodological dead ends.

Such a list will surely strike some readers as an unfair overgeneralization. And indeed, not every item applies equally to every single neuroX production. Moreover, speaking of *a* neurodisciplinary field or *the* neuroscientific turn risks obscuring not only the great diversity of topics but, more importantly, the variety of "goals and dynamics" proposed for articulating the brain sciences and the human sciences, from desirability of mutual influence to complementary interaction to far-reaching integration (Beaulieu 2012, 156). Nevertheless, other than the fact that our list results from examining a vast number of publications across all neurodisciplines, generalizing here is a matter of scale and perspective. In their introduction to *The Neuroscientific Turn,* Melissa Littlefield and Jenell Johnson (2012, 9) note that, beyond the differences among the neurodisciplines that make up the "turn," the *neuro* "signifies a hypothetical *location* (i.e., the nervous system, brain, neuron) where we should look for answers to our deepest questions about consciousness, learning, selfhood, and so forth." That is the core belief that brings all neuroX together and from which the features we have identified ultimately derive. There are in this regard no exceptions.

What Are Humanists Afraid Of?

Does criticism or repudiation of that core belief and its corollaries manifest "fear" of the consequences of placing humans among animals? We don't think so. The claim that "many from the social and human sciences [let us call them 'humanists'] react with horror to this suggestion that our specifically human social capacities have neurobiological bases" (Rose 2013a, 15) may have rhetorical appeal, but it is by and large a misrepresentation. Of the numerous critics of the neuroscientific turn we have read, heard, and personally met, none "from the social and human sciences" have either reacted "with horror" or fear being considered biological creatures. In fact, none to our knowledge deny that humans are products of evolution and that what they are and do has neurobiological bases. Their point (and ours) is rather that neuroimaging methods and the neurobiological level of analysis are not always and ultimately the most appropriate to account for human phenomena. That is why some of the most relevant critics of the neuroscientific turn in the human sciences are brain scientists.

For example, as a basic neuroscientist with experience in the area of molecular biochemistry and biophysics, Susan M. Fitzpatrick is well prepared to examine whether functional neuroimaging methodologies offer "a constructive path for the neuroscientific turn in the humanities and social sciences" and, reciprocally, whether the typical neuroX topics are "actually well suited for study by functional brain imaging" (Fitzpatrick 2012, 180). To both questions, her answer is a clear "no." (After fifteen years as vice president, Fitzpatrick became in January 2015 president of the James S. McDonnell Foundation, and her position has presumably informed the foundation's decision not to fund projects "proposing to use functional imaging to identify the 'neural correlates' of cognitive or behavioral tasks.")[6]

Now, is Fitzpatrick afraid of acknowledging the neurochemical bases of human behavior and therefore of seeing humans as biological creatures? Here too the answer is "no." Her reasons for considering the neuroX misguided are of a different order. First, as she notes, "the findings rarely reveal novel insights about brain structure-function relationship . . . but are usually reinforcing psychological findings" (Fitzpatrick 2012, 182). Indeed, most neuroimaging research (in the neuroX) superficially enhances already established findings without illuminating them further. In the worse cases—and they are numerous—the *neuro* is merely tacked onto the discussion of mental phenomena. It may well be that Bernard of Clairvaux's mystical visions "were likely mediated by the effects of sleep fragmentation on the prefrontal cortex," while Teresa of Avila's "by the inhibition of the thalamus" (Hendrix and May 2012, 116). But such speculations bring no insight into mystical experience, which is the purported object of study.

We recognize our own frustration in the impression Fitzpatrick (2012, 183) gained from reading neurodisciplinary papers, that "we learn very little of substance beyond what was already known from the cognitive psychological or behavioral studies—except for perhaps some converging evidence that behavior is accompanied by brain activity." And even such "evidence" is pointless. Discovering that believers and nonbelievers display different patterns of brain response when they listen to prayers simply confirms that listening to prayers is not the same thing for the two groups. When "beauty" is operationalized as hedonic judgments (where subjects choose between like or dislike), then a correlation between "beauty" and the "activation" of brain areas that govern pleasure and positive reward is only to be expected. Describing the neurobiological correlates of the mental and bodily effects of meditation may have scientific interest, and "mindfulness neuroscience" will perhaps uncover their "underlying mechanisms" (Tang and Posner 2013). But to say that showing how mindfulness-based stress reduction alters gray matter density "demonstrates that changes in brain structure may underlie some of these reported improvements and that people are not just feeling better because they are spending time relaxing" (Press Release 2011) is both trivial and immodest: changes in the brain necessarily and always "underlie" the effects of practices (and that includes relaxing), yet those effects do not become any more real when they are shown to have neural correlates (see Walton 2015 for a self-contented catalog of the ways "meditation actually changes our brain").[7]

Second, the primary problems with fMRI use in the neuroX are not technical and will not be solved with improved technology for the same kind of research. They are conceptual. As Fitzpatrick (2012, 183) emphasizes, "most neuroscientific-turn questions are framed assuming we already understand how neural substrates serve cognitive functions." In our list, such framing shows up as the slippage from correlations to causes and in the vocabulary of neural *substrates* and *underpinnings* that sustains it. That vocabulary, the usual one in neurodisciplinary publications (Schleim and Rosier 2009), gives a wrong idea of what imaging may contribute. Fitzpatrick (2012, 188) remarks that the interpretation of the changes in neuronal activity that neuroimaging brings to light "is meaningless without a theory of the cognitive operations involved in performing the experimental tasks, and a well designed set of tasks to test the theory," and she concludes: "Attempting the 'neuro-turn' via functional neuroimaging is most likely a wrong turn" (194).

In short, it is definitely not because humanists deny the existence of neurobiological bases of behavior or refuse to be considered biological creatures that some of them react critically to the neural turn. Is it then, as Nikolas Rose (2013a, 15) has also claimed, because they "feel that their space is being colonized, their expertise displaced"? Again, no, even though such feelings may arise in some situations. For example, in 2010 "neuro lit crit" was presented as the "next big thing in English," as "the cutting edge of literary studies," and as an approach that might "save humanities" (see here Chapter 4). Perhaps such claims generate among literary scholars a sense of "colonization" and "displacement."

However, even when those terms make sense, the bottom line has less to do with fears of oppression or suppression than with points 11 and 12 of our list, *hierarchy* and *neglect*. As for hierarchy, the only feature that makes neuro lit crit "cutting edge" is the *neuro* label. The alliance with "science" moves literary studies closer to the top of the epistemic hierarchy—as if other methods provided no genuine insights, as if we hadn't realized that reading literature may be worthwhile before brain-scanning technology existed, or as if assessments of cultural value could be brain based. The humanist's discomfort derives from a clash of views about knowledge and method. Not all humanists are doggedly attached to the "ontological primacy" of the sociocultural or build their "neuro-critique" on that presumption (Fitzgerald and Callard 2014, 8–9). Even those who are not willing to give away all reciprocity or to engage in "interdisciplinary collaboration as a practice of subjugation" (Callard and Fitzgerald 2015, 96) recognize that there are differentials of power, and anyone who has admired a teacher knows that these differentials can be productive. Rather, much more basically, the unstylish humanist has concluded, on the basis of solid evidence, that the neurosciences, and neuroimaging experiments in particular, have hardly anything to offer when it comes to understanding the complex phenomena usually associated with cultural and historical research.

The *neuro* stance toward existence and knowledge can be characterized in terms of levels of analysis and interpretation. Studies of the "seductive allure of neuroscience explanations" showed that adding irrelevant neuro-scientific information to an argument makes it more persuasive than it would be without it and that people are more likely to agree with scientific claims when they are supported by a brain image than when they are accompanied by another type of image (Skolnick Weisberg et al. 2008, McCabe and Castel 2008). After having been criticized (Farah and Hook 2013), those results have been corroborated (and the cycle is likely to continue). Superflu-ous neuroscience information was again found to make explanations of psy-chological phenomena more appealing (Fernandez-Duque et al. 2015), and evidence from neuroimaging research turns out to be perceived as more convincing than evidence from behavioral science research, especially among subjects motivated to disbelieve the evidence (Munro and Munro 2014). Perhaps most importantly, as historical and sociological studies dem-onstrate (e.g., Dumit 2004, Hagner 2009, Joyce 2008), the power of images is not purely a matter of individual psychological response but a social phe-nomenon embedded in contexts, ranging from business and technology to academia and epistemology, of which the *neuro* industry is itself a part.

NeuroX practitioners may, like everyone else, fall prey to the "illusion of explanatory depth" (Rozenblit and Keil 2002). This effect is particularly strong in connection with causally complex systems such as technological devices and natural phenomena. Studies of people's posture toward their own explanatory knowledge about an object or phenomenon have found that the best predictor of overconfidence is the ratio of visible to hidden parts—in other words, people are more convinced that they know how something works when they can easily see or visualize its components. This finding is consistent with the persuasive power of neuroimages and the conviction that they demonstrate neuronal activations. Overconfidence also reveals a "level of analysis confusion": functions are taken for mechanisms, and having knowledge at one level of explanation makes us feel that we also have it at another. We thus generate causal claims and get the impression that we grasp the mechanics of the phenomena involved. We are not saying that neuroX practitioners pretend to have full mechanistic knowledge of the processes they study but only that their threshold for acceptance of evidence and causation is such that simply affirming on the basis of correlations that a brain structure is "associated" with a phenomenon or that it is "involved" or "plays a role" in it stands in lieu of analysis or explanation.

It has been argued that functional neuroimages can support causal claims about brain function and are therefore not "merely correlational." Indeed, neuroimaging research relies on the "reasonable expectation that areas activated concurrently with task performance are likely to be causally involved in the task," even if that causal involvement "is no guarantee that the region implements the computation of interest in a given experiment." Moreover, neuroimaging experiments give information "about the influence of behavior on brain activity—which is not identical but unquestionably relevant to influence flowing the other way" (Weber and Thompson-Schill 2010, 2415). Because of the correlational nature of its results, functional imaging cannot by itself establish causal effects of brain activity on behavior. Nevertheless, if a stimulus or experimental situation S consistently elicits a pattern P of brain activation, then S must somehow be the cause of P. This means, as Colin Klein (2010, 275) has argued, that the evidence of neuroimaging does not reside in the images it produces but that its findings may nonetheless "point us to where the evidence for functional hypotheses might be."

We shall return to these matters; here we only emphasize that the neurodisciplines are not alone in their incapacity to attribute causal efficacy to the factors involved in their experiments. The problem is that their language and entire outlook suggest otherwise. By placing brain processes at the beginning of causal chains, they give those processes overall primacy and drive the metamorphosis of explicit correlation into implicit causality, of stated probabilities into implied causes. The hierarchy embodied in those methodological and epistemic moves is unwarranted, regardless of eventually "colonizing" intentions.

The second factor at the root of what has been tendentiously depicted as the humanists' fear of oppression or suppression is encapsulated in point 12 of our list: *neglect*. Again, aside from individual intentions, the objective disregard of research in the humanities and social sciences is a structural effect of the epistemological and ontological positions just sketched. The problem is not "expertise displaced" but expertise neglected—and it is particularly objectionable given the neurodisciplines' constant appeal to interdisciplinarity, collaboration, and "bidirectionality." Of course not every project and publication manifests it in an equally grievous way, but the tendency is clear.

In the life sciences at large, there seem to be fruitful two-way movements. In one direction, a "biohumanities" model proposes to go beyond passive acceptance or mere commentary of the implications of the life sciences and produces research that feeds into those sciences and our understanding of them (Meloni 2013). In the other direction, biology "becomes social" (Meloni 2014). Evolutionary theory and molecular epigenetics blur nature/nurture boundaries and place environments and cultures center stage; the neurosciences no longer see the brain as an isolated data processor but as an organ both largely shaped by the external world and designed to create social relationships. The neuroscientific turn looks like a major protagonist of such integrative motions, since it is largely made up of projects that aim at developing shared conceptual and methodological frameworks and transcending disciplinary boundaries. Yet translating these good intentions into good deeds has proven singularly arduous.

As our cases will illustrate, the main reason for that difficulty lies not (as often claimed) in the fact that the neurodisciplines "are still young" but in their internal logic, which derives from the assumption that all manifestations of the human are best understood as products of the human brain. This assumption precludes any true bidirectionality. The human sciences are crucial insofar as they provide topics and questions: For the areas we shall examine, these might be "What is beauty?" or "What accounts for intercultural differences?" But then these topics and questions are *neuro* operationalized in ways that tend to ignore the scientific production of the source disciplines and to cancel the meanings they have in domains such as aesthetics or anthropology. Instead of mutual exchange, asymmetry and imbalance emerge as the key features of the neurodisciplinary setting. This is embodied not only in how the neuroX handle the concepts and the intellectual output of the human sciences but also in the concrete interactions between humanists and neuroscientists.

Formal testimony is rare, but a study documenting from the inside a transdisciplinary fMRI lie detection experiment brings to light some of the basic psychological and interpersonal features of those interactions. The first element of asymmetry is that the humanists and social scientists quickly ended up thinking "like neuroscientists, rather than like transdisciplinary deployed neuro-collaborators" (Littlefield et al. 2014, 8). This is only natural. In spite of the ideal of bidirectionality and the discourse of transdisciplinarity, the research is a neuroimaging experiment. It is therefore the humanists, not the neuroscientists, who experience "disciplinary double consciousness." Positions and relations within "*neuro*-collaborations" are not symmetric or reciprocal: Even if by definition they take place "between neuroscientists and other scholars from a variety of disciplines," they remain focused "on the central nervous system" (9). Such a focus necessarily structures the interactions and generates a hierarchy where the *neuro* is the dominant culture. This happens even in circumstances, like those of the experiment mentioned here, in which humanist participants felt ambivalence or had a critical attitude toward the very possibility of applying neuroimaging to lie detection (Fitzgerald et al. 2014).

Within neuroscience, the "translational imperative" (Harrington and Hauskeller 2014), which demands that research be applicable in the form of products and therapies, is in practice driven by a "promise of porosity," by the expectation that, one day in the distant future, laboratory work will lead to clinical interventions (Brosnan and Michael 2014); against rampant hype in the area, John Ioannidis (2015, 39) has called for failures and negative results to be recognized "probably as the most useful outcomes that translational research efforts can offer." But nobody pursues that as a goal. *Neuro* collaborations proclaim a commitment to inter- or transdisciplinarity, and when failure arises, they urge more voguish "entanglements." Such commitment, mandatory for a research proposal to have chances of being funded, has (in the neurodisciplines) been by and large intellectually unsuccessful. In practice, it is mainly rhetoric and promises, and in the end, only the *neuro* remains.

A Word on Neuroimaging

Contrary to what any responsible neuroscientist could explain, in 2007 an official document of the American Psychological Association, the world's largest professional society in the field, affirmed that fMRI "isn't quite a mind reader, but it comes close," declared that it "produces movies starring the brain," and added that psychologists and others "aren't using fMRI just to see what lights up in people's brains as they perform different mental tasks" but also "to help answer classic questions within psychology" (APA 2007). This is merely one example in thousands of the hype and the misleading publicity that have sustained the growth in the range of application of fMRI. We do not mean to downplay the extraordinary achievement of having developed such a tool nor ignore its potential or deny the existence of many scientifically significant results. Any honest specialist would nonetheless have to concur with Bruce Rosen, one of the world's leading fMRI experts, that over twenty years after its invention, the technology has not "changed the world" in the ways suggested in the APA document and assumed and proclaimed in many other contexts (Rosen and Savoy 2012).[8]

fMRI has left a modest mark on presurgical planning, treatment evaluation, and clinical assessment; it has been widely adopted as a tool for research into mental illness but plays no diagnostic role and has not influenced psychiatric care. Although it has transformed cognitive neuroscience, where it is the most frequently used technology, the gargantuan increase in the quantity of neuroimaging papers is not universally seen as having been accompanied by a proportional contribution to the understanding of the mind; while, as we shall see, it has been extensively applied to issues related to personhood, society, and culture, its most significant impact in that connection has been to drive the emergence and justify the existence of the neurodisciplines we shall discuss here. Its major role and effects have been cultural, sociological, and economic. Neuroimages have become contemporary fetishes whose power derives from the belief that fMRI offers "windows into the mind" (the metaphor is widespread). The technology has nourished the growth of various commercial and academic industries; it has dramatically contributed to "neurologize" research into the human (and the criteria by which it is evaluated) and has been occasionally misused for worthy purposes (as when it was wielded as evidence that mental illness is organic, decisively advancing the adoption of the U.S. Mental Health Parity Act of 2008).

As far as the neurodisciplines are concerned, much more than a tool, fMRI has been a condition of existence. The main empirical form of the neuroX consists of applications of fMRI to a range of humanistic and social science issues, and insofar as the availability of this form of neuroimaging has been the engine for the rise and growth of the neurodisciplines, the neuroscientific turn has been technology driven. Nevertheless, the inherent flimsiness of the neuroX does not derive from their dependence on a particular technology but from their premises, regardless of the particular imaging methods they might favor at any point. This distinction is not always clear.

One can take measures to face the "crisis of confidence" in the replicability and reliability of published neuroscientific findings, try to diminish the impact of excessively small samples and the mind-boggling proportion of false-positive findings (Eklund, Nichols, and Knutsson 2016; Miller 2016), and in general call for vigilance with respect to the uses of neuroimaging and the interpretation of its results (Boekel et al. 2015; Boekel, Forstmann, and Wagenmakers 2016; Button et al. 2013; Rachul and Zarzeczny 2012; Whelan and Garavan 2014).[9] Reality checks, however, are not incompatible with adhering to the neuroX project and trying to solve "from the inside" problems such as the abundance of false positives, the cover-up of negative results, or many investigations' lack of validity and statistical power (see for example the "nine ideas for a better neuroscience" by Bareither, Hasler, and Strasser 2015).

These shortcomings can be overcome, but it is not because of them that the neurodisciplines are, as we said above, self-defeating. Nikos Logothetis (2008, 876–877), a specialist in the neural mechanisms of perception, director at the Max Planck Institute for Biological Cybernetics, and one of the most authoritative neuroimaging experts worldwide, has pointed out that the limitations of fMRI "are not related to physics or poor engineering, and are unlikely to be resolved by increasing the sophistication and power of the scanners; they are instead due to the circuitry and functional organization of the brain, as well as to inappropriate experimental protocols that ignore this organization." In some areas, fMRI can be certainly used "for gaining insights into brain function and formulating interesting and eventually testable hypotheses" (877). However, as Logothetis's observations about brain organization and experimental design make clear, it cannot be an adequate tool for investigating the phenomena the neurodisciplines claim to study.

While the availability of functional neuroimaging has driven the neuroX, their existence also depends on the principle (mentioned above in connection with neuroethics) that "imaging the brain provides information about the mind" (Farah 2010b, 4). Historically, such a principle has been often associated with the belief that brain images allow some form of "mind reading." The boom of optical instrumentation that began in the mid–nineteenth century gave plausibility to brain/mind-reading fictions. With the development of electroencephalography in the early 1930s, some researchers became convinced that the recorded waves would offer direct insights into mental life (Borck 2005). Five decades later, after a period of relative "iconophobia" and the spread of brain-as-computer models, the rise of digital brain imaging technologies has given mind-reading hopes a new lease of life (Hagner 2009).

As far as fMRI is concerned, the assertion that "imaging the brain provides information about the mind" implies that correlations between brain activity and the realization of a particular task during scanning capture a relationship between mental states and brain states that somehow illuminates the former. How does this work? During the control and experimental conditions, fMRI registers a hemodynamic signal known as BOLD ("blood-oxygen level-dependent"), whose intensity depends on local glucose metabolism. The precise relationship between neuronal activity and the BOLD signal is still under investigation, but it is widely assumed that the signals correlate with neuronal or local synaptic activity. It is in any case a surrogate that does not straightforwardly "reflect" that activity, and it is not clear "how it is coupled to the underlying neurophysiology, and how this coupling varies across the brain, across tasks and across individuals" (Singh 2012, 1121). The main output of neuroX studies consists of BOLD signals that correlate at a statistical level judged significant with responses to the tasks subjects carry out inside a scanner.

Through complex statistical procedures, the signals are ultimately transmuted into the familiar images, displaying the brain areas colloquially said to become "active" when subjects perform the experimental task. Even when they look like realistic reproductions of an entire brain (a gross external view, either flat or three-dimensional) or of traditional brain sections (sagittal, coronal, or transverse), fMRI images neither apply optical principles nor reproduce anything: they are "belief-opaque in a way photography is not" (Roskies 2007, 871). Their appearance results from decisions about how to generate, process, and represent numerical data recorded from a functioning

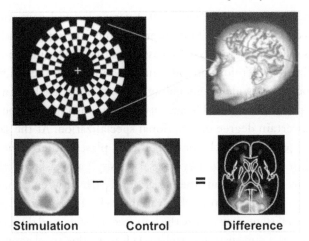

Stimulation   **Control**   **Difference**

FIGURE 2.1 Illustration of subtractive technique. Data collected during a control condition are subtracted from data collected during performance of task of interest. Fig. 1 from Roskies (2010). Courtesy Adina Roskies and Marcus E. Raichle.

brain; this data could be given the form of graphs or curves rather than vibrantly colored brains.

The areas designated as being active are obtained by "subtraction" (Figure 2.1). Schematically: BOLD signals generated during task T1 (the control condition) indicate activity in brain areas B1 and B2; these are assumed to be "associated with" the neurobiological processes that "underlie" T1. BOLD signals generated by performing experimental task T2 demonstrate increased or decreased activity in B1 and B2 and/or "activation" of new areas. The experimental result emerges from subtracting the data of T2 and T1 (or "contrasting" the conditions). Even though the otherwise or newly "activated" areas thus identified represent correlations, they are said to "underlie" or "underpin" the functions involved in the task, in the sense of constituting their anatomical and physiological "substrates" or "bases" (Schleim and Rosier 2009).

Although by the early twenty-first century the range of subtractive techniques had become broader than this "classic" paradigm, they involve the same principles and assumptions and give rise to similar criticisms (Roskies 2010, 638). Debates around these have been highly technical and theoretical (see for example Coltheart 2006; Hardcastle and Stewart 2002; Roskies

2009, 2010; Van Orden and Papp 1997). A particularly relevant point for our discussion concerns the possibility of attributing function or significance to the results. Neuroimaging has often assumed that the brain is modular and that localization therefore provides information about mental functions. Various sources of evidence, including neuroimaging itself when it reveals replicable patterns and reliable activations, demonstrate that the brain indeed has a certain degree of functional specialization. At the same time, while not equipotential, the brain is highly plastic, interconnected, and integrated. It is characterized by "causal density": any task is likely to have effects across the brain, and "there is a causal path between changes in any explanatory variable and any other variable" (Klein 2010, 269). Moreover, different neural states may realize the same mental state, and different mental states possess the same neural correlates (a feature akin to what philosophers call "multiple realizability"; Bickle 2013). And given that B1 and B2 are also active in T2, neuroimaging cannot discriminate the function of the various activations nor provide the basis for interpreting modified or new activation beyond noting an "association" between them and the experimental task. That is why it cannot answer the *So what?* question.

In defense of the subtraction method, it has been judiciously noted that neuroimaging results, hardly meaningful in themselves, are not or should not be considered piecemeal and in isolation (Roskies 2010, Rugg and Thompson-Schill 2013). Rather, they need to be interpreted by means of "triangulation," with reference to other results of the same subjects, other neuroimaging experiments, and information of different sorts, such as that obtained (for example in psychology or neurophysiology) via other methods. A multimodal approach is indispensable. It has so far played a negligible role in the neurodisciplines but should in principle be beneficial. However, would going beyond the verbs ("underpin" and so forth) commonly used to describe the role of structures and circuits with regard to the function or task of interest give the neurodisciplines the power to say something relevant about their topics?

Results obtained until now suggest that it is unlikely and that research will reinforce the problem of multiple realizability and deepen the challenge of interpreting neural correlates. For example, each person has a "connectivity profile" that distinguishes him or her "regardless of how the brain is engaged during imaging" (Finn et al. 2015). Similarly, contrary to legend, while there are sex differences in the brain, largely attributable to hormone exposure during perinatal life (McCarthy 2015), brains don't come in male and female forms; rather, most brains are unique "mosaics" of features, "some more common in females compared with males, some more common in males compared with females, and some common in both females and males" (Joel et al. 2015). Thus, while "your brain scans are quintessentially you" (Finn 2015), you are not quintessentially your brain scan.

Discussions about the uses and abuses of neuroimaging, especially in neurodisciplinary contexts, have not abated. In 2014, the Hastings Center, a major bioethics institution, devoted its *Report* to assessing functional neuroimaging. In it, Martha Farah, a cognitive neuroscientist and a leading figure of neuroethics, systematically probed the critiques and concluded that while each had a "kernel of truth," each could also be rebutted. Yes, Farah noticed, a BOLD signal is not a direct measurement of brain activity, and we do not know to which aspects of neural activity it corresponds. But the relationship between the two is strong enough to make fMRI a useful tool. Images, like graphs or maps, are indeed manufactured, but they are not fabricated. Yes, fMRI is about localization, but it is not for localization's sake, and most neuroimaging is not motivated by it. (That may be an accurate generalization across its many fields of application, but the neurodisciplines seem to be an exception.) Subtraction? Yes, neuroimaging initially assumed context-independent modules, such that "a cognitive process A will have the same neural instantiation whether it is accompanied by cognitive process B, C, D, or E" (Farah 2014, S23). But other approaches, such as the analysis of functional connectivity, should allow fMRI to transcend such limitation.

Neuroimaging research has been criticized for its use of "reverse inference," that is to say for attempts to infer the presence of specific mental processes from detected "activations" (Poldrack 2008): If there is activation in area B1, then mental process M1 must be taking place. If there were a one-to-one correspondence between brain regions B and such processes M, one could infer which Ms are taking place by identifying active Bs. However, since (as mentioned above) psychological processes involve several brain regions, and since a single brain region is usually involved in multiple processes, such inference is hardly possible. Farah (2014, S24) argued that, with appropriate precautions and information derived from manipulating processes and observing brain activation, reverse inference may be legitimate. If performance of a task T involving mental process M consistently "activates" brain region B, then activation of B may indicate that M is taking place. Although such inferences are not the goal of neurodisciplinary research, they may serve, as Poldrack (2008, 224) pointed out, to "drive subsequent behavioral or neuroimaging studies, rather than as a direct means to interpreting neuroimaging results."

Farah is right to emphasize that (as also mentioned above) many of the problems with fMRI (reverse inference, highly artificial experimental environments, small samples, low reliability, weak validity, and inflated statistical significance) are shared by other areas of scientific inquiry and that there is a difference between specifically criticizing particular applications or studies and "wholesale criticism" of neuroimaging. The latter concerns the method itself and casts doubt "on the conclusions of any research carried out with imaging, no matter how well designed and carefully executed" (Farah 2014, S28). Such an indiscriminate position would certainly be unfair, and there is no doubt that, since the neuroX wave began in the 1990s, enormous progress has been done in meeting methodological and technical challenges. Yet the congenital shortcoming of the neuroX does not reside in the quality of the design and execution of the experiments but in their adequacy to the objects they claim to be studying. We illustrate this below.

It is revealing of the self-assurance that prevails in the neurodisciplines that such a fundamental issue goes unexamined. In an article of 1999 entitled "If Neuroimaging Is the Answer, What Is the Question?" the cognitive psychologist Stephen Kosslyn doubted that mental processes can ever be better understood by observing which neural sites are "activated" when subjects perform a task. Rather, he argued, one should start with questions that inspire experimental tasks in ways that take advantage of the strengths of neuroimaging techniques. Although Kosslyn (1999) was reflecting on work done mainly since the early 1990s, his question remains as topical as ever. Over fifteen years later, and after considerable hype and groping, the neurodisciplines show evident signs of methodological and theoretical refinement. At the same time, we must still ask whether the assumptions and approaches of the neuroX are suitable for the goals, questions, and objects they define for themselves. The answer, as our case studies suggest, is that they are not.

The Neurodisciplines of Culture

Most neurodisciplines aim at capturing the commonality that underlies the heterogeneity of behaviors and experience—in other words, universal neurobiological processes, which are "modulated" by contextual factors. In contrast, the neurodisciplines of culture, such as neuroanthropology and cultural neuroscience, focus less on commonality than on difference, on that which gives cultures their specificity and on how culture is "inscribed" in the brain. Like all disciplines of the *neuro*, they attempt to draw their explanations from knowledge about the brain. But they are particularly careful to emphasize the brain's "enculturation" and the *interactions* of culture and brain. They thus offer an opportunity to examine how the notion of culture operates within a framework built to study transcultural neurobiological processes.

The editors of *The Encultured Brain*, a book that presents itself as "an introduction to neuroanthropology," state that the discipline's project is "to examine different neural systems empirically, understand how neural capacities develop, and document which biological and environmental factors shape their realization" (Downey and Lende 2012, 24). Such a project has been considered as part of an "exciting shift" toward a more "integrative" biological anthropology, insofar as it demonstrates "that anthropology can provide for neuroscience contextual examples of how enculturation can help explain differences in brain functioning, while neuroscience offers anthropology direct evidence of neuroplasticity's role in social and cultural dynamics" (MacKinnon 2014, 357). *The Encultured Brain* claims to break with earlier notions of culture:

> For a long time, anthropologists have focused on culture as a system of symbolic associations, public signs, or shared meaning. But from the perspective of the nervous system, patterns of variation among different groups also include significant unconscious, non-symbolic traits, such as patterns of behavior, automatized response, skills, and perceptual biases. This neuroanthropological framing opens more space for considering why all types of cognition may not operate in identical fashion, and how non-cognitive forms of neural enculturation may influence thought and action. (Downey and Lende 2012, 37)

In other words, culture is not only about shared representations but also about "shared conditionings of the nervous system." This may seem self-evident,

since it is unlikely that there can be shared patterns of behavior, either symbolic or automatic, in the absence of some shared brain processes. Yet for
the authors we just quoted, it is the "implications" of that principle that
appear "obvious."

For them, "the predominant reasons that culture becomes embodied . . .
is that neuroanatomy inherently makes experience material" (37). The trivial
observation that "without material change in the brain, learning, memory,
maturation, and even trauma could not happen" leads them to the apparently significant statement that "cultural concepts and meanings become
neurological anatomy" (37). All of this, as the authors say, is obvious. The
questions are whether, or in what sense, examining changes in the brain significantly adds to an understanding of culture beyond reiterating that neurobiological processes are involved, and how the notion of *culture* operates
within the conceptual and methodological framework of the *neuro*.

These issues can be explored in at least two ways. On the one hand, with
respect to research itself, we may ask: How do the neurodisciplines of culture translate their emphasis on brain-culture "bidirectionality" into concrete investigative strategies, and what are their empirical results? On the
other hand, these disciplines can be examined in regard to their implicit values and epistemologies. In spite of their insistence on the two-way processes
that embody the brain in culture and culture in the brain, they usually assert the ontological primacy of the brain and see the human groups that
constitute cultures as a "community of brains" (Domínguez Duque 2015,
292). Such a premise turns culture, however defined, into an external factor
that "shapes," "influences," and "impacts on" neural activity, function, and
processes. The spontaneous use of such action verbs is emblematic of how
the neurodisciplines of culture approach their object (Gutchess and Goh
2013; other references are given below). What is the consequence for those
disciplines and for the very concept of culture?

Neurologizing Culture

Like other neurodisciplines, those that concern culture passed in a few years
from being an informal group of scholars with common interests to having
their own name and Wikipedia article, professional institutions, journals,

societies, colloquia, educational events, blogs and websites, programs, and graduate students. Special issues of journals not specifically devoted to them highlight synergies between those new disciplines and more established fields. Thus, cultural neuroscience has been the subject of special issues of *Psychological Inquiry* (2013), *Social Cognitive and Affective Neuroscience* (2010), the *Asian Journal of Social Psychology* (2010), and *Progress in Brain Research* (2009). The *Handbook of Social Neuroscience* offers an overview (Chiao 2011), and the collective volume *Cultural and Neural Frames of Cognition and Communication* (Han and Pöppel 2011) includes several contributions from the discipline. As for neuroanthropology, calls for such an endeavor first emerged in the late 1970s; the word was used in the early 1990s and had entered reference works of anthropology by the middle of the decade (Marcus 1997, Downey 2012a). In 2012, the same year *The Encultured Brain* appeared, the journals *Anthropological Theory* and *Annals of Anthropological Practice* devoted special issues to neuroanthropology.

In turn, the term "cultural neuroscience" seems to have first appeared in print in 2007, in a chapter for the *Handbook of Cultural Psychology*. It was then defined as "an area of research that investigates cultural variation in psychological, neural, and genomic processes as a means of articulating the interrelationship of these processes and their emergent properties" (Chiao and Ambady 2007, 238). Cultural neuroscientists of course acknowledge that social factors offer more than "minimal interest" for understanding brain and behavioral processes (Zhou and Cacioppo 2010). At the same time, considering that the sociocultural level of analysis is by itself insufficient, they highlight the *interdisciplinarity* of their endeavor and the *bidirectionality* of the processes they investigate and speak of "biocultural co-constructivism" and "multiple" or "reciprocal determinism" (Zhou and Cacioppo 2010). Cultural neuroscientists hold that values, practices, and beliefs both "shape and are shaped by the mind, brain and genes" and that the study of "cultural variation in mental, neural and genomic processes" therefore constitutes a means of "articulating the bidirectional relationship of these processes and their emergent properties" (Chiao and Cheon 2012, 288; Chiao et al. 2013; Kim and Sasaki 2014).

While the notion that complex behavior "results from the dynamic interaction of genes and cultural environment" is not new, cultural neuroscience is supposed to represent "a novel empirical approach to demonstrating bidirectional interactions between culture and biology by integrating theory and methods from cultural psychology, neuroscience and neurogenetics" (Chiao and Cheon 2012, 289). It seeks "to explain a given mental phenomenon in terms of a synergistic product of mental, neural and genetic events" (289) and claims to have "potential implications" not only for psychiatry, business, and technology but also for global public policy issues in health, globalization, immigration, and interethnic ideology (Chiao 2009b, Denkhaus and Bös 2012). At the research level, cultural neuroscientists are motivated by two "still unanswered" questions: How do cultural traits "shape" neurobiology and behavior? And how do neurobiological mechanisms "facilitate the emergence and transmission of cultural traits"? (Chiao et al. 2010, 356).

Neither neuroanthropology nor cultural neuroscience represents the first attempt at broaching culture with neuroscientific tools. Since the early 1990s, cognitive neuroscience has incorporated the study of interpersonal and social behavior, and "social neuroscience" emerged toward the end of that decade (see Cacioppo and Berntson 1992 for an early use of the term); by 2005, a handbook of "key readings" had already been published (Cacioppo and Berntson 2005). The field derives from findings in cross-cultural psychology that show how social cognition and behavior depend on sociocultural context, and it combines neuroimaging, cognitive science, and social psychology to investigate the neural "representation" of social interaction and the neural "substrates" of social processes (Han and Northoff 2008, Zhou and Cacioppo 2010). The journals *Social Neuroscience* and *Social Cognitive and Affective Neuroscience* were launched in 2006, a Social and Affective Neuroscience Society "committed to research investigating the neural basis of social and affective processes" was established in 2008, and it was followed in 2010 by a Society for Social Neuroscience, intended to support "the interdisciplinary academic field devoted to understanding how biological systems implement social processes and behavior, and how these social structures and processes impact the brain and biology."[10] The journal *Culture and Brain* was founded in 2013, with a focus on "cultural differences in neural activity" and "the mutual constitution of culture and the brain" (Han 2013).

The social, affective, and cultural neurosciences largely overlap with one another as well as with neuroanthropology and transcultural neuroimaging (Domínguez Duque et al. 2009, 2010; Han and Northoff 2008; Lende and Downey 2012a); labels such as "sociocultural neuroscience" are forged to underline interconnections (Wajman et al. 2015). At the same time, these emerging disciplines are engaged in dynamics of differentiation. In particular, neuroanthropologists have emphasized the differences between their approach and that of cultural neuroscience (Domínguez Duque 2012, Lende and Downey 2012a). While, in their view, cultural neuroscience wishes above all to offer brain-level explanations, neuroanthropology aims to combine such explanations with an ethnographic perspective. It is thus "in a better position to move back and forth between the neural, the phenomenal and the cultural domains" (Domínguez Duque 2012, 22) and to test neuroscientific hypotheses "against the reality of what people actually do, say, and experience" (Downey and Lende 2012, 42). Ethnography, in short, should be able to provide "empirical access" to the ways social and cultural processes shape brain function, meaning and behavior (42). Some neuroanthropologists have expressed concern about cultural biases in research and called for an increased awareness of the historical, social, and political circumstances under which experiments are conducted (Domínguez Duque et al. 2010); others see the interface between anthropology and the neurosciences as a way of doing anthropology experimentally (Roepstorff and Frith 2012).

In short, there is a cluster of *neuro* disciplines aimed at understanding how the brain "mediates" social interactions and culture and produces emotion and cognition. The question is how and whether these questions and programmatic statements translate into research results capable of going beyond generalities such as "cultural practices adapt to neural constraints, and the brain adapts to cultural practice" (Ambady and Bharucha 2009, 342), which simply reiterate the field's premise.

Causes, Correlations, Plasticity

Insofar as neuroimaging is supposed to show "how 'deep' culture can go into the human brain" (Kitayama and Park 2010, 124), it has been the method of choice for directly studying the "encultured" brain. However, to the extent that neuroanthropology draws its main concepts and questions from cultural anthropology, it emphasizes fieldwork as its empirical basis and is as a consequence not inclined to use neuroimaging, which requires an experimental setting. That is why most neuroanthropological studies limit themselves to citing brain research and juxtaposing it to other kinds of materials, drawn directly from the study of cultural settings and situations, under the assumption that such a juxtaposition demonstrates the impact of those situations on the brain or the "interplay" of culture, brains, and experience (see for example *The Encultured Brain* [Lende and Downey 2012a] or the special issue "Neuroanthropology and Its Applications," *Annals of Anthropological Practice* 36, 2012). In short, "neuroanthropology" is mainly the name of a potential framework with seemingly little impact on concrete anthropological work. A study on the anthropology of opioid maintenance treatments for addiction may redescribe itself as "neuroanthropology" and its topic as "the neuroeconomics and neuroracial politics of opioid pharmaceuticals" (Hansen and Skinner 2012). Similarly, it is simply by attaching the label "neurocognitive" to the skills involved that ethnographies of rugby or capoeira become instances of neuroanthropology (Downey 2012b, 2012c). The renaming operation is purely cosmetic and constitutes good marketing, but it brings no methodological, empirical, or conceptual gains.

In contrast to neuroanthropology, cultural neuroscience uses neuroimaging so systematically that it is often described as "cultural neuroimaging." This is not to say that neuroanthropology would benefit from turning to neuroimaging but that imaging methods have so far been the chief way of going empirically beyond merely juxtaposing the neurobiological and the cultural. The question is whether they satisfy the stated purpose of illuminating culture.

The difference between neuroanthropology and cultural neuroscience with regard to neuroimaging is consistent with their conceptual and disciplinary roots in cultural anthropology and cultural psychology respectively. Cultural psychology is indeed cultural neuroscience's "parent discipline" (Denkhaus and Bös 2012)—but in a manner that involves little more than replacing the "mind" of the *psy* by the "brain" of the *neuro*. Indeed, the University of Chicago anthropologist Richard Shweder (1991, 72) defined cultural psychology as the study of "the way cultural traditions and social practices regulate, express, and transform the human psyche, resulting less in psychic unity for humankind than in ethnic divergences in mind, self, and emotion." If we here substitute *human brain* for *human psyche* and *neural unity* for *psychic unity* and then add "brain" to the loci of ethnic divergence, we obtain an accurate depiction of cultural neuroscience.

This field assumes that "understanding cultural and genetic *influences* on brain function likely holds the key to articulating better psychological theory" (Chiao 2009b, 290). The quest for "influences" is reinforced by the premise that "human behavior *results* from neural activity" and by the further inference that behavioral variation among cultures "likely *emerges* from cultural variation in neural mechanisms underlying these behaviors" (Chiao 2009b, 290, our emphasis; see also Chiao and Cheon 2012, 289). Though here left vague through the use of "likely," the reasoning presupposes a direction and hierarchy of causes, from genetics and the brain toward mind and culture. Neuroimaging and genomic methods for "mapping" neural processes and genes "to" neural, mental, and cultural processes produce correlations, but these are presented in a causal perspective reinforced by the belief that cultural traits constitute evolutionary adaptations (Chiao and Blizinsky 2010).

The tension between correlational results and causal claims, as well as the existence of an implicit epistemic hierarchy, undermine cultural neuroscience's calls for synergy and bidirectionality. We later examine the relevant research, but let us first take as an example the assertion that cultural values, practices, and beliefs "impact human behavior" or that the "cultural dimension" of *individualism-collectivism* (a favorite of cultural psychology) "affect[s] a wide variety of human mental processes at a behavioral level" and "modulate[s] neural and electrophysiological responses" (Chiao 2009b, 291, 295). Such statements embody a circular reasoning. On the one hand, a cultural "dimension" includes by definition mental and behavioral processes, and these necessarily correlate with some feature of brain functioning. On the other hand, the cultural dimension is itself defined, at least in part, on the basis of the mental and behavioral processes it is supposed to "affect."

Culture "influences" brain functioning, "modulates" neural mechanisms, "shapes" neural systems (291). Thus, people living in culture X may develop "distinct neural mechanisms." Yet these mechanisms may "underlie" behaviors identical to those observable in culture Y, where they correlate with other neural processes (290). Cultural neuroscience has explored these effects in connection with emotion (providing "evidence that culture influences how people infer emotional states"; 296), interpersonal perception (showing that individuals from egalitarian vs. hierarchical cultures display greater mesolimbic activity to dominant vs. facial cues; Freeman et al. 2009), and social cognition (demonstrating that cultural values rather than ethnic affiliation "modulate neural response during self-evaluation"; Chiao 2009b, 297). The discipline studies a wide range of psychological process, from visual and semantic processing (Goh et al. 2010, Gutchess et al. 2010) to fear (Chiao et al. 2008), empathy (Cheon et al. 2011), and self-representation (Kitayama and Park 2010; Mrazek, Harada, and Chiao 2014). The basic claim is always the same: culture shapes activity in some part of the brain, which in turn guides behavior.

Cultural neuroscientists consider their findings supported by the existence of neuroplasticity (the brain's ability to change as a result of experience) and its main theoretical consequence: to challenge the belief that brain functions have fixed localizations and that the brain is malleable only within strictly limited critical periods. Celebrated, as we saw in Chapter 1, as a revolutionary finding and immediately taken up by a large spectrum of interested individuals from brain fitness dealers to philosophers and psychiatrists, political scientists, and rehabilitation specialists, neuroplasticity also seems to confirm that cultural differences at the neural level reside in patterns of connectivity. Sustained engagement in cultural tasks, understood as repeated participation in routinized behaviors, results in different patterns of brain activation and functional and structural modifications (see Hanawaka et al. 2003, for Japanese abacus experts, or Maguire et al. 2000, for London taxi drivers). Brain plasticity can therefore be depicted as the feature that enables the interaction of brain and culture at the three interrelated levels of explicit values, conventions, and routines; socially shared scripts for action; and individual idiosyncrasy. It thus explains intercultural neural differences as a consequence of practice and experience.

Investigative Practices

Cultural neuroscience has followed two strategies. One, "culture mapping," involves "determining which cognitive or neural processes vary across cultures without determining whether the differences are learned or innate"

(Ambady and Bharucha 2009, 342). For example, while performing numerical tasks, native English speakers showed more activation in brain areas "associated" with language processing; native Chinese speakers showed more activation in an area "associated" with visual-spatial processing (Tang et al. 2006). The finding is hypothetically attributed to exposure to different visual patterns. Greater premotor activity in the Chinese "could be due" to the visuospatial nature of their language, whereas activation of language areas in English speakers suggests that "the retrieval of mathematical facts may be mediated by phonological processing" (Ambady and Bharucha 2009, 342–343). The second strategy, "source analysis," attempts to determine "the source or causes" of cultural mappings, including genetic commonality or difference, cultural learning "mediated by brain plasticity," and the degree of similarity between cultural environments. This strategy has been less pursued than mapping, "but new technologies promise to advance" it "quickly" (343, 344); by 2016, it seems to have remained programmatic.

In turn, neuroanthropologists see themselves as uniquely situated to explore brain-culture bidirectionality, and (as we mentioned) they take a critical stance vis-à-vis cultural neuroscience. Yet they too focus on how culture "influences" or "changes" brain function and structure and how brain areas "respond to regularities in the cultural stream of experience" (Domínguez Duque et al. 2009, 43). They too hail as "extraordinary" the fact that culture "affects" not only brain function but also brain structure (60; see also Domínguez Duque 2012, 22). Indeed, as neuroanthropologists explain, the prefrontal cortex "stands first to be modified or constituted by cultural experience as it is the structure that *lays* culture's foundations" (Domínguez Duque et al. 2009, 60, 61, our emphasis). The notion that the prefrontal cortex is *constituted by* culture while also being what ultimately *generates* it goes beyond describing the reciprocal interaction, which obtains at all levels, of body and world. It highlights the foundational asymmetry of the neurodisciplines of culture. The assertion that culture, as a complex of activities including forms of learning, "modifies" the brain is certainly substantiated by empirical observation. In contrast, except in its most diluted interpretation, the claim that the prefrontal cortex "lays" the foundations of culture formulates an ontological assumption. And this assumption translates into the way research is performed.

Let us take a frequently quoted article in the field, published in 2009 in *Human Brain Mapping* and entitled "Neural Basis of Individualistic and Collectivistic Views of the Self." Its goal was to understand how individualism and collectivism "modulate neural representations underlying social cognition" (Chiao et al. 2009, 2813). According to earlier studies, people who support individualistic values think of themselves and others as independent and as having stable personal traits, whereas those who endorse collectivistic ideals see people as interconnected and describe themselves as immersed in a social context. The authors drew on the notion of self-construal style (SCS), which has been used to differentiate Western and East Asian views of the self; they do not refer to research questioning the capacity of self-construal to reflect individual-level cultural orientation or mediate and explain cross-cultural differences (Levine et al. 2003).

Based on previous work suggesting that activity in the medial prefrontal cortex (MPFC) "reflect[s] the neural basis of self-knowledge" (Chiao et al. 2009, 2814; Kelley et al. 2002), the authors hypothesized that individualists would show a greater response for general self-descriptions, and collectivists, for contextual self-descriptions, in the anterior rostral portion of the MPFC. Twenty-four right-handed university students were recruited for the study, twelve native Japanese from Nagoya and twelve "Caucasian-Americans" from Chicago. They were shown seventy-two stimuli, in Japanese or English respectively: twenty-four general self-descriptions, twenty-four contextual self-descriptions, and twenty-four self-descriptions in italicized or roman font. In comparison with collectivists, individualists exhibited during self-judgments greater activation in the bilateral thalamus, right putamen, bilateral cuneus, right insula, bilateral cerebellum, and right superior frontal gyrus; in turn, collectivists showed during the same task greater activation in the left middle temporal gyrus. Self-construal style, as established after scanning, interacted in a statistically significant manner with "type of self judgment in neural response within the anterior rostral region of MPFC," the bilateral parahippocampal gyri, right middle temporal gyrus, and left superior occipital gyrus (Chiao et al. 2009, 2817).

The results seemed to demonstrate that "self-relevant processing within MPFC varies as a function of [self-construal style]." People who endorse individualistic values show greater MPFC activation during general self-descriptions, while those who endorse collectivist values display greater

MPFC activation during contextual self-descriptions. In both cases, in-creased MPFC activity "reflects the role SCS plays in how knowledge about the self is formed, and possibly also stored and retrieved." The researchers concluded that "knowledge self-representations of one's self . . . are cultur-ally specific at the neural level." They also speculated that greater activity within the right superior frontal gyrus may "reflect evidence of enhanced self-relevant processing in individualists relative to collectivists" and called for more research to elucidate how cultural values "affect" neural processing (Chiao et al. 2009, 2819). A meta-analysis of research in the area published between 2003 and 2014 confirmed that "East Asian cultures are associated with increased neural activity in the brain regions related to inference of others' mind and emotion regulation whereas Western cultures are associ-ated with enhanced neural activity in the brain areas related to self-relevance encoding and emotional responses during social cognitive/affective pro-cesses" (Han and Ma 2014, 293).

Such studies of the neural "bases" of individualism and collectivism are typical of the neurodisciplines in at least two ways. First, they illustrate a characteristic slippage between the establishment of statistical correla-tions (here, with culture as predictor) and the identification of anatomo-functional "bases" or "underpinnings." Second, the outcomes that could matter are predictable without neuroscience or neuroimaging. The authors of "Neural Basis of Individualistic and Collectivistic Views of the Self" point to "an intriguing aspect" of their findings, namely that participants' cultural values (individualism or collectivism) rather than cultural affilia-tion (being white American or native Japanese) "modulated" neural re-sponse during self-judgments (Chiao et al. 2009, 2819). But in the Western and East Asian contexts from which the study drew its subjects, people ad-just to various environmental demands, so that culture, as defined by ethnic or national affiliation, cannot be expected always to match individual be-havior. Its findings are therefore far from "intriguing." The main thing a study such as the one we just summarized does is to convey the assumption that culture is based on the brain and the belief that a phenomenon becomes more real or objective by virtue of having a neural correlate. Unless these assumptions are made, there is no need for neuroscience to apprehend the "dynamic nature of cultural values across individuals and cultural groups" (2819).

Cultural neuroscientists may retort that they have not simply corroborated results from the social sciences but added something essential by showing "how such dynamic cultural values shape neural representations" (2819). However, in the same way that they cannot demonstrate the neural "bases" of culturally contingent values or attitudes, they cannot show how *particular* values or attitudes shape the brain. To be sure, "cultural values, beliefs, and practices must be important for social brain functioning" (2819). This, however, is so by definition. First, because anything brained organisms do is related to brain function. Second, because given that "social brain" refers to the brain regions involved in understanding others (Blakemore 2008) and that social cognition is, in humans at least, inseparable from culturally determined ways of interacting with others, culture is necessarily "important" for the social brain. The contrary would in both cases constitute a sensational finding, if not a *contradictio in adjecto*.

Cultural Diversity as "Neurodiversity"

On the one hand, with respect to their significance for understanding culture, neuroimaging experiments recover at the end what they put in at the beginning, namely the notion that culture has "neural bases." On the other hand, the rhetoric of wonder, in which findings are always "intriguing" or "extraordinary," betrays the persistence of a dualistic attitude. Celebrating the discovery that "culture" somehow "modifies" brain function implies imagining at least two dualities: brain and person, culture and individual. Yet, as has been pointed out from inside the discipline, "it should not be surprising per se that there exists a neural difference underlying a psychological difference"—in fact, the existence of such a difference is "an axiomatic assumption" of cultural neuroscience, not an "empirical question" (Freeman 2013, 26).

The cultural neuroscientists whose study we just sketched reported on the "influence of cultural values on neural responses within MPFC during self judgments, despite the absence of differences at the behavioral level," and concluded that their results "reveal an advantage of examining cultural values such as SCS at the neural level" (Chiao et al. 2009, 2819). The "advantage" resides in the potential to discover cultural affiliation in the absence of overt behavior. Now, such inscription of cultural values "at the neural level" could mean two things. One is that culture, including beliefs, norms, and meanings, is somehow embodied in individuals, specifically in their brains, prereflexively shaping their actions (Choudhury and Slaby 2012a, Gallagher and Zahavi 2008, Noë 2009). Another is that the neural level displays a truth about humans as cultural beings that is not knowable by examining social and cultural practices. Although the programmatic statements of cultural neuroscience seem to favor the former interpretation, cultural neuroscientific practice materializes the latter.

A frequently cited study on the "neural basis of cultural influence on self representation" provides another illustration of such a perspective (Zhu et al. 2007; see also the replications: Ng et al. 2010, Ray et al. 2010). The authors used fMRI to analyze brain activity of Western and Chinese subjects while they judged personal-trait adjectives regarding self, the mother, or a public person. As others in the field, they started with the observation that whereas North Americans and Europeans tend to view the self as independent, autonomous, and separate from others, East Asians emphasize interdependence and interconnectedness. The experimental design was standard: thirteen Chinese and thirteen Western college students were scanned while asked to judge if an adjective was adequate to describe the self, the mother, and other. They were also asked (as a neutral control condition) to judge the font of the words.

The findings were said to provide evidence of a neural distinction between self and intimate persons for Westerners but not for Chinese. Thus, in Chinese subjects, "mother-judgments" generated enhanced MPFC activity (compared with "other-judgments" and the neutral condition), such that "the representation of Chinese mother" could not be distinguished "from the representation of their selves, in terms of the MPFC activity." This result was said to indicate that whereas the Chinese "use" MPFC to "represent" both mother and the self, MPFC activity in Western subjects corresponds to a "representation" of the individual self alone (Zhu et al. 2007, 1314). The data seemed to the authors significant for both anthropology and neuroscience because it suggested "that culture influences the functional neuroanatomy of self-representation" and provided evidence of an "interplay of biology and culture in shaping the mind and the brain" (1315).

The study apparently mediated between a social constructivism that downplays the role of biology in cultural and social processes and practices and a naturalistic reductionism according to which interpersonal and cultural relations arise in the brain. However, unless one holds one of the two positions and thereby engages in a form of dualism, it is hard to justify costly experiments to arrive at statements such as "culture influences the functional neuroanatomy of self-representation" or "habitual cognitive processes are accompanied by detectible [*sic*] parallel neural processes" (1315, 1314). The paradox is that a significant Cartesian bias persists behind the explicit emphasis on brain-culture reciprocal interactions.

As the authors explained, social psychology demonstrated behavioral and cognitive differences between the Western and the East Asian self. But since it did not tell "whether cultures influence the relevant neural mechanisms," it remained necessary to look for neuroimaging evidence that Western and Chinese selves effectively differ "at a neural level" (1313, 1315). Two poles were thereby joined: culture both "affects the psychological structure of self" and "shapes the functional anatomy of self-representation" (1310). The trouble with such a claim is twofold. On the one hand, correlations do not reveal relations that can be captured by verbs such as "affect" and "shape." On the other hand, the use of those verbs manifests a peculiarly abstract and mechanical view of culture. Contrary to the way they are here conceptualized, notions, attitudes, and practices connected to the self are integral parts of culture; they are among the key features that contribute to enact it, not something that a mysterious agent called "culture" shapes from the outside.

Insofar as cultural diversity is conceptualized essentially as a form of neurodiversity, the experimental setups and results of cultural neuroscience may become part of identity politics (Roepstorff 2011, 40). At the same time, by positing the existence of difference between selves "at a neural level," cultural neuroscience contributes to the downplaying of diversity within the group. In both scenarios (interethnic difference and intragroup identity), the brain is endowed with ontological primacy: The mind is what the brain does, and culture is included in the process. One of the main concerns here is not that cultural neuroscience seems to suggest that universal values do not exist (Begley 2010) but that it naturalizes cultural stereotypes in the laboratory (Choudhury 2010, Choudhury and Kirmayer 2009). There have been calls for a more nuanced consideration of socioecological factors (Cheon et al. 2013), but so far they have not been systematically implemented in experimental work, and cultural neuroscience still has to draw consequences from the complex intellectual and political histories of sampling categories such as the usual "Caucasian-American" (see Painter 2010 for an overview). Indeed, as critics have pointed out, in practice cultural neuroscience tends to classify subjects on the basis of outer appearance at the expense of behavior or sociological or cultural dimensions, has "an understanding of 'culture' and 'race' which still appeals to biology, blood and ancestry," and may therefore seem to reinforce "Western dominance in a postcolonial situation" (Martinez Mateo et al. 2012, 160; 2013, 3). Regardless of whether cultural neuroscience is really so politically incorrect, the notion of "culture" it assumes indeed functions as a proxy of "race" (Heinz et al. 2014).

From Culture to Brain

One could object that individualism/collectivism and self-representation are particularly problematic subjects or that we confined ourselves to investigations that explicitly claim to be about "neural basis" (for a synthesis, see Zhu and Han 2008). After all, research has been carried out on topics such as perceptual processing (Kitayama et al. 2003), attentional modulation (Hedden et al. 2008), language (Tan et al. 2005; Lei, Akama, and Murphy 2014), music (Nan et al. 2008), number representation and mental calculation (Tang et al. 2006), emotional processes (Chiao et al. 2008), mental attribution (Tang et al. 2006), and self-representation and self-awareness (Han and Northoff 2008); other topics, such as the default mode network; regulation and inhibition of feelings, thoughts, and actions; prejudice and dehumanization; and fundamental warmth and competence judgments (Ames and Fiske 2010), have been identified as up-and-coming research areas. Moreover, important integrative effort has been made concerning the neuroscience of intergroup and intercultural relations (Cikara and Van Bavel 2014, Warnick and Landis 2015).

The studies we chose are nonetheless representative. Let us take another one that has been justly described as "foundational"[11] for the field: Shihui Han and Georg Northoff's 2008 overview of the area, implications, and future directions of transcultural neuroimaging in relation to "culture-sensitive neural substrates of human cognition." The authors position themselves clearly from the start: "A fascinating mystery facing human beings is how the brain gives rise to the mind." Transcultural neuroimaging emerges as a way to deal with that mystery and is considered promising insofar as it "can bridge the gap between neuroscientific investigations of supposedly culture-invariant neural mechanisms and psychological evidence of culture-sensitive cognition" (Han and Northoff 2008, 646). Once again, then, the mind, and culture as its ultimate collective product, are essentially what the brain does.

The authors aptly ask whether cultural experiences modulate or determine preexisting patterns of neural activity. This is a crucial question, one common to all attempts at bridging brain and culture. But is it is relevant for understanding culture? As the authors themselves point out, even if the same brain region is "recruited" by different groups for the same task, "two cultures might have different meanings for the concepts involved in a task" (652). The significant level of analysis must therefore be that of meanings and practices.

Han and Northoff realize that the notion of culture involves complexities that cannot be studied via the usual experimental designs. They recognize, for example, that there is no such thing as a homogenous "Western" or "East Asian" culture. Research practices, however, are less nuanced. It has been remarked that cultural psychology may give the impression that "there is a very small number of cultural identities (North American vs. East or Southeast Asian) that vary principally on the dimensions of individualism-collectivism or independent-interdependent self-construal" (Cohen 2009, 194). The same applies to cultural neurosciences, whose methods and experimental designs inevitably homogenize and factorize culture. More importantly, cultural neuroscience does not take culture as its object of study but as an independent variable on which a dependent one, such as the individualist-collectivist position, rests.

We have already noted that some contributors to the neurodisciplines of culture think of their object in a subtler way. Anthropologists have suggested an experimental approach that would take into account both the anthropology of experimentation and research subjects' lived experiences (Roepstorff and Frith 2012, Roepstorff and Vogeley 2009). Together with a number of colleagues, Juan F. Domínguez Duque, "the first Ph.D. in Neuroanthropology,"[12] has criticized cultural neuroscience's "primarily psychological" concept of culture, understood as a set of variables affecting the brain. Such an approach ends up setting aside "the actual social processes by which cultural knowledge is constituted" (Domínguez Duque et al. 2010, 143, 144).

For example, individualism and collectivism cannot be reduced to a simple variable and may be integrated in the same person to handle pragmatically different perspectives on the same situation. Domínguez Duque and colleagues would like to reduce the projection of the investigator's cultural values and beliefs onto the groups analyzed and to situate the circumstances under which the experiments take place. For them, therefore, neuroanthropology is a sort of self-reflexive radicalization of cultural neuroscience, one in which "research and analysis techniques from cultural (and more broadly, social) neuroscience are integrated into and embedded in ethnographic research" (Domínguez Duque 2012, 25). In a similar vein, Suparna Choudhury (2010) proposes to approach cultural neuroscience from the angle of "critical neuroscience." To do that, she suggests attending to the conceptualization of culture in the design and interpretation of experiments, taking into account the historical contexts of the phenomenon under scrutiny, considering the meanings that experimental categories may have in different cultures, and identifying how cultural biases and beliefs may influence the design and results of experiments (see also Choudhury, Nagel, and Slaby 2009; Choudhury and Slaby 2012b). It is hoped that such suggestions will inspire fruitful synergies between cultural neuroscience, neuroanthropology, and critical neuroscience (Lende and Downey 2012b, 411).

As for the concept of culture itself, an alternative to the psychologism of cultural neuroscience emphasizes that culture is socially created and transmitted and should be understood as "shared structures of meaning" through which people interact with one another (Domínguez Duque et al. 2010, 139; Domínguez Duque 2012). Such criticism of the notion of culture implicitly used by the "first generation" of cultural neuroscientists, as well as the emphasis on the contested and evolving nature of the concept, are accompanied by proposals to incorporate an anthropological understanding of culture into experimental settings. Those laudable goals, however, are not specific to the *neuro* in "neuroanthropology" or "neuroethnography." Rather, they can be achieved by complementing various qualitative and quantitative methods with critical theory and reflexive ethnography and by "historically, socially and politically contextualizing the circumstances under which enquiry takes place" (Domínguez Duque et al. 2010, 144).

In a similar perspective, the German scholars Ruth Denkhaus and Mathias Bös point out that most criticisms of cultural neuroscience have already been raised with regard to its mother discipline, cultural psychology. They propose to replace the "entity conception of culture" underlying the homogenizing and essentializing tendencies of cultural neuroscience with a notion of culture as "patterns of representations, actions and artifacts that are distributed or spread by social interaction" (Denkhaus and Bös 2012, 445). Reference to "actions and artifacts" implies that culture is not in people's heads but is simultaneously in the individuals, their brains and minds, and the world they inhabit (450). The authors thus criticize the assumption that culture is "stored in people's brains" (Ames and Fiskes 2010, 72). Han et al. (2013) have also underlined the constitutive, rather than merely modulatory, role of context. While in the modulatory context-dependence model, neuronal and cultural influences interact but remain separate and independent, the notion of constitutive context-dependence implies no clear-cut separation between the biological domain of the brain and the social domain of culture. In this model, brains are "biosocial" and culture is "sociobiological" (353). Some cultural neuroscientists have proposed to redefine culture as that which is manifest in "the direct dependence of the brain's neural activity" on context (Northoff 2013b, 95), and others would like to integrate factors such as socioeconomic status, unemployment rate, residential mobility, or population density in their definition of cultural influences as a way to address within-nation variation (Ng, Morris, and Oishi 2013).

Such views, however, do not alter the basic assumption that neuroscience provides "the most fundamental perspective yet available" on how people appropriate culture (Domínguez Duque et al. 2010, 140). That is why the declarations of intention about the co-construction of brain and culture have not had a major impact on how experimental and fieldwork are conducted, nor have they prevented neuroanthropologists from claiming that "the shared webs of signification that make up culture are primarily the product of the activity of the PFC [prefrontal cortex]" (Domínguez Duque et al. 2009, 60). Such a statement gives expression to the foundational creed of the field's epistemology and thought style.

A 2015 overview asserts that, from a functional connectivity analysis showing that neural connections between MPFC and bilateral temporoparietal function (said to be "implicated" in perspective taking) "were much stronger for Chinese than for Danes during the judgment of social attributes of the self," it can be concluded

> that the Chinese self is constituted by a more integrated, or holistic, representation of both direct and indirect appraisals. In comparison, the Western self appears more one-dimensional in the sense that it is defined largely on the basis of the first-person perspective alone. (Kitayama and Huff 2015, 6)

Such an inference, however, is fallacious. This is not because it implies the questionable existence of a homogeneous, perfectly self-consistent Western or Chinese self. In fact, the study itself reports that in Asian American individuals, who have multiple cultural identities, brain response patterns depend on which "cultural frame" is made salient (10), but it does not establish that this cannot happen in allegedly monocultural persons. The inference is fallacious because the nature of a self cannot be inferred or even hypothesized from the existence of certain "neural connections." However, such an inference exhibits the ultimate implicit goal of much neurodisciplinary research: to diagnose and classify on the basis of brain data, thus saving the trouble of engaging in apparently messier and less objective human science research. (We shall see in Chapter 3 that a similar ambition characterizes some areas of psychiatric neuroimaging.)

Culture?

What, then, other or beyond a "product" of prefrontal cortex activity, is *culture* for the neurodisciplines of culture? The concept was notoriously capacious by the time that the anthropologists Alfred Kroeber and Clyde Kluckhohn (1952) enumerated over 150 definitions, and it has remained so (Shweder 2001). In *Primitive Culture*, Edward Tylor (1871, 1) defined "Culture or Civilization" as "that complex whole which includes knowledge, belief, art, morals, law, custom, and any other capabilities and habits acquired by man as a member of society." Since then, many others have followed more or less his lead, seeing in culture "the complex of values, customs, beliefs and practices which constitute the way of life of a specific group" (Eagleton 2000, 34). Different emphases are also to be found, with a range and overlap of meanings, as illustrated in Raymond Williams's (1985, 91) observation that "in archaeology and in *cultural anthropology* the reference to *culture* or *a culture* is primarily to *material* production, while in history and *cultural studies* the reference is primarily to *signifying* or *symbolic* systems."

What exactly goes into "culture" has been long discussed. Kroeber and Kluckhohn identified different types of definitions (descriptive, historical, normative, psychological, structural, and "genetic" in the sense of developmental) and came up with a lengthy list of the conceptual elements that went into them, from acts and activities to feelings, language, and traditions. They themselves proposed a very encompassing definition but recognized that there is no way to proceed normatively. Nonetheless, two things emerge. One is that students of culture tend to characterize their object as "the organization of human experience and action by symbolic means" (Sahlins 2000, 158). The other is that those organizations and means are not static and do not form systematic and homogenous totalities. Early twentieth-century anthropologists sometimes did regard culture in that way, producing what Marshall Sahlins (159) critically called "anthropology-cultures." In that framework, it was always possible somehow to dig out the authentic native who reflected the culture. Indeed, as James Clifford (1988, 338) noted, the very idea of culture "carries with it an expectation of roots, of a stable, territorialized existence" (see also the neuroanthropological point of view in Roepstorff, Niewöhner, and Beck 2010).

Such bounded, functionally integrated, coherent units operating as a consistent totality probably never existed, and if they did, they certainly no longer do in the context of what has been famously characterized as "locally lived lives in a globally interconnected world" (Gupta and Ferguson 1992, 11). Two points about this context are especially relevant here. One is the possibility of internal contradiction. The debate around Margaret Mead's 1928 *Coming of Age in Samoa* is in this respect illuminating. Mead offered the image of a harmonious society with a liberal attitude towards sexuality. Her book had a huge social impact and became the bible of an entire generation. Then in 1983, Derek Freeman published *Margaret Mead and Samoa: The Making and Unmaking of an Anthropological Myth*, where he argued (on bases that were later questioned) that Mead was misled by native informants and ignored evidence contrary to her depiction of Samoan life.

Details aside, a major lesson of this controversy is that Samoan culture contains paradoxes and contradictions, which are, as Nancy Scheper-Hughes (1984, 90) put it, "culturally structured but never actually resolved." Aggressive and harmonious behavioral poles and values may function in the same individuals and groups depending on circumstances. Mead therefore captured *a* Samoan truth, not *the* Samoan truth. Anthropologists have given up the idea "that everything in a society must adhere to a single configuration or pattern" and no longer "think of 'culture' as a single integrated reality" (90, 91). Yet whenever cultural neuroscience draws on tools such as the Self-Construal Scale, it assumes exactly such a view, according to which any one factor (being "independent" or "interdependent") must correlate with some basic principle or attitude considered defining of the culture (such as individualism or collectivism). Like neuroaesthetics trying to establish the neural correlates of beauty but incapable of taking into account the fact that the same stimulus may be judged both ugly and beautiful, cultural neuroscience can only identify the supposed neural correlates of isolated factors, under the postulate that these correlates represent the cerebral embodiment of culture. The definition of "culture" as "factors that affect the biological and psychological processes that shape beliefs and norms shared by groups of individuals" illustrates precisely that (Hyde et al. 2015, 76).

The second consequence of lives being "locally lived" in an "interconnected world" is that cultural difference is not a basic given that correlates with being or belonging to some form of "people" (Western, Asian) but is rather "a product of a shared historical process that differentiates the world as it connects it" (Gupta and Ferguson 1992, 16). Cultural neuroscientific practice assumes separate and discrete cultures, which it juxtaposes in its experimental designs. It thereby participates in the processes whereby differences are constructed. This is in itself unproblematic and, perhaps, an inevitable artifact of studying culture. The problem and the challenge lie deeper, and they apply to all the neurodisciplines of culture: their assumption that culture is essentially, both ontologically and causally, a byproduct of the brain does not equip them well to deal with cultural phenomena—while at the same time it gives them a powerful tool for shaping culture itself.

Varieties of Neuroaesthetic Inquiry

As both aesthetics and neuroscience, neuroaesthetics is pursued in a variety of contexts and from different methodological perspectives; such diversity, as that of other *neuro* fields, is expected to generate and enable the testing of "foundational hypotheses" across domains (Skov and Vartanian 2009b, 1). The official story of how neuroaesthetics came about follows a familiar pattern: first, "the possibility of equating an aspect of mind to brain processes is introduced and hotly debated"; models and hypotheses are then advanced, followed by experimental work; the area receives a decisive impulse from the advent of noninvasive neuroimaging techniques; and now, "just as the rise of the experimental study of consciousness and moral psychology has increased neuroscience's relevance to these fields of enquiry," a similar trend is foreseen for aesthetics (2–3). Insofar as mechanisms involved in aesthetic judgment also intervene in the case of objects that are not considered artworks, the relevant neural processes are likely to be common to different functions, and that is why, it is said, neuroaesthetics should be seen "as a basic part of the larger neuroscience program" (5).

The word itself seems to have been coined in the late 1990s by Semir Zeki, a specialist of the primate visual brain at University College London. A decade later, neuroaesthetics was depicted as coming of age "as a field of study in its own right," devoted to exploring the "neural foundations of aesthetic appreciation" (Cela-Conde et al. 2011, Chatterjee 2010, Nadal and Pearce 2011). The new discipline might be undergoing "growing pains" (Chatterjee 2012), but that has not prevented it from being outstandingly productive.[13] Moreover, in 2013 a major institutional step was accomplished with the establishment of a Max Planck Institute for Empirical Aesthetics in Frankfurt. While this institute does not identify itself with "neuroaesthetics," it includes a neuroscience department, and its self-presentation as an interdisciplinary enterprise that "focuses primarily on the foundations of aesthetic evaluation, perception and experience" largely matches the discipline's profile.[14]

Neuroaesthetics has been characterized as studying "the neural bases for the contemplation and creation of a work of art" (Nalbantian 2008, 357) in the visual arts, literature, music, dance, theater, and film, but it is actually both narrower and broader. Narrower because, as we shall see, it tends to concentrate on issues of appreciation, which it operationalizes as subjects' liking or disliking stimuli or judging them for their "aesthetic" qualities; broader because it should be made to encompass two areas of study that place today's brains—those of the scanned subjects—in long-term temporal perspectives.

On the one hand, there is evolutionary aesthetics, which generally considers art, and more specifically aesthetic preferences, as adaptations that evolved to enhance reproductive success (see the concise discussion in Davies 2009). More precisely, the "exhibitory form of art" is said to have roots in "courtship rituals by animals where genetic qualities of the displaying animal are assessed by the potential mate" (Zaidel 2013, 229; see also Schaeffer's [2010] interpretation, based on evolutionary costly signaling theory, or Menninghaus's [2008] claim that the evolutionary perspective confirms Kant's theory of aesthetic judgment). Discussing these conjectures would lead us too far away from our topic. Suffice it here to say that the main relevance of evolutionary aesthetics for neuroaesthetics lies in the belief that the alleged adaptations have become inscribed in neuronal function (such that, for example, they determine humans' supposedly innate preference for savanna-like landscapes; Falk and Balling 2010). Neuroaesthetics, in fact, has been said to belong in a wider field of "bioaesthetics" aimed at understanding representation, emotion, and creativity from a neuroscientific and evolutionary point of view (Fitch, von Graevenitz and Nicolas 2009; Skov 2006).

On the other hand, there is "neuroarthistory" (Onians 2008a). The book that launched this barbarous term was advertised as "a fascinating account" of "one of the newest and most exciting fields in the human sciences," and its author was praised by his university for "cracking the real Da Vinci code."[15] He did that thanks to the concept of neuroplasticity, which allegedly opens the way to explaining why art looks the way it does. This, it is claimed, will for the first time in art history allow access to artists' minds. For example: the reason why the animals painted in the Chauvet-Pont-d'Arc Cave look so "strikingly naturalistic" (a questionable assertion) is that paleolithic humans observed them intently; the neural networks of a contemporary Australian bark painter were particularly attuned to parallel lines because as a child he admired his father's expertise in using fibers to make fish traps and was later exposed to parallel lines through op art; a similar reasoning explains why Florentine Renaissance artists made more use of lines, while Venetians, more use of color. In such a perspective, the answer to the question "What might neuroarthistory add to the discussion of Du-champ?" is that the objects Duchamp used in his readymades

> had become so familiar as valuable devices that viewers would necessarily have enjoyed seeing them. They might not have placed them in the category of art, but the response they evoked was one at the centre of artistic experience, an unconscious pleasure, a pleasure enhanced by the additional references associated with title and text and context. It was such a neurally-based response that Duchamp unconsciously exploited. (Onians 2008b)

The word "nothing" is a more straightforward way of answering that question, and, as a reviewer noted, "it is disturbing that these often ludicrously tendentious ideas . . . are being advanced not by some mad autodidact on a park bench but by a serious academic" (Tallis 2008b, 19).

While neuroaesthetics has some methodological roots in the neurobiology of perception, it should be properly considered as part of a broader neuroscience of art, which could itself be situated within "empirical aesthetics." Since its beginnings in the nineteenth century, empirical aesthetics has used mainly the methods of experimental psychology to restore *aisthesis*— perception and sensation—at the heart of aesthetic experience, and has devoted itself mostly "to the discovery of a finite set of universal laws that govern people's interactions with objects" (Vartanian 2014, 8; see also the other chapters in Tinio and Smith 2014, part I; as well as Lichtenstein, Maigné, and Pierre 2013 for the French context). Although these disciplines have in principle a range of concerns that encompasses human responses to non–art objects too, questions related to art predominate in their actual productions, and they see their most important cultural significance in the answers they can provide to questions such as "What is art?" "What is beauty?" "How does aesthetic judgment work?" and, in the case of neuroaesthetics, "How do they all emerge from the brain?"

The neuroaesthetic mode of inquiry into these questions, as well as the answers provided, presuppose that art is essentially an extension of the brain's cognitive and adaptive functions. Both reflect a "truth" Zeki (2002, 54) declared to be "axiomatic," namely "that all human activity is dictated by the organization and laws of the brain; that, therefore, there can be no real theory of art and aesthetics unless neurobiologically based." That is indeed the assumption of the entire field. From it follows neuroaesthetics' most typical hermeneutic device, which consists of considering artists as neuroscientists *qui s'ignorent,* as individuals whose main task is to explore the brain with the tools of their trade. What accounts for the impact of Michelangelo's *Pietà?* The fact that the sculptor "instinctively understood the common visual and emotional organization and workings of the brain," which in turn "allowed him to exploit our common visual organization and arouse shared experiences beyond the reach of words" (Zeki n.d.; specifically on the widespread idea of the artist as neuroscientist, see also Cavanagh 2005, Ramachandran and Hirstein 1999, Zeki 2000).

In all fairness, the notion of the artist as neuroscientist also appears more metaphorically in connection with artistic artifices known to be consistent with the physiology of vision. For example, shadows must be darker than their immediate surroundings, but they can have the wrong color or shape yet function as recognizable shadows; given some basic cues, transparent material can be recognized in spite of gross deviations from the optics of refraction, and almost any reflection (as in mirrors) will be accepted as such; figures need not be complete in order to be meaningfully understood. These observations, however, were made without using neuroimaging and initially without reference to neuroscientific results at all. Most importantly, they were made in contexts where the main goal was not to explain art but to use artworks as stimuli for studying the visual system. It has, for example, been argued that finding statistical regularities (in amplitude spectrum and fractal dimension, among others) in paintings' spatial structure may "grant insight into the range of spatial patterns that humans find compelling" and that art, both representational and nonrepresentational, replicates the "basic statistics" of the world (Graham and Field 2007, 149, 150). The goal here is to investigate the visual system through the analysis of artworks; the shared and divergent statistics of art and natural scenes "may provide new tools for uncovering the coding strategies of the visual system" (162).

The same approach is at work when the neurophysiologist Margaret Livingstone examines the reasons for the elusive quality of the Gioconda's smile (it is more apparent in the low-spatial-frequency range and therefore more visible as a smile when one is *not* looking at her mouth) or claims that Rembrandt and many other artists were stereoblind, a condition that might have been an asset for flattening three-dimensional scenes onto a canvas (Livingstone 2000, Livingstone and Conway 2004). The accuracy and relevance of such interpretations have been questioned. The point, however, is that their object is not primarily art but "the biology of seeing"—how certain effects of artworks depend on the neurobiology of luminance, stereoscopy, shading, and other processes—as well as how art may enter into research about the visual brain.

This is very different from, say, suggesting that impressionism is "effective" because of, or thanks to, the amygdala, a bilateral brain structure involved in emotional learning and memory consolidation. Research has shown that the amygdala responds more strongly to a blurry version of faces expressing fear than to the same stimuli depicted in sharp detail (Vuilleumier et al. 2003). But does this mean that the impact of impressionist works results from their connecting "more directly to emotional centres than to conscious image-recognition areas" (Cavanagh 2005, 305)? In one case, artworks are used as stimuli to explore the functioning of the visual system; in the other, information about the functioning of a brain structure is extrapolated to explain in the simplest possible terms a complex historical and experiential phenomenon.

It is one thing to acknowledge that experience of artworks begins with the automatic analysis of perceptual stimuli; it is another to affirm that contextual, educational, biographical, and other nonbiological factors are nonessential because they only "modulate" neurophysiological processes. Such an assertion would simply get rid of art and aesthetics. A German neuroanatomist, for example, has proposed a "universal model of aesthetic perception based on the sensory coding of natural stimuli," on the assumption that cultural, historical, and social factors are variable and that "being variable [they] cannot be relevant in the search for a universal theory of aesthetic experience" (Redies 2007, 100). The model boils down to the idea that artworks induce a particular "resonance" based on the evolutionary adaptation of the visual system to natural scenes and that sensory stimuli are "more or less aesthetic" depending on the degree of "neural resonance" they induce (106).[16] Even when "aesthetic" is, as here, unduly employed as a synonym of "likeable," those "variable" factors remain constitutive of human's relation to art.

As the definitions and conjectures we just sketched illustrate, neuroaesthetics oscillates when it comes to specify its object. It seems to distinguish implicitly between the *aesthetic relation* and the *artistic relation* or experience.[17]

Partly following Kant, the French literary theorist Gérard Genette finds at the bottom of the aesthetic relation a "disinterested" attitude in which we pay attention to and judge an object independently of considerations about usefulness, focusing on its appearance rather than on its function. We can engage in this kind of relation with any object (including natural ones), but any object we approach aesthetically can also be treated in other ways (as a commodity, as a piece of decoration, as an instrument of power). From a biological and evolutionary standpoint, the aesthetic relation is seen as part of the natural behavioral repertoire of the human species. In the artistic relation we recognize in addition that objects are imbued with an aesthetic intention, but of course, such objects too can also be handled in nonaesthetic ways and, inversely, numerous nonartistic objects have become art (see for example Fraenkel 2007, on the case of prehistoric cave drawings). As Genette (1999, 11) puts it, "it is not the object that makes the relation aesthetic, but the relation that makes the object aesthetic."

In practice, neuroaesthetics focuses on responses to artworks, with the response generally operationalized as hedonic judgments concerning preference or degrees of liking. It thus systematically reduces the aesthetic relation to those judgments, which it uses as a proxy for the response to art. Its neglect of the specificity of the artistic relation is embodied in its very methodology, which assumes that subjects lying in a scanner can engage in an aesthetic and specifically artistic relation to the small-size copy of an artwork displayed on a screen for a few seconds and for an explicitly scientific purpose. Although some neuroaestheticians worry about the prodigious lack of ecological validity of studies in their field (Chatterjee 2010, Lacey et al. 2011, Nadal and Pearce 2011), the adjective "aesthetic" continues to be identified with "hedonic" and to be applied indiscriminately to perception, assessment, preference, and experience.

Beauty

As mentioned, neuroaesthetics is varied, ranging from investigations into the neuronal processing of lines, color, or kinetic contours to research into the neural correlates of aesthetic judgment. We shall first look at a fundamental concern of aesthetics, *neuro* and otherwise: beauty, with a focus on the contributions of the neuroaesthetics pioneer Semir Zeki, long the most visible figure of the field.

Zeki is a prolific author and blogger, fond of somersaulting over the abyss that separates single-cell measurements from the biggest names of the Western canon. He is of course perfectly aware that there is such a thing as culture and that, whatever its definition, it does not exist exclusively as encephalic mass. He recognizes, for example, that the correspondence between Mondrian's or the constructivists' preference for straight lines and the existence of brain cells that selectively respond to straight lines of different orientations and the fact that kinetic art seems "admirably suited for stimulating the cells in V5 and anticipated artistically the physiological properties of motion-selective cells" (Zeki 2001, 51) do not mean that the aesthetic feelings aroused by a Malevich or a Calder are attributable exclusively to the individual activity of certain neurons. Rather, the correlation between the existence of certain types of artwork and certain types of neurons shows that an artwork's "constituent elements" are "a powerful stimulus for these cells" and that a brain deprived of the appropriate neurons "will not be able to appreciate" the art in question (Zeki 1998, 14). In other words, constructivist or kinetic art "would give no aesthetic feeling in the absence of those cells" (Zeki 2000, 100). Details aside, this apparent insight boils down to saying that one cannot enjoy painting without a visual system or music without an auditory one (or their functional equivalents).[18] It therefore adds nothing to aesthetics. And since those perceptual systems are involved in nonaesthetic functions, an analysis such as Zeki's "tells us nothing about Picasso and Cézanne that does not apply equally to Häagen Dazs and McDonalds" (Hyman 2006). For all their superficial references to Plato, Kant, or Schopenhauer, speculations à la Zeki (for example about the origins of Dante's, Michelangelo's, and Wagner's art in their having "formed an ideal of love in their brains" [Zeki 2002, 62]) leave out so much of relevance that their contribution to aesthetics amounts to nil (Ione 2003). What, however, do neuroaestheticians do in the lab?

In 2004, Zeki and Hideaki Kawabata, a professor at Keio University in Japan, published in the *Journal of Neurophysiology* an article entitled "Neural Correlates of Beauty." This brings us right to the heart of the neuroaesthetic enterprise: The topic is beauty, the methodology is brain imaging, and the main product is a correlation. The purpose of the study was to find brain areas that would respond specifically to one category of painting (for example, the portrait) as well as brain areas that would be "consistently active" across subjects when they perceive a painting they judge beautiful or ugly. This strategy, the authors explained, allowed them to circumvent the question of how what an individual regards as ugly or beautiful is conditioned by culture, education, and inclination (Kawabata and Zeki 2004, 1699). As it turns out, since these factors cannot be switched off, the authors do not really circumvent them but just ignore them—and with them, most of what makes something be "art" or "aesthetic," "beautiful" or "ugly."

Four categories of painting were retained: abstract, still life, landscape, and portrait. Each of the ten twenty- to thirty-one-year-old subjects ("5 females") was shown three hundred paintings on a computer monitor and was instructed to give each a score for ugly (1–4), neutral (5–6), or beautiful (7–10). Of these, 192 were selected (those with scores 1–2, 5–6, and 9–10), and they were randomly shown to each subject, who again had to rate them while inside the fMRI scanner. The procedure generated a two-factor 3×4 event-related design: one factor was three different response conditions (beautiful, neutral, ugly); another, the four genres of painting.

The study revealed a functional specialization of the visual brain, particularly for faces and landscapes, regardless of aesthetic judgment. A still life produced the greatest change in the visual cortex V3 area, and landscapes in the parahippocampal place area. Second main result: diverse contrasts (that is, subtractions in the sense of fMRI methodology) revealed activity in the medial orbitofrontal cortex, the anterior cingulate gyrus, the parietal cortex, and the motor cortex. With responses to the neutral stimulus acting as a baseline, the authors also found that, regardless of painting categories, signals increased for *beautiful* judgment at the orbitofrontal cortex and at the motor cortex for *ugly* judgment. Moreover, aesthetic judgment correlated with brain structures "known to be engaged during the perception of rewarding stimuli" (1702)—something to be expected since, at least in the lab, beauty is associated with pleasure, and reward is pleasurable by definition. But no separate area was "specifically engaged" when stimuli were judged to be ugly. The conclusion was that, since both beautiful and ugly stimuli "modulate activity in the same cortical areas . . . it is the modulation of activity within those areas that correlates with the judgment of a stimulus as being beautiful or not" (1704). Such a result, the authors emphasized, does not tell "what constitutes beauty in neural terms" but rather addresses the "Kantian" questions of "what are the conditions implied by the existence of the phenomenon of beauty (or its absence) . . . and what are the presuppositions that give validity to our aesthetic judgment. In aesthetics," they declared, "the answer to both questions must be an activation of the brain's reward system with a certain intensity" (1704).

This conclusion is consistent with the postulate that "the almost infinite creative variability" of art "arises out of common neurobiological processes" (Zeki 2001, 51). Does such commonality illuminate artistic processes beyond the fact that art making and aesthetic response share mechanisms with making and responding in other domains? Do these processes differ in artists and nonartists? The late Robert Solso, the author of *Cognition and the Visual Arts* and *The Psychology of Art and the Evolution of the Conscious Brain*, conducted MRI scans of a portrait artist while making thirty-second drawings of faces inside the scanner. When the results were compared to those of a psychology graduate student with no formal art training, they revealed decreased activity in the artist's fusiform face brain area (FFA), frequently associated with face identification, and increased activity in his right middle frontal area, usually associated with "more complex associations and manipulations of visual forms" (Solso 2000, 82). The lower FFA level of activation indicated that the artist "may be more efficient" than the novice at processing facial features, and the involvement of the right frontal part of the brain suggested that he engaged in a "'higher order' interpretation" and an "abstracted representation" of the perceived face (83; also Solso 2001).

What have we learned? A visual artist can be expected to be more efficient than a novice at processing visual stimuli, and morphological and physiological brain changes always accompany doing and learning. The hippocampus of taxi drivers grows larger as the drivers spend more time in the job (Maguire et al., 2000), meditation has been shown to increase thickness in brain regions associated with attention and processing sensory input (Lazar et al. 2005), pianists' brains display increased gray matter density and white matter integrity (Han et al. 2009)—and so on for the rest of us who are neither cabbies nor Buddhist monks nor pianists. Of course empirical details are important, and they may be useful to detect, diagnose, and treat any number of conditions attributable to a brain lesion. Art, however, is not in the brain as a lesion is in the brain, and the commonality neuroaestheticians claim to discover does not illuminate the differentiated aesthetic reactions individuals can have vis-à-vis a Mondrian and a Malevich: you can find the one beautiful and the other ugly even though perceiving both necessitates the firing of neurons specialized in straight lines. The fact that perceiving a Mondrian and a garden fence activates the same cells reveals nothing specifically about hedonic preference and even less about the aesthetic relation.

So does neuroaesthetics say anything about beauty that would be significant for aesthetics? Kawabata and Zeki refrained from theorizing beauty or venturing normative definitions. But almost a decade later, Zeki and another Japanese collaborator finally concluded that beauty is "some quality in bodies that correlates with activity in the mOFC [medial orbitofrontal cortex] by the intervention of the senses" and that such definition could give rise to standards of judgment: "A painting by Francis Bacon may be executed in a painterly style and have great artistic merit but may not qualify as beautiful to a subject, because the experience of viewing it does not correlate with activity in his or her mOFC" (Ishizu and Zeki 2011, 8–9, 7; for more on Bacon along the same lines, see Zeki and Ishizu 2013).

Whether Bacon or Beauty, these studies miss their putative objects. They postulate dichotomy as the basic structure of judgment, thus excluding the possibility that you might admire and enjoy Rubens's or Bacon's composition, palette, and brushwork but be displeased by the looks of their characters—not to say anything of the aesthetics of ugliness (Eco 2007) or of the fact that one can find an object or person simultaneously superb and hideous. Yet from the neuroaesthetician's point of view, the problem with the study by Kawabata and Zeki was that it did not deal with "aesthetic judgment per se" (Jacobsen et al. 2006, 276). That is why researchers at the Institute of Psychology of the University of Leipzig and the Max Planck Institute of Human Cognitive and Brain Sciences, also in Leipzig, proposed to identify "the neural correlates of genuine aesthetic judgments of beauty" (281) through the use of geometrical shapes rather than paintings. Subjects were asked to describe each of four stimuli as beautiful or not and as symmetric or not. Since earlier studies found that aesthetic judgment is often ruled by symmetry, it was expected that "differences between the brain correlates of aesthetic judgment and symmetry judgment should be solely due to differences of judgment processes per se" (276–277). The use of abstract black-and-white patterns was intended to factor out elements, such as attitudes and memory, said to interfere with aesthetic judgment.

As anticipated, symmetry was the most important predictor of judgments; symmetric and regular pictures were generally found to be more beautiful than the others. Some brain areas were differentially activated. Aesthetic judgments elicited activation in the right frontomedial cortex; symmetry judgments, bilateral activation in the dorsal premotor cortex, the superior parietal lobule, the intraparietal sulcus, the left ventral premotor cortex, left fusiform gyrus, and the visual cortex. While symmetry had no significant influence on BOLD signal changes, *beautiful* judgments led to higher signal changes than *not-beautiful* judgments in the brain areas "specifically engaged" in aesthetic and symmetry judgments. Moreover, aesthetic judgments "recruited" areas that partly overlapped with brain networks involved in social and moral judgments. In conclusion, judgments of beauty were found to "trigger" activation in a brain network "that generally underlies evaluative judgments, and hence share neural substrate with, e.g., social and moral judgments," and brain activations during aesthetic judgment can therefore not "be reduced to an assessment of symmetry but are actually due to a particular mode of judgment" (284).

Again, it is difficult to detect here a contribution to aesthetics. Neither a certain preference for symmetry nor the fact that beauty is irreducible to symmetry ranks as a novel insight. In addition, to the extent that aesthetic judgment is *judgment*, it must involve brain areas engaged in judgment. The ecological validity of the experimental tasks is yet another issue: shouldn't the "per se" in "aesthetic judgment per se" include the memory, attitudes, and other factors the experiments were designed to exclude (but naturally couldn't)? And does the experiment succeed in this? Isn't someone familiar with and liking or disliking Malevich (or op, kinetic, constructivist, or geometric art in general) likely to react to a geometric pattern differently from someone who isn't? And wouldn't the judgment be different depending on whether the pattern measures 200 × 200 cm and hangs in an art gallery with a famous signature on one corner or is ten times smaller and displayed for two seconds on a screen inside a loud and cramped scanner? Surely the frictionless plane is an important epistemic device. But neuroaesthetics goes in this respect well beyond the Galilean conceptual model, and it is not

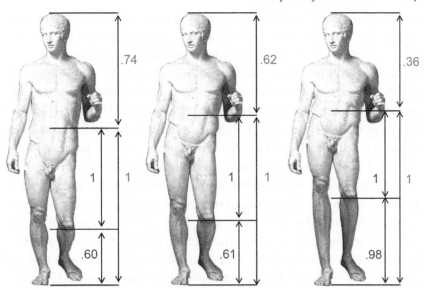

FIGURE 2.2 Example of canonical and modified stimuli. The original image (*Dory-phoros* by Polykleitos) is shown at the center of the figure. This sculpture obeys canonical proportions (golden ratio=1:1.618). Two modified versions of the same sculpture are presented on its left and right sides. The left image was modified by creating a short legs:long trunk relation (ratio=1:0.74), the right image by creating the opposite relation pattern (ratio=1:0.36). All images were used in behavioral testing. The central image (judged-as-beautiful on 100%) and left image (judged-as-ugly on 64%) were employed in the fMRI study. From Di Dio, Macaluso, and Rizzolatti (2007). Courtesy Dr. C. Di Dio.

enough to be asked to decide in a few seconds whether something is beautiful for the response to be germane to aesthetics.

Other neuroaestheticians may retort that the shortcomings of Jacobsen and colleagues is that they stayed too close to Kawabata and Zeki and hence too far away from aesthetic judgment "per se." An attempt at getting closer, carried out in Italy, examined "brain response" to images of classical and Renaissance sculptures manipulated so as to alter the golden ratio of the originals (Figure 2.2).

The purpose of the study was to find out whether there is "an objective, biological basis for the experience of beauty" or whether such experience is "entirely subjective" (Di Dio, Macaluso, and Rizzolatti 2007, 1). The question can be translated into whether "objective parameters intrinsic to works of art are able to elicit a specific neural pattern underlying the sense of beauty in the observer" (6). The authors hypothesized a positive answer, namely that humans are endowed "with species-specific mechanisms that resonate in response to certain parameters present in works of art" (6). The German team had chosen symmetry; the Italians preferred the golden ratio. Both symmetry and the golden ratio have for centuries figured prominently in empirical and philosophical inquiries into art and aesthetic experience; the fact that the debate about their role and status still goes on suggests the extent to which they do not represent merely factual matters but are dense condensations of values and worldviews. More down to earth, neither the Germans nor the Italians explained in which sense the "objective parameters" they claimed to investigate would be "intrinsic" to artworks, since they are absent in some artworks and present in non-art artifacts as well as in nature.

The Italian team used fifteen sets of three images, each including fifteen originals and fifteen modified pictures (seven with long trunks and short legs, eight with short trunks and long legs); twenty sculptures represented male bodies and ten, female bodies. The stimuli, thirty for each of six separate fMRI runs, were presented, in a randomized order and for two seconds each, in three conditions: *observation*, in which subjects were asked "to observe the sculptures as if they were in a museum," *aesthetic judgment* (subjects were asked if they liked the image), and *proportion judgment* (whether they found the image proportionate).

Two types of analysis were carried out. One contrasted brain activations in reaction to canonical vs. modified stimuli. This was supposed to uncover "the neural responses to objective beauty parameters"; the hypothesis was that the golden ratio would elicit enhanced activity in areas mediating pleasure and that signal increase would be particularly strong during the observation condition, "where brain response to the artworks was not interfered with by additional cognitive requests" of judgment (2). (Asking the subjects to look at the stimulus *as if in a museum* was obviously not considered a significant request, cognitively or otherwise.) A second analysis, aimed at brain responses "related to the overt subjective appreciation of the stimuli," contrasted brain activations obtained during the presentations of judged-as-beautiful against judged-as-ugly images.

The behavioral results showed that canonical images were evaluated more positively and modified ones more negatively. In MRI analysis, the results of viewing canonical and modified conditions were first taken together and then contrasted with the rest across all three conditions (observation, aesthetic judgment, and proportion judgment). This contrast revealed "activations" in several areas. Especially significant for the authors was signal enhancement in the insula during the observation condition. The insula is one of the most beloved brain structures of the *neuro* industry. It is involved in motor control and homeostasis, as well as in interoception and visceral states associated with emotional experience, in addition to self-awareness and sense of agency. It seems to play such a crucial role in combining information about bodily states into higher-order cognitive and emotional processes that, as a *New York Times* science writer put it: "The bottom line is that mind and body are integrated in the insula" (Blakeslee 2007).

Di Dio and her colleagues attributed the weaker insula activation effect in the aesthetic and proportion conditions to the explicit request to judge, which might have "diverted the volunteers' attention resources towards a specific cognitive demand, thus lessening the natural neural response within the insula" (Di Dio, Macaluso, and Rizzolatti 2007, 4). From this it seemed to follow that "the positive emotional *feeling* elicited in the viewer by the canonical images was determined by a preferential coding of these images, relative to the modified ones, by various cortical areas and by a concurrent, *joint* activation of the anterior insula" (6). Thus, since the golden ratio "determined brain activations different to those where this parameter was violated," the question about the existence of "objective beauty" was answered positively (6). The authors acknowledge that it would be "too reductive" to imagine that the sense of beauty "occurs because" of insula activation; joint activation of many areas and circuits is needed. In sum, could artworks "ever become a permanent patrimony of humankind without a resonance induced by some biologically inherent parameters?" (8). The answer is obviously "no."

However, like any catalog of humanity's essential attributes or components (from variations on the Western canon to enumerations of features characterizing human nature), the list of those parameters is negotiable, debatable, and likely to reflect particular interests and specific circumstances. Moreover, the fact that our brains may be predisposed one way or another toward certain qualities, such as proportion, symmetry, or scale invariance, does not help us better understand aesthetic experience as *aesthetic* and as *experience*. Common perceptual mechanisms are surely involved in looking at Andy Warhol's 1964 sculpture *Brillo Box* in a gallery and at James Harvey's original Brillo boxes in a supermarket. Harvey's representative complained about Warhol's use of his client's design, yet he despondently admitted, "What's one man's box, may be another man's art" (Gaddy 2007). While neuroaesthetics does not aim to analyze such a situation, it should at least have the means of taking it into account. All it has done so far, and all it can do with its methods and conceptual frameworks, is to correlate known facts (we react emotionally to artworks) or some narrowly operationalized aspects of the artistic relation (especially appreciation) with the activity of various brain areas, which are then said to "be involved," "associated with," "underlie," "contribute to," or somehow or other "reflect" those facts or aspects. Such are "the insights from neuroimaging" for understanding the "experience of art" (Nadal 2013).

Empathy

Although beauty has been a central topic in aesthetics, the neuroaesthetics of beauty has brought us to the point of wondering if, in spite of its name, the new discipline is about aesthetics at all. With David Freedberg, we enter a different world—one that promises a more sophisticated treatment of art as well as smarter ways of linking neuroscientific knowledge and the aesthetic relation. Indeed, Freedberg, formerly the Pierre Matisse Professor of the History of Art at Columbia University, and since 2015 the director of the Warburg Institute, had, before turning his attention to the brain, published widely on Dutch, Flemish, French, and Italian seventeenth-century art (including painting, drawing, and print), iconoclasm, the intersection of art and science, and, to a lesser extent, contemporary art. Freedberg's interest in the neurosciences relates directly to historical events he explored in his seminal *The Power of Images*.

In that book, Freedberg (1989) wished, as Ernst Gombrich (1990) noted in a sharp review, "to lead the response to art back to our elementary reactions." The author's turn to neuroscience fulfills that original desire and explains the power of images on the basis of the neurobiological mechanisms of automatic empathetic reactions. The concern for the universality of response and its transcultural psychobiological roots drives Freedberg to consider that power as an immanent property and to look *in images themselves* for the principle of their efficacy, instead of considering them as configurations that, in order to have power beyond perceptual primitives, must encompass the spectator and historical and cultural forces (Prévost 2003). Christian devotion, for example, can be inspired not only by the more or less realistic depiction of religious scenes and characters but also by abstractions such as pictures of the Sacred Heart or the monogram of Christ. The metaphoric or metonymic sign may have a cognitive, affective, and existential power equal to that of the relevant related object, and that is why, as Gombrich (1990) put it, "there is no line one could ever draw between sacred images, words, or signs." In all cases, it is context that makes those objects sacred.

As a historian of iconoclasm, Freedberg (1985) knows this; as a neuroaesthetician, however, he attributes the "energy" of the Buddhapada, the highly revered stone footprints that constitute early aniconic representations of the Buddha, to the activation of motor neurons in the viewer's mirror neuron system (Freedberg 2009c). Like the Christian abstractions just mentioned, these sculptures are objects of devotion and carriers of specific powers by virtue of cultural meanings (whether apprehended at the highest theological level or at that of mere behavioral imitation, with all degrees in between). Determining if motor neurons are activated when we look at the trace of Buddha's foot does not change that and (though scientifically interesting) does not add or subtract anything from an analysis of its power: If everyone's motor neurons are activated when contemplating the sacred footprints but only some of us feel religious reverence (or indifference, aesthetic wonder, historical respect, or fundamentalist hatred), then those neurons don't play the key role in how we experience the Buddhapada. Freedberg's position reflects his rejection of the "standard social science model" that, in his view, blocks understanding of "the relationship between the cultural construction of responses and those aspects of response that pertain to our human nature" (Freedberg 2009c; see also 2007, 2008). The solution consists of shifting the investigative and interpretive balance to focus on "how culture modulates biology" and on how "neurology informs history" and to look for "biological and psychological invariances across cultures" (Freedberg 2009c; 2007, 17, 21).

For Freedberg as for others, the neural turn will renew the humanities and make them more meaningful; in aesthetics and art history, it will counter "intellectualizing views of art" (Freedberg 2009c) and make up for the "elimination of the emotional, the empathetic and the realm of non-cognitive corporeal response" said to characterize most of twentieth-century art history and criticism (Freedberg and Gallese 2007, 199). In those fields, Freedberg (2007, 23) explains, emotions "were felt to be too random, too embarrassing and too incidental to the transcendental value of art." While the anthropological and art historical "orthodoxy" allegedly refuses to analyze emotional responses independently of their cultural and historical contexts, neuroscientific research since the 1990s has corroborated insights about empathy as an embodied emotion that were first formulated by several late nineteenth- and early twentieth-century philosophers, psychophysiologists, and art historians (Freedberg 2007, 27–29; 2009a, 87; 2009b, 70; Freedberg and Gallese 2007, 198).

Isn't Freedberg setting up a straw man to boost his cause? He singles out the philosopher Arthur Danto (1924–2013) as an instance of the "intellectualizing" stance according to which aesthetic responses are "purely a matter of the way in which the concept of art is considered" (Gallese and Freedberg 2007). Danto, however, proposed no such thing. His point was rather that telling artworks from other things requires that they be constituted as "artistic" in virtue of theories and that it is therefore these theories that make "art" possible (Danto 1964). After all, there is no evidence that the painters of Lascaux or their contemporaries believed that they were making art, yet there now is such a thing as prehistoric art; before the German psychiatrist Hans Prinzhorn and among others the artists Paul Klee and Wassily Kandinsky in the early twentieth century, the drawings, paintings, and sculptures of the mentally ill, of children, and of indigenous peoples were not "art." During the scandal that followed Australian aborigines' painting over ancient "rock art," a Ngarinyin leader declared, "Some told me recently that 'rock art is dead.' If 'Art' was dead, that would not matter to we Aborigines. We have never thought of our rock-paintings as 'Art.' To us they are images" (Mowaljarlai et al. 1988, 691). And these images, depicting the cloud and rain spirits known as Wandjina, must be repainted in order to perpetuate the presence of those spirits and "stimulate the energies that bring increase and renewal" in nature (692). The relationship to the images is largely determined by their function and status.

Pop art radicalized such connections in the making of the aesthetic relation. Danto described its challenge as being that of "indiscernible counterparts that may have radically distinct ontological affiliations" and asked, "Why is *Brillo Box* art when the Brillo cartons in the warehouse are merely soap-pad containers?" (Danto 1981, 4; Danto 1993). There might be, as he acknowledged, an aesthetic innate sense, yet responses will differ depending on how objects are classed, and the differences will be "as deep as those between bodily movements and actions, between a person and a zombie, between a divinity and an idol" (Danto 1981, 100). In short, if the aesthetic response to artworks involves processes that response to non-art things does not, then the processes involved in reacting to non-art cannot be what defines art or aesthetic response (see also Danto 1997, chap. 5). This is so much so that preclassification of artworks (for example, as fakes or originals, Leder 2001) actually affects people's preferences. Neuroaesthetics holds the diametrically opposite assumption, namely that the ultimate grounds for our response to objects lies in object properties to which our brains automatically respond by virtue of their basic physiology. All the rest comes as a supplement.

For Freedberg, the key is to be found in mirror neurons and on the supposed neural substrates of empathy and embodiment. Together with Vittorio Gallese, the co-discoverer of mirror neurons, he has elaborated a "theory of empathetic responses to works of art that is not purely introspective, intuitive or metaphysical but has a precise and definable material basis in the brain" (Freedberg and Gallese 2007, 199). How would that work?

Mirror neurons constitute a system of visuomotor cells that fire not only when an organism performs an action but also when it observes a similar action performed by another organism, conspecific or not (Rizzolatti and Craighero 2004, Gallese 2009). They were discovered in the early 1990s in the premotor area F_5 and subsequently in the inferior parietal lobule of macaque monkeys; mainly by inference from studies using electroencephalography, magnetoencephalography, and brain imaging, they are said to exist also in humans (Gallese 2007, 2008; review by Rizzolatti and Fabbri-Destro 2010). Although this has been doubted and major methodological difficulties persist, the most reliable conclusion so far is that "changes in the BOLD signal during action observation seem to be consistent with the existence of a mirror neuron system in humans, but they cannot yet furnish conclusive proof" (Kilner and Lemon 2013, R1060; Caramazza et al. 2014), and the discussions around mirror neurons and their functions continues (for example as a *Brain and Behavioral Sciences* forum on the target article by Cook et al. 2014).

As soon as they were discovered, mirror neurons were hailed as providing the bases of language, "theory of mind" (the ability to attribute mental states to others), imitation, empathy (and therefore morality), art, social cognition, as well as social life and intersubjectivity in general (hence the hypothesis that mirror neuron dysfunction underlies autism). Major doubts have been formulated concerning the functions attributed to mirror neurons (e.g., Borg 2007 in connection with intentional attribution or "mind-reading," Hickok 2009 with regard to action understanding, or Jacob 2008 with respect to the representation of an agent's prior intention; Rizzolatti and Sinigaglia 2010 reply that the brain observation-action execution circuit indeed gives the observing individual a grasp of another individual's motor goals and intentions). In fact, the controversy—which empirical evidence alone is unlikely to settle—has not subsided and in any case does not appear to affect neuroaestheticians' convictions.

A widespread way of understanding the role of mirror neuron systems is to say that they *simulate* ("mirror") observed actions, whether performed or depicted, and that such "embodied simulation" is the foundation of our capacity unconsciously to make sense of the actions, emotions, and sensations of others. That is why such simulation may function as the basis for an approach to aesthetic response (Freedberg and Gallese 2007). Looking at an artwork incites in spectators (or rather, in their brains) the simulation of the action depicted or embodied in the work; the action can be that of the figures represented but also the artist's creative motoric gesture. Thus, Michelangelo's marble *Prisoners* activate in spectators the brain areas corresponding to the muscles that seem to be exerted in the sculpture itself. When we contemplate the singing angels of Hubert and Jan van Eyck's Ghent Altarpiece, completed in 1432, "it is hard," Freedberg (2009b, 67) claims, "not to want to imitate them, even to wrinkle one's brows with the apparent difficulty of singing whatever it is they are singing." Research has established that the observation of buccal movements enhances the motor excitability of the respective somatotopic areas in the brain, thus accounting for the imitative capacities of newborns; such findings, in Freedberg's view (2009b, 76–78), add "scientific context" to much older commentaries about the vivid naturalism of the Eyckian angels.

Similarly, Freedberg asserts that, in front of a panel of Luca della Robbia's *cantoria* in Florence Cathedral (1431–1438), which displays beautiful singing angels, "the desire for some form of emulation may well up within the observer. . . . Luca's art is so remarkable that it seems to encourage its beholders, somehow or another, to participate in the movements he so vividly depicts" (72). Freedberg's lyricism is effective, but it appeals to his experience and to a generic universal beholder rather than to empirical demonstration. Caravaggio's *Incredulity of Saint Thomas* at Sanssouci in Potsdam, showing the apostle placing his finger in the resurrected Christ's wound, gives rise to "empathy for tactile sensations" (201). And Goya's *Disasters of War* provoke physical reactions in the same body parts that are depicted as mutilated in the pictures—a physical empathy that "easily transmutes into a feeling of empathy" (Freedberg and Gallese 2007, 197) and thereby opens the way for a moral response. (See also Freedberg 2008 on Rubens's *Peasant Dance*.) Together with such investigations as Paul Ekman's on the facial expressions of emotions or Peter Lang's and others' on the affective, facial, visceral, and behavioral reactions of looking at pictures, mirror neuron research seems to add empirical substance to Elaine Scarry's or Susan Sontag's classical essays about responses to the pain of others and even provides the basis for an "embodied narratology" (Gallese 2011, Wojciehowski and Gallese 2011).

These conjectures do not apply to figurative art alone. Viewing Jackson Pollock's action paintings or Lucio Fontana's cut canvases too provokes "embodied empathetic feelings" in response to the "visible traces of the artist's creative gestures, such as vigorous modeling in clay or paint, fast brushwork and signs of movements of the hand" (Freedberg and Gallese 2007, 199). Freedberg and Gallese take the artists' drippings and tears as "visible traces of goal-directed movement" (202). Yet a machine, or a chimp, or one of us could have randomly made those marks, or they could result from Mike Bidlo's calculated effort to create a *Not Pollock* that looks exactly like an existing original. It is an empirical matter whether our brains respond in the same way to Pollock-looking images when "they" know and don't know how the images were made or whether empathic simulation in front of Fontana's slashed canvases is verified in subjects who lack familiarity with sharp objects and stretched surfaces. Of course, our visual processing systems may respond to a Bidlo as to a Pollock, just as we may like or dislike certain objects independently of whether they are labeled "art." Our brains perhaps make us feel gestures that did or did not take place. But perception, awareness, and previous experience of context (including materials, size, position, authorship, and categorization) contribute decisively to the status of the traces supposed to give rise to empathic response, and this status in turn plays a crucial role in shaping our cognitive and emotional relations to them.

In experiments using various methods (EEG, transcranial magnetic stimulation, eye tracking, and event related potentials), Freedberg, Gallese, and associated researchers have reached results they see as empirical support for the "embodied simulation" thesis (Battaglia, Lisanby, and Freedberg 2011; Massaro et al. 2012; Sbriscia-Fioretti et al. 2013; Umiltà et al. 2012). They have shown, for example, cortical sensorimotor activation during the perception of static abstract-art images or corticomotor excitability effects that arise only during the observation of original works of art (as opposed to a copy or to a photograph showing a gesture they consider to be the same as that depicted in the original). Like most studies in neuroaesthetics, these articles combine highly technical methodologies with surprising weaknesses in such basic aspects as choice of stimuli and control situations. Putting these aside, their consequences for aesthetics are of two sorts.

On the one hand, these experiments, again like many in neuroaesthetics, imply the existence of biologically grounded criteria for evaluating artistic quality. This was obvious in the inquiries into beauty sketched above, whose aim, as has been justly observed, is "to extract rules that would lead to a practical definition of beauty, connecting features of objects and neural activity" (Conway and Rehding 2013, 1). Although Freedberg, Gallese, and their collaborators do not say it, they suggest the same goal when commenting on the correlation between looking at (the reproduction of) an original and a cerebral effect they interpret as evidence for embodied simulation: "Since observation of the photograph [of a gesture] did not significantly affect corticomotor excitability, we assume that this effect, in the case of the painting [depicting that gesture], must be a consequence of the artist's skill in giving the illusory impression of movement" (Battaglia, Lisanby, and

Freedberg 2011, 4). This opens the way for a dubious application of reverse inference: demonstrate cortical activation, and you have a proof of artistic quality.

On the other hand, precisely because of their focus on automatic unconscious or preconscious processes, the experiments we just mentioned differentiate between *aesthetic experience*, where embodied simulation is "an important component," and *aesthetic judgment*, which, as "the explicit aesthetic rating of an object according to culturally and socially determined aesthetic canons," represents "the most cognitive aspect of the relation established with works of art" (Massaro et al. 2012, 15). In the *aesthetic relation*, however, judgment and experience (even in the simplified senses given here) are not only continuous, as the quoted formulation suggests, but constitutionally interdependent.

Mirror-neuron-based embodied simulation theory thus brings us back to the question "What does neural hermeneutics contribute to the understanding of art and of the aesthetic and artistic relations?" It proposes mechanisms that may be universal as well as necessary for perceiving and responding to artworks. Indeed, we experience visceral sensations when we encounter tormented bodies in art. But we also feel them before early modern instructions for judicial torture or present-day journalistic reports and photographs of atrocities or gruesome accidents. In all cases we may be moved to disgust, compassion, and indignation; we may feel prompted to act or paralyzed with fear or despair; we may also be curious or admiring, find the object praiseworthy or contemptible; and we may experience any combination of the above, and more. In all these cases, the visceral-mirror component of our reactions (the supposed "embodied simulation") is conceivably primary and automatic, but its relevance ends where questions of aesthetics barely begin. (Automaticity itself has also been questioned: de Vignemont and Singer 2006 suggest that empathy requires an appraisal related to the kinds of emotions involved, the relationship between empathizer and target, features of the empathizer, and situative context.)

Studies of the "empathic brain" (Keysers 2011) may identify the regions "involved" in the human capacity to understand other people's sensations, intentions, and emotions. If anything, however, those regions' activity can be no more than one condition of the aesthetic relation, and what the research contributes is nothing more "than a further 'implementation' story about our acknowledged abilities to respond to visual representations" (Davies 2014, 11). At bottom, the question is whether the activity of mirror and "canonical" neurons is constitutive of aesthetic response.[19] In response to David Freedberg and Vittorio Gallese, two cognitive scientists and philosophers of perception argued that the authors' proposal for a neural basis of empathic responses to art was "open to the charge of irrelevance to the issues of aesthetic experience and what constitutes artworks" (Casati and Pignocchi 2007). They emphasized that if witnessing real-life scenes that correspond to artistic depictions arouses "relevantly similar" neural responses, then neuronal activation "is not sufficient for aesthetic appraisal or judgments that something is an artwork." Gallese and Freedberg (2007) replied that "mirror and canonical neurons are crucial elements in aesthetic response." For them, aesthetic judgment requires the simulated embodiment and empathetic engagement that follow upon visual observation via mirror neuron activity; such processes "might be precognitive" and "not always" informed by cognition and culture. Moreover, insofar as "artistic skill" lies in artists' eliciting emotional and felt motoric responses in the spectators, the "intellectualizing" (for example, Danto's) view of aesthetic response must be mistaken.

The two sides do not even seem to understand "constitutive" and "crucial" in the same way. It is nonetheless telling that neuroaestheticians underline that the responses they study apply "in the case of lesser known—and sometimes everyday—images" (Gallese and Freedberg 2007). Indeed, making those responses and their mechanisms play *the* essential role in thinking about art implies that the differences, perceptual and other, between a Mondrian, the Parthenon, and a grid (or a picture and its reproduction, or a passport photo and the Veronica) are irrelevant for understanding aesthetic response. Not only does neuroaesthetics neglect materiality, and with it size, texture, color, and scale; it also does not differentiate between the picture of a hammer and the real hammer one can grasp, the depicted person and the real person with whom one could interact. It disregards the radical difference between the "affordances," the distinct action possibilities offered by things and by their images.

Perceiving an artwork may neurobiologically prepare the spectator for actions or interactions, but these will not be necessarily realized; art thus transcends the actual enactive possibilities related to what it represents or conveys, but, by the same token, it opens up new, different ones (Gallagher 2010). There is no need to deny the existence of perceptual primitives or of bottom-up automatic processes to realize that neuroaesthetics disregards distinctions that are essentially relevant for the theory, practice, history, and reception of art and in general for the aesthetic relation. We concur with the philosopher David Davies (2014, 12) that empirical evidence of the kind provided by the new discipline might "inform" aesthetics but that "most of the significant philosophical issues cannot be resolved by appeal to this work." For neuroaesthetics, such an objection is not even comprehensible, since its entire project assumes that aesthetics was for the most part misguided until it began to take the brain into account.

In short, aesthetic experience begins where neuroaesthetics ends; or, as the philosopher Alva Noë (2015, 361) put it, in neuroaesthetics "art isn't explained; it is explained away." Indeed, neuroaesthetics requires us to give up the very concept of "art," to impoverish it to the point of considering its productions as a primitive form of brain imaging, and beauty as the automatic outcome of the arrangement of visual stimuli (Cappelletto 2009, 151, 152). More generally, the discipline moves art into an epistemic framework that excludes the notion of intentionally produced works (Fimiani 2009). Maybe this was the direction Freedberg (1989, 437) pointed to in *The Power of Images* when he complained that our perception is clouded by "the compulsion to establish whether an object is art or not." There are occasions indeed when that compulsion and the discourse that surrounds it are obstacles to both feeling and understanding, yet if *art* is to remain a meaningful notion and the *aesthetic relation* a meaningful experience, cortex without context simply won't do.

THREE

Cerebralizing Distress

From our exploration of the neurodisciplines that deal with culture and cultural productions, we concluded that "cortex without context won't do." By this we meant to sum up the observation that the methodologies that require leaving out or are incapable of taking into account contextual factors turn out to miss the objects and processes they claim to be studying—objects and processes that are intrinsically contextual. But if there is an area where the role of context has been the focus of debate, it is the understanding and management of mental distress in all its forms. (We shall retain the term "distress" even though, as we shall see below in connection with neurodiversity, not all diagnosed people agree that they suffer or that their suffering can be attributed to the diagnosed condition.) Contested contexts here include the entire range from the genetic to the biographical and the familial to the ethnic, economic, and sociopolitical. While the role of these environments in mental distress is widely recognized, the discussion hinges on their relative weight and on how best to understand their interactions. The "cerebralizing" of psychological suffering has long been at the heart of that discussion. While disagreements have frequently involved stark dichotomies between nature and culture or reductionistic views of different types (for reduction can be as much culturalist as genetic or neurobiological), we have here chosen to examine the ambivalent modes in which the *neuro* serves a variety of often contrary claims and purposes.

The Engines of Cerebralization

In an article published in *Nature* in 2008, Steven Hyman, a Harvard professor of neurobiology and NIMH's director from 1996 to 2001, acknowledged that "despite the disease burden attributable to neuropsychiatric disorders, and despite significant research, their mechanisms of pathogenesis and precise genetic and non-genetic risk factors have remained stubbornly out of reach" (Hyman 2008, 890). Immediately following his rather bleak assessment, Hyman claimed that "this parlous state of affairs is finally beginning to improve, in part through the application of new genomic technologies coupled to advances in neuroscience." This "glimmer of light" announced a "new dawn" in the diagnosis and treatment of "neuropsychiatric disorders" (893). We could quote dozens of claims characterized by the same structure: first comes a strongly pessimistic observation about the "current" situation, then a declaration of hope in future breakthroughs in understanding pathogenesis. These breakthroughs depend on the underlying belief that psychological distress is essentially a state of the brain and must be ultimately understood and explained as such, a belief also expressed by the common use of *brain disorder* and *neuropsychiatry* to refer to what used to be called *mental disorder* and *psychiatry*.

Hyman defines "mental disorders" as a "diverse group of brain disorders" principally affecting "emotion, higher cognition and executive function." In fact, for him, the expression "mental disorders" is an "unfortunate anachronism" dating back to an era when the conditions thus named "were not universally understood to reflect abnormalities of brain structure, connectivity or function." Widespread as it is, such conviction is, as just mentioned, invariably accompanied by the equally general acknowledgment that identifying precise neural abnormalities underlying those disorders has "stubbornly defied" research efforts (Hyman 2007, 725). At least since the 1990s, such ambivalence has been a central feature of the cerebralization of distress and, therefore, too, of how it has impinged on self-identity and self-understanding.

There is by now considerable anthropological and sociological work on these matters. Some of it deals with how neuroimaging, as a major vector of cerebralization, has contributed to shape subjectivities and has been integrated into the discourses and practices not only of patients but also of parent groups and health professionals (e.g., Borgelt et al. 2012; Buchman et al. 2013; Cohn 2010, 2012; Dumit 2003, 2004; Eijkholt, Andersson, and Illes 2012; Illes et al. 2008). In her groundbreaking ethnographic exploration of mania and depression in American culture, the anthropologist Emily Martin examined the spread of brain-based vocabularies in psychiatry and their impact upon issues of self-identity and self-identification (Martin 2007, 2010). The case of alcoholism (and we shall mention others) illustrates how the tendency to map personhood and illnesses onto the brain by way of a certain "folk neurology" coexists with and may preserve rather than upset older notions (Vrecko 2006).

The subjectivation processes at work in the area of mental disorders instantiate a phenomenon we noted when discussing the notion of the *cerebral subject* in Chapter 1: Neuroscientific ideas do not necessarily transform self-understandings in any radical manner but combine with existing perceptions and sometimes reinforce current norms. Thus, addiction understood as a brain disorder turns out to strengthen rather than weaken the appeal to individual responsibility. Maintaining a healthy brain implies a "way of life characterized by autonomous, responsible citizenship," for whose attainment an actively exerted willpower is more important than passively taken medication (Netherland 2011, 172).

In short, the cerebralization of psychological distress is no straightforward affair, and ambivalence is one of its central features. At the level of individual and group experience, interpreting mental illness as a brain condition can be liberating, but it can also generate new stereotypes and mechanisms of exclusion; it can inspire new socialities but also erect identitarian barriers. At the scientific level, it promises to be the source of progress in diagnosis and treatment, yet even its protagonists acknowledge over half a century of few advances and many failures. In this chapter, we shall explore these dynamics by way of two cases, one focused on scientific research, the other on the making of collective and individual subjectivity: the neuroimaging of depression and the claims for autism as a form of "neurodiversity." Before this, however, we must sketch some important elements of the broader contexts to which both belong, namely the emergence of the "pharma-psych nexus," the globalization of mental health, the logic of biomarkers, and the crisis of the biological model.

Pharma-Psych

The expression *pharma-psych nexus* (Williams, Katz, and Martin 2011) has been used to capture the spread of psychopharmaceutical products that target the brain's chemistry. This includes, among others, the commercialization of selective serotonin reuptake inhibitors (SSRIs) for depression and anxiety-related disorders and of psychostimulants like methylphenidate (well known under the trade name Ritalin) for ADHD as well as the use of these and other substances, such as modafinil (indicated for the treatment of narcolepsy), for recreational and enhancement purposes. In his books *The Antidepressant Era* (1997), *The Creation of Psychopharmacology* (2002), *Let Them Eat Prozac* (2004), *Mania: A Short History of Bipolar Disorder* (2008), and *Pharmaggedon* (2013), the psychiatrist and historian David Healy has critically examined the collusion between medicine and the pharmaceutical industry, particularly in the domain of mental health, with depression as a major case (see also Bentall 2009, Greenberg 2010, Kirsch 2009). Healy and others have demonstrated how extensively the production of evidence in psychiatry has been co-opted by economic and marketing considerations. Pharmaceutical companies largely draw on biased ghostwriting, make sure that only positive results are published while reframing or concealing the negative outcomes of clinical trials, and exaggerate the effectiveness of medications (Angel 2004; Dumit 2012; Goldacre 2013; Gupta 2014; Healy 2004, 2008; Kirmayer and Raikhel 2009). Insofar as the drug-based approach has fueled the expansion of mental illness to its current epidemic proportions, the system sustains itself (Whitaker 2010).

The pharma-psych drive is not merely a matter of economics and medicine but also of professional ethics. The pharmaceutical industry's funding of biomedical research and education generates conflicts of interest that medical doctors and researchers frequently prefer not to disclose. Companies use material incentives to increase prescription rates and encourage the adoption of new drugs over available generics; most American physicians have accepted gifts yet tend to downplay their influence (Armstrong 2012; Gibbons et al. 1998; Grande 2010; Green et al. 2012; Grande, Shea, and Mitchell 2009; Hodges 1995; Wazana 2000). This situation has led to a significant weakening of public trust and to intense discussions about how best to regulate this area of the medical profession (Grande 2010).

As reflected in hundreds of online forums, trust has also been shaken by other factors. One is increasing awareness that the spread in the use of a particular medication enlarges diagnostic boundaries and even generates new diagnostic categories. Depression, for example, has expanded to encompass sorrow, sadness, and shyness—states that, even when intense or prolonged, do not necessarily signal mental illness (Frances 2013, Horwitz and Wakefield 2007, Lane 2007). Another concerns medication. The discovery in the 1950s of the antipsychotic and antidepressant effect of certain synthetic compounds (chlorpromazine was the first) and the subsequent introduction of prescription psychotropic drugs gave rise to the claim that mental illness is caused by a "chemical imbalance" in the brain (Whitaker 2010). As far as depression is concerned, the "imbalance" view received support from the fact that SSRIs have antidepressant effects in some patients. In fact, neither the cause of those effects nor the modes of action of those drugs are known. Many call the imbalance theory a "myth," and it is clear that it should be seen at the minimum as a metaphor (Moncrieff 2008). Yet it has been uncritically conveyed by the media and successfully promoted by both psychiatrists and the pharmaceutical industry, for whom it has had a huge marketing value (Lacasse and Leo 2005, Leo and Lacasse 2008).

Contrary to the assertions of pharmaceutical advertising about the action of particular medications, psychiatric drugs lack specificity and have overlapping effects that do not correspond neatly to specific symptoms, disorders, or neurotransmitters. Pharmaceutical companies, however, have covertly and intentionally committed the "therapeutic fallacy," suggesting that the drugs they advertise are supported by a causal theory about the targeted psychopathology. The theory, however, looks valid mainly because the marketed drug improves certain symptoms. Recent debates about antidepressants probe three possible explanations for the effectiveness of these medications: They are effective because their active component has specific and targeted psychodynamic action (this is most commercially interesting claim), the placebo effect is responsible for the medications' effectiveness, or the drugs involve some unknown mechanism of action that provokes a nonspecific altered mental state alongside with a placebo effect (Gupta 2014, 59).

The psychiatrist and bioethicist Mona Gupta notes, "All three interpretations are plausible, but none is self-evidently true or false" (59). Now, if that is the case, then the psychiatric community has the prerogative to determine which is most likely. Professional and financial interests tend to balance the choice toward the first explanation, which presupposes specific antidepressant effectiveness. Hence, as the historian Edward Shorter (2013, 4–5) sharply put it,

> Today, with the ubiquity of the diagnosis of depression, we have the idea that low mood and an inability to experience pleasure are our main problems; we see ourselves as having a mood disorder situated solely in the brain and mind that antidepressants can correct. But this is not science; it is pharmaceutical advertising.

Globalization

Psychopharmaceutical marketing has also contributed to the globalization of psychiatry and the high prevalence of depression, as documented by research in India (Ecks 2013, Ecks and Basu 2009, Sumeet and Jadhav 2009), Japan (Applbaum 2006, Kirmayer 2002, Kitanaka 2011), Brazil (Béhague 2009; Biehl 2005, 2006; Leibing 2009) and Argentina (Lakoff 2005, 2006). While ethnographic studies tend to corroborate the existence of a global psychopharmaceutical hegemony (Good 2010), the distribution of spending on pharmaceuticals is strongly asymmetric and is determined by economic incentives (Petryna and Kleinman 2006). In the area of mental health, the result is overdiagnosis and overmedication in the richest countries and dismal neglect in the poorer ones (Kleinman 2012).

Such an imbalance in the distribution of resources must be placed in the framework of the discussion concerning the contribution of mental disorders to the global burden of disease (GBD) as measured in Disability-Adjusted Life Years (DALYs, or number of years lost due to ill health, disability, or early death). Neuropsychiatric conditions, including common ones such as depression and anxiety, addiction disorders (alcohol and substance abuse), and psychoses and dementia, account for up to a quarter of all DALYs and up to a third of those credited to noncommunicable diseases, with high variability among countries and income levels (Prince et al. 2007, 2014). Depression is considered to be the major contributor to the GBD and, together with anxiety disorder, accounts for between one-quarter and one-third of all primary healthcare visits worldwide (Prince et al. 2014, 103). The high burden imposed by mental disorders according to epidemiological estimates coexists with the secondary place of mental health in global health agendas and policies. The Global Mental Health (GMH) movement, which was most visibly launched by the British medical journal *The Lancet* in 2007, highlights the "treatment gap" between the need for and the availability of mental health services, especially in low- and middle-income countries. To overcome this gap, the World Health Organization (WHO) initiated in 2008 the "Mental Health Gap Action Programme" (mhGAP) (Cohen, Patel, and Minas 2014; Hanlon, Fekadu, and Patel 2014; Patel 2012; WHO 2008).

These proposals have been accompanied by controversy, for example over the validity of diagnostic instruments across different countries and the reliability of epidemiological estimates of the global prevalence of mental disorders (Mills 2014; Summerfield 2008, 2012; Watters 2010). However, disputes about technical aspects ultimately concern the conceptual framework that merges mental distress with neurological disorder, the fundamental assumption that mental illnesses are essentially disorders of the brain. Inherent in the cerebral localization of mental distress is the epistemic hierarchy we noticed in the previous chapter: It is believed that only the discovery of neurobiological causes will satisfy the ambition to "define true madness" and that thereby will the "real" contribution of mental disorder to the GBD be established (Rose and Abi-Rached 2013, 130).

Such convictions about causality (to which we shall return) are relevant for the quest, present in the GMH movement, to reconcile biological universality and cultural particularity. Different cultures have different beliefs about the meaning of *mind* and mind-body relationships, but it is accepted that brains are basically the same across the entire human species. The 2001 World Health Report states: "Mental disorders are not the exclusive preserve of any social group; they are truly universal. Mental and behavioural disorders are found in people of all regions, all countries and all societies" (WHO 2001, 23). The universality of disease is in this case sustained by a universal neurobiology, which justifies introducing in different cultural contexts trans-cultural intervention packages and modes of diagnosis while at the same time acknowledging variation at the level of the expression and triggers of psychopathology (Cohen, Patel, and Minas 2014; Patel 2012; WHO 2013).

Since the mid-1990s, however, and at least within the GMH movement, the universality of disease has become dissociated from the globalization of nosologies and even from the global use of the very notion of "mental disorder." Primary care workers in developing countries are uncomfortable with that notion and "contend that the use of symptoms to diagnose mental disorders, without consideration of context . . . essentially flags non-clinically significant distress" (Jacob and Patel 2014, 1433). Thus, noticing, for example, that in low- and middle-income countries "very few patients report feeling depressed" and that most interventions targeting depression avoid the use of the label, two major actors of GMH have advocated not just dimensional approaches to distress but the abandonment of prevalent international classifications in favor of new bottom-up taxonomies that would be elaborated independently of specialist perspectives (Jacob and Patel 2014; compare with Patel and Winston 1994).

Biomarkers

At the research level the assumption that mental distress involves brain anomalies fuels the quest for biomarkers that can distinguish better between normality and pathology, grasp etiological factors, and aid in the development of treatments that will be effective because they hit the targeted abnormalities. Yet even the staunchest advocates of the neurobiological approach acknowledge that biomarkers remain "stubbornly out of reach" (Hyman 2008, 890). In 2002, in a contribution to the preparation of the fifth edition of the *DSM*, the American Psychiatric Association's *Diagnostic and*

Statistical Manual of Mental Disorders (released in 2013), a group of prominent biological psychiatrists remarked that psychiatry had "thus far failed to identify a single neurobiological phenotypic marker or gene that is useful in making a diagnosis of a major psychiatric disorder or for predicting response to psychopharmacologic treatment" (Charney et al. 2002, 33). Over a decade later the situation has not changed.

Thus, an article from 2011 on the challenges involved in searching for autism biomarkers concluded that "despite huge advances in the basic scientific understanding of autism, comparatively little has been achieved to date with regard to translating the resulting evidence into clinically useful biomarkers" (Walsh et al. 2011, 609–610). And in 2014, the Emory University neuropsychiatrist Helen S. Mayberg, a major figure in the field of depression neuroimaging, pessimistically admitted that "the claims of clinics that they can reliably use structural or functional brain scans" for diagnostic and treatments goals "is without medical or scientific support." Even worse, such claims are "beyond the scope of current research and give false hope to patients and their families" (Mayberg 2014, S34).

The reason for such failures reside, partly at least, in the categories for which the biomarkers are being searched, which are those provided by the *DSM* and the *ICD* (the WHO's *International Classification of Diseases*).[1] Possibly there are no biomarkers for the symptom clusters these classifications identify as diagnostic categories. Hence the NIMH initiative, launched in 2011, to turn away from *DSM* categories and develop "Research Domain Criteria" (RDoC) aimed at transforming psychiatric diagnosis via the convergence of genetics, neuroimaging, and cognitive science (Insel et al. 2010; Insel 2013; Kapur, Phillips, and Insel 2012).

RDoC represent a new angle in the search for neurobiological markers but not a radical new departure. In fact, they maintain intact the established neurobiological view of mental disorder, with its focus on discrete biological mechanisms at the expense of a more integrated "ecosocial" approach (Kirmayer and Crafa 2014). In RDoC, biomarkers will no longer be coupled with *DSM* categories, but mental illnesses will remain defined as "biological disorders involving brain circuits that implicate specific domains of cognition, emotion, or behavior" (Insel 2013). The problem is that elucidating the etiology of those disorders demands "trust in the brain, not in *DSM*" (Rose 2013b, 10), yet trust in the brain has so far produced virtually no results of clinical or diagnostic utility. Various neuroimaging findings have been said to correlate with learning and performance in children and adults, criminality, health-related behaviors, and responses to treatments, and it has been claimed that insofar as they may function as neuromarkers, they may contribute to personalize practices in those domains (Gabrieli, Ghosh, and Whifield-Gabrieli 2015). Yet the situation remains as Nikolas Rose and Joelle Abi-Rached (2013, 138) described it, namely, "Each of the pathways that neuropsychiatry has attempted to trace through the brain seems to run, not into the bright uplands of clarity, but into the murky, damp, misty, and mysterious forest of uncertainty." Perhaps the reason is that RDoC are based on the brain-disease model of mental health just at a time when the "bio-bio-bio" model (Read 2005; Read, Bentall, and Fosse 2009), which combines neurobiology, genetics, and pharmacology, has come under attack at the epistemic, ontological, sociomoral, and cultural levels.

Crisis of the "Bio-Bio-Bio" Model?

What are the grounds for criticism of the "bio-bio-bio" model? First, recent and in principle better-targeted medication has not worked as anticipated. The new generations of antipsychotics are not more effective than older and (by now) much cheaper drugs like Thorazine, a trademark of chlorpromazine. The new drugs, moreover, have been related to sudden cardiac death, cardiovascular risk, weight gain, and the development of diabetes (Álvarez-Jiménez et al. 2008; Foley and Morley 2011; Luhrmann 2012; Ray et al. 2009; Weinmann, Read, and Aderhold 2009). The pharmacological disenchantment matches the failure to identify genetic and neurobiological biomarkers and is reinforced by evidence about the role of culture in the prevalence and prognosis of disorders such as schizophrenia (Luhrmann 2007, 2012; but as Cohen and Gureje 2007 document, producing and assimilating evidence about these matters in ways that are not heavily determined by interests and preconceived views is as difficult in relation to culture as in relation to biology).

Second, psychological therapies have made a comeback. The cerebralization of psychiatry has conflicted with psychodynamic, mainly psychoanalytic approaches, around issues of efficacy, diagnostic validity, and

prevalence. Current debates around GMH and the articulation of universality and particularity in mental disorders constitute the most recent chapter
of this ongoing tension, which seems to be entering a new phase. Indeed, as
we shall see below, the recent retreat from prevalent contemporary classifications is at bottom a move toward "denosologizing" mental illness altogether,
that is, giving up diagnostic categories as we know them, to focus instead on
dimensions that can be variously combined and treated in context-sensitive
ways at the behavioral and psychological level.

Given the evidence that the efficacy of antipsychotics has been
overestimated—and their toxicity underestimated—as well as emerging data
regarding alternative treatment options, it has been argued that patients
should be given more choice concerning medication and therapy. A 2012 editorial in the *British Journal of Psychiatry* argued that nonadherence and dis-
continuation of medication by some psychotic patients may "represent a
rational informed choice rather than an irrational decision due to lack of insight or symptoms such as suspiciousness" (Morrison et al. 2012, 83). The
authors emphasized the significance of evidence-based alternatives to antipsychotic medication, mainly psychosocial interventions. Studies show the
efficacy of cognitive behavioral therapy (CBT) in reducing psychotic symptoms compared to other psychological methods (Turner et al. 2014) and
conclude that it "seems to be a safe and acceptable alternative for people with
schizophrenia spectrum disorders who have chosen not to take antipsychotic
drugs" (Morrison et al. 2014, 1395).

While CBT has been recommended in the United Kingdom for new
cases of schizophrenia, long-term psychotherapy has become standard in
some parts of Scandinavia (Balter 2014). Contrary to widespread skeptical
or negative perceptions about psychodynamic therapies, assessments using
randomized control trials support their effectiveness (Bhar and Beck 2009;
Fonagy et al. 2015; Leichsenring and Klein 2014; Leichsenring and Rabung
2008, 2011; Rosenbaum et al. 2012; Shedler 2010; Thoma et al. 2012). Carried
out in the context of a generalized turn to evidence-based practices in insurance and healthcare policy making, these studies are encouraging for
both CBT and psychodynamic therapies and contribute to the credibility
of psychological approaches to severe mental disorders at a time when the
search for biomarkers and the use of antipsychotics as a first option seem to
have stalled. As far as depression is concerned, CBT has become less and

less effective, its effect size falling by half since 1977 (Johnsen and Friborg 2015).

Yet the evidence-based conditions under which psychotherapy is being validated have themselves raised objections. Thus, the emphasis on probabilistic outcomes has been criticized as a threat to the psychotherapeutic focus on the specificity of each patient's experience (McKinley 2011); the generalized use of treatment-as-usual (TAU) as a control condition is problematic, given that in each case TAU encompasses diverse treatments and that its composition varies in ways that affect assessment outcomes in uncontrolled manners (Löfholm et al. 2013); and the real-world *effectiveness* of depression treatments that demonstrated *efficacy* in strict randomized control trials criteria has for the most part not been assessed (Balt 2014, Blais et al. 2013).

In short, the therapy wars are not about to end (Burkeman 2015), and the fact that they likely never will points to a further consideration relative to criticism of the bio-bio-bio model since the late twentieth century. Advocates of psychotherapy do not defend a purely psychological understanding of mental disorders, and the backlash against the model does not deny the etiological role of genetics or neurobiology. Rather, it reflects the emergence of a more systematic focus on the interactions among biological, social, and cultural factors. Epigenetics has come to provide not only empirical data but also a model, insofar as it concerns the study of changes in the regulation of gene activity and expression that do not depend on gene sequence but are heavily influenced by the environment (Carey 2012; Meloni 2013, 2014a, 2014b; Rose 2013b). The epigenetic approach has profound implications for research, mental health services, and prevention, as it replaces the earlier focus on genetic predispositions and inborn susceptibility or vulnerability and opens ways for showing how an adverse social environment "gets into the mind" and "under the skin" (Hyman 2009) and affects mental health (Toyokawa et al. 2012).

For example, while there is evidence of a relationship between childhood trauma and subsequent psychosis, understanding it requires integrating biological and psychosocial paradigms, and that is likely to be done largely via the identification of epigenetic processes (Larkin and Read 2008; Read, Bentall, and Fosse 2009). In the case of schizophrenia, nutritional intake (an environmental factor) can affect epigenetic processes associated with the disorder. The study of survivors of the Dutch "Hunger Winter" of 1944 and of the Great Chinese Famine of 1959–1961, both of which implied prenatal food deprivation, revealed a twofold increase in the cumulative risk of schizophrenia in the birth cohort. The effect of such deprivation on the *IGF2* gene, which provides instructions for making a protein that plays an essential role in prenatal development, offers a plausible epigenetic mechanism for the environmental roots of schizophrenia (Toyokawa et al. 2012). Epigenetic differences linked to susceptibility to psychiatric disorders might arise through exposure to stress-related factors during critical developmental periods, and various models propose to explain how epigenetic regulatory mechanisms contribute to behavioral phenotypes in schizophrenia and depression, drug addiction, and fear-related anxiety disorders (Dudley et al. 2011). Models are also being developed for gene-environment interactions accounting for the well-documented effects of early life stress (childhood abuse, neglect, and loss) as a risk factor for the later development of depressive disorders (Heim and Binder 2012).

In short, by aiming to give equal weight to genetics and the environment, the epigenetic trend instantiates a sort of social turn in the biological sciences (Meloni 2014a) and, in any case, represents the breakdown of the bio-bio-bio model as far as the understanding of psychic distress is concerned.

Finally, there have been new departures with regard to the moral and political effects of the pure biomedical approach. As long as it was believed that reframing mental illness as brain disease reduced stigma, the bio-bio-bio model could be perceived as nourishing acceptance, diversity, and human rights. Biological explanations seemed to exempt individuals from responsibility for their disease (Corrigan et al. 2002, Lopez-Ibor 2002). Claiming that mental disorders are diseases "like any other," antistigma campaigns espoused cerebralization in the belief that public acceptance of biological causality would inspire more tolerant attitudes (Cheek 2012). Even scholars who debunked the chemical imbalance theory of mental disorders saw in it a "convenient" way of helping to destigmatize psychiatric illness (Angell 2011).

It turns out that this destigmatizing effect has been exaggerated and that the biological conception of mental illness has sometimes even provided new grounds for intolerance (Angermeyer and Matschinger 2005; Bennett, Thirlaway, and Murray 2008; Phelan 2005; Read and Harré 2001; Schnittker 2008). The idea that individuals are not responsible for their disorders and that, therefore, their brain is to be "blamed" can promote further stigma. Neurobiologization can heighten the perception that mentally ill individuals are dangerous, precisely because they lack control and appear unpredictable. In addition, it contributes to erect boundaries between "healthy" individuals and mental health sufferers, now seen as biologically different. Thus, 42 percent of people interviewed in a Canadian survey would no longer socialize with a friend with mental illness, and 55 percent would not marry someone suffering from a mental disorder (Cheek 2012). Other research suggests that attributing mental illness to genetic or biological underpinnings increases public stigma and social distance and that "people who are the intended beneficiaries of stigma reduction campaigns . . . may internalise a stigma-reduction message while the society around them fails to do so" (Buchman et al. 2013, 71). If the biomedical model of illness ever had a moral justification, it is also now largely gone.

The situation we just sketched is complicated. Different factors drive both the cerebralization of distress and its critique, and they may interlock in various ways. It is therefore more productive to map that complexity than to reproduce dichotomies that are not confirmed in the field. For example, it may seem that the cerebralization of mental distress goes hand in hand with the reification of nosological categories—that, to put it bluntly, neuroimaging and the *DSM* express the same epistemic outlook on mental illness. However, as we saw, while constituting a move away from *DSM* categories, the Research Domain Criteria program pursues the search for neurobiological markers and strengthens the brain model of mental distress. Such ambivalence, we propose, is an essential feature of the cerebralization of psychic distress and therefore also of the processes whereby human experience and views about personhood sometimes come to embody ways of "being brains."

Depression

Perhaps more than any other psychiatric condition, depression remains torn between biomedical and psychological accounts, between neurobiological causes and contextualized explanations. Though generally understood as involving an array of factors, from genetic predispositions to environmental circumstances, it has turned out to be impossibly challenging to bring these factors together. The stakes of the challenge are considerable, since (in 2010) major depressive disorder (MDD) ranked as the second leading cause of disability worldwide and as the eleventh leading cause of global burden of disease (Ferrari et al. 2013). Depressive disorders have therefore become a global health priority, and the WHO recurrently calls for coordinated global action.[2]

At the same time, there have been intense debates over whether depression is overdiagnosed and antidepressants overprescribed (Reid 2013, Spence 2013) as well as over the efficacy of antidepressants (better than placebo? only in severe cases? Fournier et al. 2010, Gibbons et al. 2012, Kirsch et al. 2008, Turner et al. 2008). The difficulties of seeing clearly in this domain are compounded by the fact, mentioned above, that drug manufacturers withhold negative data and that only positive data tends to be published. A significant turn occurred around 2011, when major companies including Novartis, GlaxoSmith-Kline, AstraZeneca, Pfizer, Merck, and Sanofi decided to stop investing in research on drugs for brain disorders and redirected their efforts toward genetics (see Tracy 2016 on the "neuro funding rollercoaster"). The decision was motivated by commercial considerations: Because many generic psychiatric drugs are available, because new medications do not work better than the older ones, and because most candidates aimed at new brain targets fail after years of clinical trials, companies concluded that there are better chances of identifying genetic biomarkers than neurobiological ones (Abbott 2011). This crisis joins the weakening of trust sketched above, rooted in the practices of industry and the failure to globalize psychiatric diagnoses and classifications effectively.

Again, however, the situation is complex. In the overall picture, the calls for abandoning the bio-bio-bio model and for elaborating local idioms of psychic distress coexist with programs that variously pursue the model as a means to revamp classification so as to make it truly universal and fully transcultural. Our focus here will be on neuroimaging as a major player in this context. As elsewhere, the uses of neuroimaging in the field of depression and the claims that are made for its significance encapsulate epistemic and moral as well as social and psychological mechanisms involved in bringing about cerebral subjects.

Just Like Diabetes?

Even a superficial look at the successive editions of the *Handbook of Depression* (Beckham and Leber 1985, 1995) demonstrates that, like most psychological and psychiatric entities, depression is not a single thing and that no single approach may be considered simultaneously necessary and sufficient for understanding and treating the condition.[3] In the *Handbook*'s most recent edition, thirty chapters address four main areas (Gotlib and Hammen 2014). Part 1 reviews "descriptive aspects" such as the epidemiology, course, outcome, and assessment of depression as well as issues in methodology, classification, and diagnosis (for example, the relations between personality and mood disorders or the comparison of unipolar and bipolar depression). Part 2 moves from the genetics of major depression to the interpersonal and social environment of the condition, dealing along the way with the contributions of neurobiology and affective neuroscience as well as with depression and early adverse experience, children of depressed parents, and the cognitive aspects of depression. Part 3 examines depression in specific populations (with a chapter on understanding the condition across cultures), and Part 4 considers prevention and treatment, not only pharmacological but also cognitive, behavioral, and psychosocial.

Obviously, neuroimaging is used in only a few of these areas. Given the vastness of the field of depression, the condition's high degree of comorbidity with other psychiatric disorders, the heterogeneity of the category, and the diversity of possible approaches, it is one element in a wider framework of investigative, therapeutic, and economic practices and interests. Neuroimaging, however, is not just one more approach in research and assessment. As we shall see, it is endowed with a certain methodological and epistemic primacy and thereby with the authority to prove that, by virtue of actually being brain disorders, mental disorders, depression included, are "just like diabetes."

Depression is of course an organic illness. It is in general reasonable to admit that mental disease is in important ways "like any other medical disease." To begin with, since "all diseases involve the self," the self-affecting aspects sometimes said to be unique to mental disorders are in fact not exclusive to those conditions (Hofmann 2015). But is depression *just like* an organic illness? Maybe yes, in the trivial sense that it is a biochemical state with potentially discoverable causes. Nevertheless, in addition to being neurobiological and having a cause, depression is a state with contents and "reasons," and it can be judged to be warranted or unwarranted, desirable or undesirable, meaningful or meaningless. You may accept that your depressive symptoms are neurochemical, but if you are told that they are "just like diabetes," you might feel that they are not acknowledged as being also warranted and meaningful (Arpaly 2005).

Less phenomenologically, imagining that depression *is just like diabetes* involves a fundamental confidence in the possibility of discovering biomarkers that will enable diagnoses of the condition according to purely biological criteria. Such redefined criteria might ultimately contribute to "denosologize" psychiatry in the following sense: Presently, categories such as "major depression disorder" are defined on the basis of syndromes, or collections of behavioral *signs* (what is observed) and *symptoms* (the patient's complaints). A denosologized psychiatry would focus on symptoms that, instead of being linked to particular conditions envisaged as discrete entities, would be shared by several conditions (as currently defined) and correlate with *dimensions*, such as aggression, anxiety, or mood. An instance, provided by Herman van Praag, a Maastricht University psychiatrist who has long criticized his field's "nosologomania," is the stress-inducible, "anxiety/aggression-driven depression" (van Praag 2005; on denosologizing, van Praag et al. 1987; van Praag 2000, 2010). For van Praag (2008, 31), the reason why half a century of intensive research has failed to elucidate the biology of depression is that "insufficiently specified diagnostic constructs" will not turn out "to be caused by specific, well-definable pathological processes."

Though not always so strongly formulated, the trend in biological psychiatry moves toward dissolving current nosographic categories and toward identifying the neurogenetic factors involved in depressive symptoms (e.g., Scharinger et al. 2011), diagnostic biomarkers, and biomarkers that will make it possible to prognosticate the illness's evolution and predict treatment efficacy and clinical response. Genetic findings and neural circuit maps link different syndromes or distinct subgroups within syndromes. This is the Research Domain Criteria perspective sketched above. In contrast, diagnosing mental disorders on the basis of clinical observation and patients' reports is seen as implying that the syndromes embody "unique and unitary disorders" and thus as undercutting the possibility of identifying illnesses linked to pathophysiology.[4] The assumption here is that clinical heterogeneity maps onto biological heterogeneity and that the only way out of the nosographic mess is to replace the examination of clinical symptoms by the identification of biomarkers.

Biomarkers are to be understood in terms of vulnerability and susceptibility, risk and probability; moreover, since they are based on groups, their predictive power as risk factors for individuals is low (Singh and Rose 2009, Walsh et al. 2011). Neuroimaging depression research is essentially about the identification of such biomarkers, which in its case take the form of patterns of neural activation that systematically correlate with a diagnosis (major depression disorder, bipolar disorder), with particular symptoms, or with treatment outcome. The neuroimaging of depression thus looks like the neuroimaging of any other "brain disorder." But there are some significant differences.

In schizophrenia, as we mentioned, social and experiential indicators, such as adversity, stressful life events, or childhood abuse and trauma, have been correlated with chances of developing the disorder; conversely, psychological and social interventions play a role in its management. Nevertheless, more than biopsychosocial models, which emphasize factor interdependence, it is the diathesis-stress model, according to which a stressor may trigger an initial illness episode in persons with a genetic predisposition (*diathesis*), which seems to have become the predominant framework for thinking about the condition (see Jones and Fernyhough 2007 for a discussion of the neural version of this model). In spite of the epigenetic turn and the awareness that culture matters, schizophrenia remains depicted primarily as a brain disease.

The diathesis-stress model is also central in depression research. The etiology of depression, both unipolar ("major" depression) and bipolar (the former "manic depression"), is generally thought to include a significant genetic component in the determination of risk, and the condition correlates with changes in neurotransmitter systems involving serotonin, norepinephrine, and dopamine. Nevertheless, while giving considerable weight to biological factors, depression studies tend to underline the interdependence of a multiplicity of risk and etiological mechanisms. It seems more difficult to turn depression into a purely organic illness than it has been to isolate the purported neural correlates or "signatures" of schizophrenia or autism spectrum disorder (on the former, see Cabral et al. 2013 and Hart et al. 2013; on the latter, Ecker et al. 2010 and Deshpande et al. 2013, both accompanied by considerable media coverage misleadingly suggesting that henceforth diagnosis can be made on the basis of brain scanning).

Cultural and historical factors hint at the sources of the difficulty. While there is debate on whether depression overlaps with melancholy and on how much continuity there might be between psychiatric categories and the *melancholia* that the Western tradition links to genius and to a superior manner of being in the world, depression sometimes retains the dark luster of the ancient black bile and is often accompanied by an exceptionally penetrating reflexivity.[5] The comparative literature scholar Matthew Bell (2014, xi) insightfully notes:

> One distinctive feature of Western culture is the high status that it has accorded to self-consciousness. Melancholia, or at least the psychological symptoms of melancholia as reported from Hippocrates right down through Western history, depends upon the West's peculiarly introspective culture. The psychological symptoms of melancholia are, to put it crudely, a disorder of malignant self-consciousness.

Certainly some depressed people associate their distress to a large spectrum of causes and reasons, from the random to the meaningful, from the reductively genetic to the deeply psychoanalytic. Nevertheless, in diverse, often contradictory ways, personal accounts by hitherto unknown patients, movie stars, famous writers, diagnosed academics, or mental health professionals have contributed to the modern persona of the depressive and the public image of the condition.

Such autobiographical narratives neither counterbalance nor contradict neurobiological explanations (Dumit 2003). Nevertheless, the evocation of contexts, moments, relationships, and inner lives gives depression cultural resonance as well as meanings that function as a kind of causal interpretation. For the authors of depression memoirs (admittedly a minority of the diagnosed population) such elucidations make more existential sense than the demonstrations of biological psychiatry. Self-reflexive depressed persons may be fascinated by brain scans and acknowledge that depression is biological (Buchman et al. 2013, Cohn 2010, Martin 2010). However, as autobiographical writings show, they wish primarily to understand contextual and relational factors that neuroimages and correlations can hardly reveal and illuminate. While organic explanations of autism or schizophrenia may satisfy the persons concerned (patients or caregivers), they seem intrinsically insufficient to those directly or indirectly touched by depression. For them, depression is not just like diabetes.

Neuroimaging Depression

In 2005, an article in the *New York Times* noted that brain scans, long celebrated as "snapshots of the living human brain," had been counted upon to illuminate the mystery of mental illness but that the promise had not been fulfilled (Carey 2005). The neuroscientists' response, expressed in that article by Steven Hyman, was that those who oversold the technology forgot that "the brain is the most complex object in the history of human inquiry." For him, the key consisted of pursuing the same line of research. Since that is indeed what happened, it is appropriate to ask what kind of progress has been made.

Meta-analytic analyses of neuroimaging publications, which seek to identify consistent patterns and results across a large number of studies, appeared both before and after the *New York Times* asked, "Can Brain Scans See Depression?" In 1998, Wayne C. Drevets, who later became senior investigator at the Neuroimaging Section of the NIMH Mood and Anxiety Disorders Program in Washington, D.C., reviewed the contributions of functional neuroimaging to knowledge of the pathophysiology and "anatomical correlates" of major depression (Drevets 1998, 341). He hoped that such neurocorrelational studies would "ultimately localize specific brain regions for histopathological assessment, elucidate anti-depressant treatment mechanisms, and guide pathophysiology-based classification of depression" (342). At the time, Drevets noted that the capabilities of neuroimaging to determine diagnosis or guide treatment had not yet been established. Functional imaging seemed nonetheless a promising approach: The fact that some depressive symptoms could be experimentally induced in nondepressed subjects opened the way for depressed-control comparisons of the changes in cerebral blood oxygenation and glucose metabolism "associated with" depression.

However, the exact nature of the association remained nebulous. For example, nondepressive conditions sometimes present in depressed patients can affect functional brain imaging measures; regional blood oxygenation or metabolic differences between depressives and control subjects "may thus reflect either the physiological correlates" of depression "or pathophysiological changes that predispose subjects to or result from affective disease" (Drevets 1998, 342). In short, as a 2008 review of biological vulnerability factors in early-onset depression put it, the quest for the "neurobiological roots" of the condition is obscured by the fact that, when assessing differences in brain function or activity between patients and controls, "it is unclear whether we are measuring causal factors making an etiological contribution to the illness, or, conversely, consequences or associated factors of the illness" (Nantel-Vivier and Pihl 2008, 105).

What are these authors saying? On the one hand, their language remains ambiguous: Is *may* freely conjectural or more or less rigorously hypothetical? On the other hand, it conveys ambiguity concerning the nature of the results. The language avoids causal connectives, employing *predispose* and *result* in the context of a speculative remark, yet at the same time, it suggests a capacity to detect and measure causal factors.

On its first page, the review we just quoted explains that the "putative biological, psychological, and environmental etiological mechanisms" of pediatric depression are "intrinsically linked, interactive, and complementary." Starting with the second page, however, it becomes clear that the analyzed research concerns "biological correlates" supposedly pointing the way to a better understanding of "etiological roots" (Nantel-Vivier and Pihl 2008, 103–104). The authors claim that, by studying pediatric populations, they "significantly decrease the likelihood of the occurrence of confounding factors and can therefore more clearly investigate causative neurobiological forces by getting closer to their etiological roots" (105). One of the main goals of "disentangling the neurobiological factors" is to develop a "biological etiology" and, on that basis, a taxonomy of illness that will yield "more homogenous diagnostic categories" (106). But if some factors are "confounding," then they are not "intrinsically" linked to the others. In fact, the purpose of the study is to isolate the "forces" to which causal efficacy can be attributed, that is, the neurobiological ones. As far as we can tell, such ambiguities in language, as well as the slippage from correlation to causation, are commonplace in neuroimaging depression research and characterize the field of psychiatric neuroimaging as a whole (Boyce 2009).

The same can be noted about the prevalent attitude vis-à-vis the variability of research results. The clinical heterogeneity of depression and the anatomical differences across individuals are major sources of variability; such heterogeneity, as Drevets (1998, 343) explained, also implies that "diverse signs and symptoms may exhibit distinct neurophysiological correlates." "Localization," he wrote, "is now limited as much by the anatomical variability across individuals as by the spatial resolution of imaging technologies" (345). At the time, a related source of confusion came from the fact that imaging results did not differ significantly between subjects with primary depressive syndromes and those whose similar syndromes derived from neurological conditions such as Parkinson's or Huntington's disease (353).

The two chief explanations for the inconsistency of the data (low spatial resolution and the secondary nature of the symptoms) were placed on the same level. Yet, while imaging resolution can improve, as it indeed has since the 1990s, variations in anatomy and brain circuitry are not limitations to be overcome. It is nevertheless hoped that they will cease being an obstacle when the clinically based nosography that still frames neuroimaging studies is replaced by a "pathophysiology-based classification." The stated hope is to refine "our understanding of the anatomical correlates" of depression (358), with the ultimate goal of integrating imaging, neurochemical, and anatomical data so as to move from physiological correlates to anatomo-pathological localizations. At the same time, the data Drevets reported seemed to support a "circuitry model in which mood disorders are associated with dysfunctional interactions between multiple structures, rather than increased or decreased activity within a single structure" (355). A vocabulary of localization thus coexisted, and still does, with an emphasis on brain circuitry.

In 2002, a shorter overview of depression neuroimaging noted the lack of a "general theory" to integrate the findings about functional abnormalities in the amygdala and hippocampus and reached circular conclusions of confounding generality: Since the medial prefrontal cortex is connected to areas where neuroimaging uncovers structural and functional abnormalities, dysfunction in this region may be fundamental to depression. . . . These results thus support a neural model of depression in which dysfunction in regions that modulate emotional behavior may result in the emotional, motivational, cognitive and behavioral manifestations of depressive disorders. (Erk, Walter, and Spitzer 2002, 67)

The recurrent *may* is the hopeful expression that the cause-and-effect connections here envisaged as *possible* will turn out to be true. The ambiguous, evocatively rather than assertively causal language is the same as in Drevets, but Erk and colleagues add an element of self-evidence, since dysfunction in regions that modulate emotion necessarily affect emotion. Insofar as the nosography of depression includes emotional signs, depression necessarily involves brain areas implicated in emotion.

A Quest for "Objectivity"

Also in 2002, an extensive review was coauthored by Richard J. Davidson, the high-profile director of the Laboratory for Affective Neuroscience at the University of Wisconsin–Madison. As a scientist with noticeable media presence and a well-publicized connection to the Dalai Lama, Davidson has been described as "a veritable rock star in the world of neuroscience" (Smith 2009) and was one of the world's hundred most influential people in *Time*'s 2006 ranking.

One of Davidson's best-known messages is that meditation alters the brain. The observation is trivial, since any human activity whatsoever involves and affects the brain. It could be scientifically interesting to know what exactly appears to be altered. In 2003, Davidson and colleagues reported increases in left-sided anterior activation, a pattern associated with positive affect, as well as increases in antibody titers following influenza vaccination in meditators compared with a nonmeditators control group (Davidson et al. 2003). A decade later, Esch (2014) reviewed the effects of meditation and mindfulness that can be detected in the brain as functional and structural alterations, especially in areas related to attention and memory, interoception and sensory processing, and self-regulation, including control of stress and emotions.

While the results are far from surprising and don't really require neuroscience, Davidson's ultimate purpose is to demonstrate that meditation can be put to useful social and psychological uses, such as reducing stress for all or making life easier in maximum-security prisons. Similarly, Tania Singer, the director of the Max Planck Institute for Human Cognitive and Brain Sciences in Leipzig, wishes her neuroimaging research into compassion and empathy to inspire a more peaceful world (Kupferschimdt 2013). In a review of "social influences on neuroplasticity," Davidson and McEwen (2012, 693) write:

> It has also been claimed for thousands of years that specific forms of mental training can produce robust beneficial and enduring effects on behavior. The rigorous investigation of such effects and the neural mechanisms responsible for producing them has only recently become a serious focus of neuroscientific study. The findings that we discuss underscore the structural plasticity of emotional circuitry in response to both acute and chronic stress, particularly alterations of spine density and dendritic length and branching in hippocampus, amygdala and prefrontal cortex.

The modern confirmation of ancient wisdom, lyrically celebrated as "a confluence of streams and a flowering of possibilities" or more soberly as "the convergence of science and the contemplative traditions" (Kabat-Zinn and Davidson 2011, 3) is surely worthwhile for those engaged in the growing enterprise of mindfulness neuroscience (Tang, Hölzel, and Posner 2015) but does not call for spending hundreds of thousands of dollars on brain scans. The empirical results add pieces to our knowledge of the brain, and it is probably relevant to investigate kindness, compassion, and well-being with the same tools that have been used to study hostility, aggression, and suffering. However, the effects of meditation, empathy training, or cognitive therapy do not become more real because they are shown to have neural correlates, nor does knowing that experiential factors shape neural circuits help promote positive social behavior.

Davidson declares that the best way to study the mind is to study the brain (Redwood 2007). Yet neither the neurosciences in general nor neuroimaging in particular can tell us anything about the psychological or social effects of meditation. That is why, when asked about "the link between compassion for others and a sense of personal happiness," Davidson referred to psychological, not neuroscientific data, citing the well-known experiment "in which participants were given $50 to spend. Half were instructed to spend it on themselves, half to spend it on others. Those who bought gifts for others reported feeling happier after the exercise" (Smith 2009). Illustrating claims for neuroscience by discussing psychological rather than neuroscientific results is a widely shared strategy among neurocultural actors—and one through which they involuntarily reveal the limitations of their own cause (see for example Frith 2007 and the critique by Tallis 2007).

Davidson's 2002 review of affective neuroscience perspectives on depression focused on research about the representation and regulation of emotion in the brain (Davidson et al. 2002a, almost identical to Davidson et al. 2002b). It illustrated a growing emphasis on the brain circuitry "underlying" mood, emotion, and affective disorders and how it coexisted (as it still does) with a focus on brain structures (prefrontal cortex, anterior cingulate cortex, hippocampus, and amygdala). It also illustrated the ultimate goal of the majority of such studies, namely to redefine depression subtypes without relying "on the descriptive nosography of psychiatric diagnosis" but "on a more objective characterization of the specific affective deficits in patients with mood disorders" (Davidson et al. 2002a, 546). In other words, the goal is to deconstruct complex processes into "elementary constituents that can be studied in neural terms" and "examined with objective laboratory measures" instead of self-reports (546).

The heterogeneity of mood disorders is one of the "crucial issues" that the neurologizing of clinical concepts aims to resolve. Symptoms are broadly similar, but the proximal causes can be extremely varied, and even "the underlying mechanisms may differ" (547). Indeed, symptoms come in clusters whose specific features "are likely mediated by different neural circuits despite the fact that they culminate in a set of symptoms that are partially shared" (547). Since descriptive phenomenology does not yield a "clean separation of underlying neural circuitry," one should move beyond it, "toward a more objective, laboratory-based parsing of affective processing abnormalities" (547).

The claim to "objectivity," here identified with what happens in a laboratory, bolsters the ultimate goal of reevaluating the relationships between etiology and nosography by defining symptom clusters "that may arise as a consequence of dysfunctions in specific regions" and thus of offering "suggestions for different ways of parsing the heterogeneity of depression in ways that more directly honor the circuitry of emotion and emotion regulation in the brain" (547). Depression types and symptom profiles "should vary systematically with the location and nature of the abnormality" (565). Thus, the "delineation of brain-based illness models . . . is seen as a promising strategy for redefining our depression nosology" (Mayberg 2007, 729), and neural markers of "at-risk individuals may prove to be more sensitive predictors of subsequent depression and sensitivity to treatment than the clinical predictors we have at present" (Keedwell 2009, 97). From a developmental viewpoint, "identifying depression subtypes based on age of onset and neurobiological characteristics may provide us with more etiologically consistent and uniform diagnostic categories" (Nantel-Vivier and Pihl 2008, 111). We have provided many quotations to show how the usual language, floating between the normative and the expectant (*should*), the permissible, the possible, and the hoped-for (*may*), contrasts with the methodological and empirical technicalities of the research, implicitly favors biological causality over integrative models, and conflates "objectivity" with laboratory research and anatomical description.

A Desire for Causality

In the early 2000s, Davidson and his colleagues' expression "may arise as a consequence of" was as far as they advanced toward understanding the causal mechanisms of depression. In connection with the prefrontal cortex, for instance, they observed that some types of depression "may be caused" by abnormalities in the circuitry that implements positive affect–guided anticipation; similarly, anatomical differences in the brains of patients with mood disorders "might account" for some of the detected functional differences (Davidson et al. 2002a, 548, 550). The existence of hippocampal-dependent Pavlovian conditioning (in the form of an association between places and fear responses) "has important implications for our understanding of the abnormalities that may arise as a consequence of hippocampal dysfunction" (556).

Davidson, however, noted: "Whether hippocampal dysfunction precedes or follows onset of depressive symptomatology is still unknown" (557). "We do not know," he added, if any of the discussed functional and structural abnormalities "precede the onset of the disorder, co-occur with the onset of the disorder, or follow the expression of the disorder" (565). Such remarks highlight the limits of neurocorrelational research, which is by definition unable to fulfill its own stated goal of differentiating between causes and consequences. By the end of the decade, neither the updated version of the same review (Davidson, Pizzagalli, and Nitschke 2009) nor any of the brain-related articles in the new *International Encyclopedia of Depression* (Ingram 2009) offered a different view or evidence of progress toward the longed-for knowledge of causes and causal mechanisms.

While the scientific literature invariably underlines progress in knowledge of the brain structures said to "subserve" or be involved in depression, it also acknowledges persistent ignorance about causality and localization. In 2008, for example, an article in *Current Directions in Psychological Science* reviewed the status and unresolved issues in neuroimaging and depression. It summarized neurocorrelational research, assessing the role of several brain structures in major depression, and concluded that heightened activity in the limbic structures engaged in emotional experience and expression dampens activation in the dorsal cortical structures involved in affect regulation. The article devoted different sections to distinct structures or systems (the amygdala, the subgenual anterior cingulate cortex, and the dorsolateral prefrontal cortex) and pointed out that identifying "the patterns of functional connectivity that characterize the depressive neural network" was still a challenge for future work (Gotlib and Hamilton 2008, 161).

As the authors made clear, the fact that "neural abnormalities" accompany depression was known before the advent of neuroimaging. But they also recognized that determining the timing of those abnormalities, as can be done by means of activation patterns (for instance, greater-than-normal amygdala reactivity to affective stimuli during a depressive episode), has so far not illuminated their actual connection to the disorder. The results concerning the temporal relation between neural activation and depression as well as the etiological role of neural dysfunction "are complex and do not cohere to tell as clear a story as we would like" (162). Indeed, anomalies can be present in a diagnosed person's brain or precede the onset of the disease "without being involved in its development" (162).

As in earlier literature, the findings discussed in the 2008 *Current Directions* article "underscore the fact that 'depression' refers to a heterogeneous group of disorders that are not carved at their neurobiological joints in *DSM-IV*"; hence the desire to define depression subtypes and symptom profiles "that are related systematically to neural functional and structural abnormalities" (162). In other words, one should go beyond correlations, establish causal links, and amend the nosology of depression on the basis of the disorder's neural substrates. The goal of deconstructing present diagnostic entities in that way is widely shared among researchers in psychiatric neuroimaging (Abou-Saleh 2006). A more recent overview notes again that "the current classification of depression is essentially clinical and aetiological and pathophysiological factors do not play a significant role"; it also comments that thanks to the development of operational criteria, diagnosis has become "reasonably reliable" but that "doubts about validity can be resolved only by a better understanding of pathophysiology" (Cowen 2013, 11).

It is revealing that the metaphor of "parsing" is applied to the *heterogeneity* of depression. It implies that depression *should* not be heterogeneous—or not in the present manner—but, rather, that it should be reconceptualized so as to facilitate its breakdown into clear-cut brain-based nosographic types and components (for example, patterns of brain activation that correspond to individual differences in severity, accompanying symptoms, or treatment response, though some studies also seek biomarkers to differentiate established categories, such as major depression and bipolar disorder; see Kempton et al. 2011).

The main research operation always consists in *correlating*, but the ultimate aim is to *relate causally*. Hence the problem of what to do with the observation (one among dozens of similar ones) that positive correlations between increased functional connectivity in the amygdala network and Geriatric Depression Scale scores in elderly patients with amnestic mild cognitive impairment "suggest" that connectivity in those areas "is related to the degree of depression." It seems impossible to go beyond hazy general conclusions—in this case, that there is an "interactive neural mechanism" between the dysfunction of emotional processing (supported by the amygdala) and cognitive and memory functions (Xie et al. 2008, T259). Although the predominant "functional connectivity" strategy aims at extracting patterns of covariance, it is assumed that the "activity changes in different locations influence one another" (Mayberg 2007, 729).

The same language that is used when neurobiological interactions and associations are inferred from statistical covariance characterizes a more recent application in psychiatric imaging research, namely diffusion tensor imaging (DTI) studies of white matter hyperintensities. White matter hyperintensities appear on magnetic resonance images as ultrawhite patches that indicate injury to axons. DTI produces neural tract images on the basis of the diffusion of water in tissue (such as the axons in white matter). The variation of diffusion along different spatial directions provides information about diffusion anisotropy (the direction preference of the diffusion process); the results are couched in terms of "fractional anisotropy" (FA), that is to say in degrees of anisotropy (from 0 for isotropic, or homogenous in all directions, to 1 for fully anisotropic). The technique is used to investigate tissue structure and connectivity between regions or points in the brain. While DTI is different from fMRI and other imaging technologies, its basic goal—to correlate pathologies with cerebral locations and circuits—continues to illustrate the assumptions, promises, and limitations of the neurocorrelational logic.

In the field of depression, white matter hyperintensities have been found consistently in elderly unipolar patients. A DTI study of 2009 established that, in comparison with controls, patients with major depressive disorder tend to show lower FA values in the left sagittal stratum; the implied structural changes "may contribute" to the previously detected dysfunction in the limbic-cortical network in depressive patients (Kieseppä et al. 2009, 5). Another meta-analysis of MRI studies of brain volume in MDD observed that some of the areas "involved in" emotion regulation and stress responsiveness exhibit volume reduction. The authors concluded that the integration of MRI and DTI measurements "may improve our understanding of the neural circuitry involved in MDD" and that their own meta-analytic results "strongly suggest that studying brain structure in MDD will contribute to understanding the pathogenesis of this disease" (Koolschijn et al. 2009, 11, 13). They do not explain, however, how pathogenesis can be inferred or demonstrated without some sense of causality or at least temporal direction (see Smith 2015 for numerous other references on white matter hyperintensities).

A 2008 meta-analysis of structural imaging studies remarked that after twenty-five years of scanning bipolar patients and generating over seven thousand MRIs, brain regions "affected in" the disorder remained ill-defined. Given the number of studies considered, significant findings were surprisingly few. There are in fact only three, all "regionally nonspecific." First, bipolar disorder is "associated" with lateral ventricle enlargement and (second) with increased deep white matter hyperintensities; third, lithium use is "associated" with increased total gray matter volume. Conclusion: "There may be genuinely limited structural change in bipolar disorder, or between-study heterogeneity may have obscured other differences" (Kempton et al. 2008, 1026). Regarding major depressive disorder ("unipolar" depression), meta-analytic studies are just as inconclusive: "we still lack information concerning the extent to which structural and functional changes co-occur in a depressed brain," and the "essential neural correlate characteristics for the phenotype of a depressive episode" are still to be discovered (Sacher et al. 2012, 142, 146–147). The high inter- and intrastudy heterogeneity, and the fact that individual investigations are chronically underpowered (that is, they have a small probability of detecting a statistically significant effect), mainly because of excessively small samples, are crucial for understanding such limited achievements. Nevertheless, however much they may be explained by deficient sampling, the occurrence of false positives and false negatives, insufficient control of intervening variables (such as medication), or discrepant nosologies, it is likely that the results also express a variability that is a characteristic feature of the objects and phenomena studied rather than a methodological artifact. (See also Fitzgerald et al. 2008, as well as Hasler 2010, who highlights "the most limited overlap of findings" from functional imaging.)

Neuroimaging depression research has kept looking for treatment-specific biomarkers capable of predicting an individual's improvement in response to a particular treatment and nonresponse to an alternative treatment. For example, it has been suggested that neural response to emotional stimuli in visual cortical areas might be a useful biomarker for identifying patients who will respond favorably to scopolamine (Furey et al. 2013). Similarly, in 2012 another study confirmed earlier suggestions that decreased reactivity to negative words in the subgenual anterior cingulate cortex (sgACC) predicts outcome in cognitive therapy for depression (Siegle et al. 2012; cf. Greicius et al. 2007). And research published in 2013 found that insula hypometabolism is associated with good results for cognitive behavioral therapy and poor response to escitalopram (a selective serotonin reuptake inhibitor), and insula hypermetabolism, with remission to escitalopram and poor response to the same kind of therapy (McGrath et al. 2013).

Functional, structural, and postmortem studies suggest that sgACC abnormalities are the most solid finding in connection with MDD. This was probably to be expected, given the role of sgACC as a crossroads in a network of structures involving the control of mood, memory, appetite, and sleep. These findings led Helen S. Mayberg, already introduced as a leading figure in depression neuroimaging, to try deep-brain stimulation (DBS) of sgACC as a treatment (the pioneer study was Mayberg et al. 2005; for recent reviews, see Anderson et al. 2012 and Schlaepfer et al. 2014). The apparent discovery of a "depression switch" (Dobbs 2006) received glowing media coverage, with most journalists preferring to ignore Mayberg's failure to disclose her financial ties to medical technology manufacturers (Bass 2010). The hype was dampened in 2013, when the U.S. Food and Drug Administration suspended a trial because it failed the "futility analysis," which monitors whether an experimental treatment has reasonable chances of being shown to be significantly better than the control treatments (Horgan 2014).

While treatments for such devastating conditions as depression are to be welcomed, the cautionary value of the DBS story can be extrapolated to the entire field of predictive neuroimaging and the goal of revamping nosology on purely neurobiological bases. The identification of diagnostic biomarkers is supposed to help redefine bipolarity "in terms of different underlying pathophysiological processes that are likely to include abnormalities in neural circuitry" (de Almeida and Phillips 2013, 115). It is hoped that, in combination with genetics (and taking environmental factors into account), neuroimaging will reveal "neural predispositions" that increase the probability of developing some form of depression (Northoff 2013a). Patients should in the future be "managed" according to "algorithms" based on brain states rather than on clinical examination and patient or professional preference (McGrath et al. 2013). As we saw above, this goal has been a response to the limited efficacy of antidepressants, which is in turn widely attributed to the heterogeneity of the condition and rests on the conviction that "depression" likely refers to multiple diseases, "each with a distinct neurobiology"

(Holtzheimer and Mayberg 2011, 4). However, as meta-analytic studies demonstrate, not only have even the most apparently foundational neuro-imaging results been challenged, but the dominant region-of-interest approaches ignore activity (and hence potential anomalies) in regions currently not considered "of interest" for the study of depression, thus sorely weakening the significance of the obtained results (Hamilton et al. 2012).

Again "Just Like Diabetes"

Awareness of these limitations has reinforced the view that abnormalities in neural networks rather than in discrete brain structures underlie psychiatric disorders. It has also contributed to move psychiatric neuroimaging research toward resting-state models (Broyd et al. 2009) and to come in line with emerging approaches to brain connectivity (Price and Drevets 2010) and the concurrent transformation of fMRI research (and brain science in general) into a big-data worldwide endeavor (Lohmann et al. 2013, Thompson et al. 2014). Launched by Marcus Raichle in 2001 (Raichle et al. 2001, Raichle and Snyder 2007), the notion of a "default mode" of brain function has come to describe a "resting state" characterized by very slow neural oscillation (see Callard and Margulies 2011 for a history and larger significance of these notions). The resting state is the "state" of large-scale networks that are active when the subject is awake but not focused on the external environment; their activity is therefore driven neither by tasks nor by external stimuli. Neuroimaging studies of the relationship between the default mode network and mental disorder began in the early 2000s and have shown, for example, that the network is functionally overactive in schizophrenia and hypoactive in Alzheimer's disease (Buckner et al. 2008).

Resting-state research has also gained momentum in the field of depression neuroimaging. A 2012 review of sixteen resting-state fMRI studies published between 2005 and 2011 described various default mode network "abnormalities" in major depression (Wang et al. 2012; see also Veer et al. 2010, not included in the review, as well as the meta-analysis by Alcaro et al. 2010). What is supposed to be their role? The most ambitious resting-state model of major depressive disorder (Northoff et al. 2011) does not aim at "denosologizing" the category. Rather, it preserves *major depressive disorder*

(MDD) in all its heterogeneity—at the level of its symptoms, the affects it encompasses (anxiety, sadness, grief, panic, pain), the bodily systems it involves (from the vegetative and endocrine to the cognitive), the neuro-anatomical regions observed to be "abnormal" in the condition, and the biochemistry pertaining to each of those systems and regions. It then seeks to correlate those different levels, mustering a vast amount of neuroanatomi-cal, psychopathological, and biochemical information to turn major de-pression into a specific brain system–network disorder.

MDD turns out in this model to be characterized by a subcortical-cortical imbalance, with resting-state hyperactivity in some regions and hypoactiv-ity in others. Certain subcortical and cortical regions are hyperactive in the resting state, while others (especially cortical) show hypoactivity. Such ab-normal resting-state patterns "may strongly impact the neural processing of external stimuli" in the regions concerned, and that "may enable and pre-dispose the occurrence" of major depression symptoms (7). Higher affective and cognitive functions are "highjacked [*sic*]" by subcortical primary-process emotional systems (1, 11). For example, depressive hopelessness arises by way of a "psychopathologically specific" relationship with resting-state ac-tivity in the ventromedial prefrontal cortex (VMPFC). On the one hand, in depressed individuals, elevated resting-state activity in the perigenual anterior cingulate cortex (PACC) and the VMPFC has been found to cor-relate with high scores on a self-report inventory known as the Beck Hope-lessness Scale. On the other hand, in "healthy" subjects, PACC and VMPFC are associated with the slowing of time in subjective perception. The "ab-normally elevated" VMPFC resting-state activity therefore "seems to impair anticipation and hence one's experience of extending hopes into the future" and "to block the ability of MDD patients to project hope into the future, thereby promoting hopelessness and ultimately helplessness" (10).

The authors of the model acknowledge that such causal pathways are speculative. For us, the most revealing feature of their work is that they of-fer as primarily predictive and etiological a neuroanatomical model very largely based on correlational neuroimaging data. For example: "One could expect that elevated resting-state activity in these regions *would lead* to an increased self-related processing and hence to abnormally increased personal concerns in MDD patients" (11). A study actually shows that, in MDD patients as compared to "healthy" subjects, increased self-focus in connection with negative emotional stimuli correlates with significantly lower signal intensities in various subcortical and cortical regions (Grimm et al. 2009). But here we remain in the domain of correlations, not of factors "leading" anywhere.

The authors of the model we just summarized believe that if social factors known to be associated with the onset of depression were shown "to impact either the resting state-level itself or the degree of rest-stimulus interaction," then the model could become neurosocial (Northoff et al. 2011, 14). They also realize that abnormal resting-state activity is likely to be a necessary rather than a sufficient condition of depression, acting as a "neural predisposition," a "susceptibility marker," a "risk factor" (14). However, in spite of its integrative purpose, the model ultimately accounts for depression in neuroanatomical terms and depicts it as "just like diabetes." And it does so explicitly: Low insulin, the authors explain, "metaphorically corresponds" to the abnormally elevated resting state; like abnormally high blood sugar, which interacts with biochemical mechanisms in diverse bodily systems, such a state has "comparable effects on diverse brain-mind subsystems"—psychopathological effects comparable to those of diabetes, like becoming blind or suffering gangrene (15).

The neurobiologists who, working with laboratory mice, induced a network and behavioral chronic stress phenotype and then reversed it by stimulating the prefrontal cortex-to-amygdala circuit did not hesitate adventurously to attribute a causal nature to the "central brain mechanism underlying MDD" they apparently identified and to announce treatments directly targeting the relevant network interactions (Hultman et al. 2016, 449). In contrast, and quite understandably, authors engaged in neuroimaging depression research abstain from speaking directly in terms of causes. Abnormalities "play a role," are "involved in," "impact on," or "may contribute" to mental disorder; functional and anatomical differences or the activation of brain structures do not reveal the cause of depressive symptoms but only have "temporal relations" with their expression or are "significantly positively associated" with them. In the end, research cannot decide "whether particular brain changes in depression are a consequence of symptoms or due to underlying neural vulnerabilities," which are themselves at the beginning of an etiological chain (Graham et al. 2013, 424). There is much to commend in a cautious attitude toward causal connections. Yet the intentionally imprecise language not only reveals ambivalence regarding causality but is also symptomatic of a historical situation. Although the existence of a link between brain chemicals and mood disorders has been known since the 1950s, when drugs that alter those substances were found to relieve these disorders, in 2017 it is still unknown if changes in neurotransmitter levels cause depression or the other way around; the same is acknowledged in connection with volumetric, anatomical, and neuroimaging data.

As we have noted, explicit discourses about physiopathology and diagnosis are less about causes than about biomarkers. These are being sought for at different levels: some concern "predisposition" and hence the probability of developing a disorder under particular conditions; others, valuable for diagnosis, are expected to indicate, with a high level of probability, that the patient suffers from such and such pathological condition; yet others bear on treatment and thus on the likelihood that an individual will respond or not to a given pharmacological or psychological therapy. The overall goal of the shift toward biomarkers is to underpin diagnoses with pathophysiological evidence and to allow disorders to be reconceptualized (eventually in a denosologizing perspective) according to biological criteria. As already mentioned, biomarkers reflect correlational results at the population level and lack predictive power for the individual. But the explicit acknowledgment that this is so constantly clashes with the way in which goals and results are presented and discussed.

As far as causality is concerned, the situation is essentially the same with regard to neuroimaging, including resting-state results relevant to ADHD (e.g., Posner et al. 2013), schizophrenia (e.g., Arbabshirani et al. 2013), and autism (e.g., Deshpande et al. 2013). The slippage from neuroimaging correlations to etiological causation observed in all these areas has an equivalent in the diagnostic realm. In the field of depression, as in ADHD, schizophrenia, and autism, neuroimaging has been advertised as capable of becoming a diagnostic tool, and *Time* has even announced, "Brain Scans Could Become EKGs [electrocardiograms] for Mental Disorders" (Khamsi 2013). Everything leads the public not to notice that scans are said to correlate with those disorders only *after* a clinical diagnosis has been made. The hope of bypassing the challenges and apparent messiness of the clinic and of automating diagnosis by way of brain scans drives the entire field of psychiatric neuroimaging; the probabilistic rationale of the biomarker is inconsistent with the expectations raised by those who, in principle, adhere to that rationale.

Moreover, as often explained, psychopathology "is increasingly viewed from a circuit perspective in which a disorder stems not from circumscribed anomalies in discrete brain regions, but rather from impairments in distributed neural networks" (Posner, Park, and Wang 2014, 3). Yet, in the same way that the shift to biomarkers coexists with a desire for causality, the emphasis on neurocircuitry has in most cases barely altered the localizationist logic that drives the research. Connectome-based imaging studies, also correlational, have identified "disrupted topological organization of large-scale functional and structural brain networks in depression," but these "pathologic networks *associated* with depression" cannot be said to be anything beyond "*potentially* valuable biomarkers" (Gong and He 2015, 223, our emphasis). The establishment of meaningful links is left for an unspecified but intensely publicized future.

Finally, the denosologizing effect of the quest for biomarkers, which is expected to break down categories such as depression, is in tension with keeping depression as a category in Global Mental Health and in "burden-of-disease" calculations; turning away from *ICD*- and *DSM*-based diagnoses clashes with the claim that depression, as defined by these classifications, is a major cause of disability worldwide. GMH advocates propose shifting tasks to communities and local primary healthcare settings with appropriately trained lay workers. There are indications that this mode of intervention is effective for depression (e.g., Patel et al. 2011), and it is praised for bringing together bottom-up consultation processes with a top-down evidence-based approach.

The question whether and how mental illness is universal, as well as the globalization of mental illness and mental health, have given rise to acrimonious controversy (see, for a recent exchange, Miller 2014; Summerfield, 2012, 2014; White 2013).[6] Independently of local knowledge systems, the misery of people who would be diagnosed as suffering from brain disorders according to *ICD* or *DSM* is undeniable, and the way they are treated in many cultures around the world fully warrants the indignation of the medical anthropologist and cross-cultural psychiatrist Arthur Kleinman (2009), who depicted the current situation as a "failure of humanity." At the practical field level, making up for such a failure must be a top priority. That, however, does not eliminate the difficulty, which global mental health actors recognize, of somehow integrating into GMH the science we have sketched here. Indications of useful synergies between GMH and clinical neuroscience have so far remained at a general level—for example, reinforcing what is known about the impact of social and financial deprivation on mental well-being (Stein et al. 2015).

The challenges are considerable, since the neurobiological outlook on psychic distress, which we have explored via neuroimaging approaches, embodies notions of objectivity and a desire for causality that are difficult to reconcile with phenomenological and first-person understandings. We have considered such a situation by focusing on attempts at revamping nosography and diagnostics on the basis of experimental work and have thus remained in the framework of specialized knowledge production and assessment. We deal next with the other end of the circle, with contexts in which *neuro* idioms and bits of neurobiological information are incorporated into the lives of individuals and groups. Especially in the case of the autism spectrum, such incorporation serves mainly but not exclusively to depathologize and redescribe the diagnosed conditions. But seeing autism as a way of being rather than an illness is not to everyone's liking, and the neurobiologization of the spectrum gives rise to camps that defend conflicting forms of subjectivity and sociality, thus highlighting again the constitutive ambivalence of cerebralizing processes.

Neurodiversity

In an often-quoted article of 1998, significantly entitled "Thoughts on Finding Myself Differently Brained," the autistic self-advocate Jane Meyerding wrote that she "was surprised to find [herself] moving into the realm of neurology." Since the 1990s, indeed, autism advocacy has organized itself largely around "neurology" or, more accurately, as a *neurodiversity* movement. So far the movement has been dominated by people diagnosed with Asperger syndrome and other forms of high-functioning autism (although some prominent self-advocates, such as Amanda Baggs, do not speak and define themselves as "low functioning").[7] Asperger as a formal diagnosis has disappeared from the *DSM-5* and is subsumed as the high-functioning end of a new "Autism Spectrum Disorder." Individuals at that end of the spectrum believe that their condition is not a disease to be treated and, if possible, cured, but rather a human specificity that must be respected as such.[8] Their being unlike "neurotypicals" derives in their view from a brain "wiring" that is different but not abnormal. Such identity claims manifest what the activist Judy Singer (1999) has called "neurological self-awareness." Indeed, autistics' identitarian claims have gone hand in hand with the cerebralization of their condition. As we shall see, the "person-first language" generally supported by the disability rights movement is not always well received within autism self-advocacy groups, for whom the expression "person with autism" suggests that the condition is not constitutive of the individual.[9] The *neuro-* prefix and a usually imprecise *neuro* vocabulary serve to construe autism as a positive attribute and to demonstrate the legitimacy of the autistic experience. Cerebralization, which as we saw is driven by a quest for causality and "objectivity," thus sustains subjectivation.

Autism as a Biosocial Phenomenon

The emergence of the term "neurodiversity" and the corresponding movement in the late 1990s should be analyzed within a broad perspective. On the one hand, it belongs in the history of disability movements (Charlton 2000; Corker and French 1999; Corker and Shakespeare 2004; Davis 1995, 2002; Shapiro 1993). On the other hand, it instantiates the extensive and diversified societal impact of neuroscientific knowledge and practices. The neurodiversity movement is historically connected to a turn away from psychoanalysis and toward a neurobiological and genetic understanding of autism. Especially in the United States, from the 1940s to the mid-1970s, psychoanalytic explanations were paramount both in psychiatric theory and clinical practice (Nadesan 2005). The later shift was embodied in pro-cure and anti-cure discourses, both expressed in neurodiversity advocacy groups and in parent and practitioner groups favorable to behavioral and psychopharmacological therapies (Chamak 2008; Silberman 2015; Silverman 2008a, 2008b, 2012).

Other roots of the neurodiversity outlook are to be found in the antipsychiatry movement as well as in the emergence of psychiatric consumer/survivor/ex-patients groups (Graby 2015). From a nosological perspective, the categories comprised under the umbrella of "neurodiversity" (autism is the main one, but see Armstrong 2010 and Hendrickx 2010) are included in the *DSM*, overlap with learning disabilities, and historically fall somewhere between psychiatric diagnosis and disability, between mental illness and mental retardation (Eyal et al. 2010, Graby 2015). Autistic self-advocates frequently associate themselves explicitly with the Deaf rights movements, and some of them have been inspired by the American "independent living" movement (Silberman 2015).

Disability studies and the disability rights movement share a commitment to the social model of disability and the rejection of the medical model (Oliver 1990, Shakespeare 2006, Wendell 1996). The social model, which distinguishes between *impairment* and *disability*, has been criticized for downplaying the importance of impairment and consequently mixing up disabled and nondisabled people. However, underlying the pursuit of a barrier-free world is the assumption that discriminations are not attributable to individuals' impairments but result from society's failure to accommodate them. Similarly, autistic self-advocates do not reject impairment labels (for they consider their autism to be neurologically real) but reclaim them from medicine and turn them into the basis of positive identities and into the justification for claiming rights and compensations. Impairment has thus become a "difference to be expected and respected on its own terms in a diverse society" (Cameron 2008, 24), and its biological nature has allowed activists to redescribe mental disorder as a sui generis form of human diversity or even of human consciousness (Boundy 2008).

As mentioned, the neurodiversity movement also draws on antipsychiatry and the "survivor's" groups, including radical strands such as Mad Pride and its appropriation of traditionally derogatory terms such as "psycho," "crazy," and "nut." These movements wish to deconstruct stereotyped and stigmatizing representations in science, medicine, and public culture at large (Rowland 2015). While they developed as "revolts from below" (Crossley 1998), they historically followed the "revolt from above" that began inside professional psychiatry with Robert Laing and David Cooper in the United Kingdom, Thomas Szasz in the United States, and Franco Basaglia in Italy. Laing's *The Divided Self* was published in 1960, Cooper's *Psychiatry and Anti-Psychiatry* in 1967, and the Mental Patients' Union appeared in Britain in 1973. Antipsychiatry politicized the psychiatric field, opened up a space for the expression of diagnosed persons, and offered them alternative discourses about madness and normality. Although users were involved in the antipsychiatry movement from the beginning, "survivor's" groups emerged a decade later. While the former belongs in terms of style and tactics to the counterculture of the early 1960s, the latter were more politicized and, influenced by Marxism, located patient politics in the context of revolutionary action and a general class struggle (Crossley 1998, 2006).

In spite of many overlaps in inspiration and rhetoric, the neurodiversity movement differs from antipsychiatry and patients' crusades in its interpretation of the ontological status of mind, brain, and body. While neurodiversity advocates admit that their conditions are neurologically real and therefore represent physically based differences, the earlier movements tended to reject the idea of fundamental and materially real divergences between them and "normal" individuals (Graby 2015, Jones and Kelly 2015). Thus, in each camp the brain ended up fulfilling opposite ideological functions. The neurodiverse see the brain as the somatic seat of legitimate identities, different from the "neurotypical"; antipsychiatry identifies it with the reductionism and pathologizing impulse of biological psychiatry. In short, neurodiversity movements share the depathologizing rhetoric of antipsychiatry and user/survivors movements but at an ontological level have more affinities with the disability rights movements, which instead of downplaying impairment redescribe it in positive terms.

The first associations for the parents of autistic children appeared in the mid-1960s. The National Autistic Society was founded in London in 1962. In 1965, Bernard Rimland, author of *Infantile Autism: The Syndrome and Its Implications for a Neural Theory of Behavior*, together with the pioneer activist mother Ruth Sullivan and other parents, established the Autism Society of America.[10] Similar groups soon mushroomed in other countries (Chamak 2008, Chamak and Bonniau 2013, Dekker 2006, Shapiro 2006, Wing 1997). The rise of the Internet in the early 1990s was a major turning point for both parent and self-advocacy groups. One of the earliest online parents lists, the Autism and Developmental Disabilities List (AUTISM List), promoted applied behavioral analysis (ABA), a form of cognitive behavioral therapy, as a treatment for autistic children. Diagnosed adults who felt that both experts and families misunderstood or ignored them resisted the emphasis on curing. Australian and American activists then formed Autism Network International (ANI) in 1992, supplemented since 1994 by Autism Network International Listserv (ANI-L). The first issue of their newsletter *Our Voice* came out in 1992, and the first autistic retreat (called Autreat) took place in 1996 (Bagatell 2010, Chamak 2008, Orsini 2012, Silverman 2008a, Silberman 2015).

Although nonautistics may adhere to ANI, all decision making is done by autistics alone. The motto "By autistics for autistics" captures the network's core values and expresses the principle of the disability movement at large, "Nothing about us without us" (Charlton 2000, Shapiro 1993). Their goal is to fight the vision of autism conveyed by professionals and families who share an "obsession" with a cure, which ANI considers not merely disrespectful of the autistic way of life but an attempt to erase legitimate difference. Hence the strength of the anti-cure attitude in this camp (Sinclair 2005).

In the opposite camp stand organizations such as the National Alliance for Autism Research (NAAR), founded in 1994, and the Cure Autism Now Foundation (CAN).[11] The latter was created in 1995 by the parents of an autistic child and brings together families, medical doctors, and scientists to support biomedical research and education. NAAR and CAN have now merged with Autism Speaks.[12] CAN is a major target for autism movement activists, who accuse it of demonizing autistics and frightening their families, promoting narrow viewpoints on the disorder, and neglecting autistic adults' life experiences.

The growth of self-advocacy movements and their enhanced exposure in the media has intensified the political clash between the anti-cure and the pro-cure.[13] One of the most controversial issues concerns applied behavior analysis, the above-mentioned therapy that employs learning theory to improve "socially significant behaviors."[14] For many parents, ABA is the only way of helping their autistic children make some progress toward establishing visual contact and performing limited cognitive tasks. In contrast, for autism militants, ABA represses autistics' natural modes of expression (Dawson 2004). In the United States and Canada, the debate has reached the courts: While parents fight to obtain governmental support or make health insurance companies pay for the therapy, which is extremely expensive, neurodiversity advocates maintain that autism is not an illness and that attempts to cure it violate autistic rights (Baker 2011; Dawson 2004; Orsini 2009, 2012). The latter position may provide reasons for refusing to subsidize treatments, but the most adamant partisans of neurodiversity are willing to take the risk. For them, the search for therapies manifests denial and intolerance toward differences and enacts eugenic and genocidal policies; in 2004, some went as far as petitioning the United Nations for recognition as a "minority social group" that deserves protection against "discrimination" and "inhuman treatment" (Nelson 2004). Their position is one of the logical consequences of cerebralization, which in this case acts as a normalizing rather than as a pathologizing mechanism. Thus the only "distress" they claim to suffer is not caused by a pathology but by society's lack of acceptance.

The biosocial field of autism is not entirely structured by such radical polarizations, but it must nonetheless face the question whether mental disorders are necessarily harmful and whether people who have "symptoms" that don't make them suffer or put them at an increased risk of experiencing future distress or impairment should be considered mentally ill (Cooper 2015). Under *DSM-IV*, such people were not diagnosed; in *DSM-5* they are. The high-functioning end of the autism spectrum is a paradigmatic case, since some people (who would have formerly been diagnosed as having Asperger syndrome) meet the diagnostic criteria but are not unhappy and function well in society. Treatments are acceptable to alleviate discomforts they share with millions of undiagnosed individuals. Thus, the prominent neurodiversity advocate Temple Grandin, a high-functioning autistic who provided the title for Oliver Sacks's *An Anthropologist on Mars* (that is how she says she feels around neurotypicals), is not against medication. However, as is clear from her autobiography *Thinking in Pictures* (1995), she wishes to limit it to secondary symptoms such as anxiety and not to autism itself. Judy Singer, another activist, thinks that drugs are acceptable as long as they aim to relieve suffering, not change individuals' personality. Fernando Cotta, the president of the Brazilian Autistic Pride movement, concurs that respecting autistics is not incompatible with medication; for example, if an autistic "has attention problems, he can take something that can help him, just as somebody who has the flu takes an anti-flu medicament" (Lage 2006).

In short, some self-advocates insist that autism itself should not be treated but have a pragmatic attitude toward medical interventions.

Neither do all parents oppose self-advocacy movements, nor do all autistic adults favor neurodiversity.[15] The latter sometimes find it difficult to harmonize their identities in the autistic communities and in the neurotypical world, and this tension can become an important source of anxiety and suffering (Bagatell 2007). Moreover, some autistic adults do want to be cured, but they seem to represent a largely silent population. "Most persons with an autism-spectrum disorder have never expressed their opinions on someone's blog and never will," affirms Jonathan Mitchell, who suffers from a mild autism spectrum disorder, blogs against neurodiversity, and notes that "the neurodiverse often reach a vulnerable audience, as many persons on the spectrum have low self-esteem. Neurodiversity provides a tempting escape valve" (quoted in Solomon 2008).[16] Sue Rubin, a low-functioning autistic who is the subject of the documentary *Autism Is a World*, emphasizes that whereas high-functioning autistics tend to be against a cure, low-functioning autistics generally hold the opposite position. For her, "the thought of a gold pot of a potion with a cure really would be wonderful." She writes:

> As a person who lives with autism daily and will not live a normal life, I find people who are high functioning and saying society should not look for a cure offensive. They have no idea what our lives are like. Killing autism lets me enjoy a life with great friends and allows me to go to college, but I must never let down my guard or autism will take over. I don't want any more children to live, as I must, in this constant state of war. (Rubin 2005)

For persons such as Rubin, it is the advocates of neurodiversity who are insensitive and lack respect.

Finally, the relationship between parent and self-advocacy groups differs considerably depending on national context. Whereas it can be highly conflictual in the United States, the United Kingdom, and Australia, in France autistic self-advocacy remains under the influence of parent associations (Chamak 2008, 2014; Chamak and Bonniau 2013). Thus, while the biosocial field of autism seems to be extremely polarized at first glance, under closer inspection it appears to be more complex and to provide room for a variety of nuanced positions. It is therefore more appropriate to characterize it as including discourses, individuals, and groups that, while antagonistic in some respects, in others overlap or support one another, rather than to characterize it as a clash of homogeneous groups holding sharply antithetical positions.

Autistic Cultures and Neurodiversity

The term "neurodiversity" is generally credited to Judy Singer, a sociologist diagnosed with Asperger syndrome, who used it in a 1999 article titled "Why Can't You Be Normal for Once in Your Life? From a 'Problem with No Name' to the Emergence of a New Category of Difference" (the title echoes Betty Friedan's 1963 *The Feminine Mystique*, whose first chapter identified American women's dissatisfaction and yearning as "The Problem That Has No Name"). The term also appeared in Jane Meyerding's 1998 "Thoughts on Finding Myself Differently Brained," and Singer herself wrote, "I am not sure if I coined this word, or whether it's just 'in the air,' part of the zeitgeist." As explained, "neurodiversity" proclaims that some features usually associated with illness are in fact only atypical or "neurodivergent" (Harmon 2004a, 2004b, 2004c). Disability studies scholars see the rise of neurodiversity as a critique of the dominant discourse of dependency and abnormality, a celebration of difference, and an assertion of pride that, beyond the circle of the disabled, their families, physicians, and caretakers, reaches into the domain of public health and educational policies (Corker 1999, Swain and Cameron 1999). "If you do not believe there is a disability, if you do not believe there is anything that needs to be 'cured' or genetically prevented—that disability is indeed little more than a social construction— then you will likewise be freed from the need for a cure" (Cheu 2004, 209).

These ideas and the social forms they animate exist chiefly by way of the Internet. "Deaf culture" (Padden and Humphries 2006) has in this connection inspired the development of "autistic culture."[17] An autistic self-advocate states it explicitly: "Much like the deaf community, we autistics are building an emergent culture. We individuals, with our cultures of one, are building a culture of many" (Prince-Hughes 2004, 7; see also Davidson 2008). The web has become the privileged vehicle for advocacy and networks, enabling "what was thought impossible, to bind autistics together into groups" (Singer 1999, 67). That is why it seems to be for them what sign language is for the deaf or Braille for the blind (Blume 1997a). Self-advocates thus craft themselves as a "new immigrant group on line, sailing to strange neurological shores on the Internet" (Blume 1997b).

Cyberspace has turned into a vehicle and territory for new forms of "biosociality." Prominent among the phenomena sustained by websites and blogs is the emergence of a specific self-advocate vocabulary for categorizing persons (Bagatell 2007): *Aspie, Cousin* (someone who is not clinically autistic but still similar enough to autistic people to be part of their culture), *Neurotypical, Autistic* or *Autie* (preferred to the politically correct "person with autism"), or *Curebie* (derogatory term for those who wish to cure autism). Websites also recommend fictional and science literature; various online support organizations, blogs, and chat rooms facilitate interaction among autistic individuals, provide clarifications on symptoms, enable the sharing of experiences, and help their users make friends or find partners (Chamak 2008; Jurecic 2007; Silverman 2008a, 2008b). All of this combines to promote awareness and empower a community that (at the initiative of Aspies for Freedom) has since 2005 celebrated on July 18 its own Pride Day.[18]

Websites like Proudly Autistic include a marketplace where one can purchase T-shirts, tote bags, mouse pads, stickers, postcards, and greeting cards proclaiming "No more 'Trained Seal' Treatments!" (against ABA), "Not Being Able to Speak Is Not the Same as Not Having Anything to Say," or "I Am Autistic. What's Your Excuse?"[19] As Nancy Bagatell (2007) shows in the case of Ben's "coming out" as an autistic, these objects may function as powerful "tools of identity." Ben's trajectory recalls the gay and lesbian coming out, which can be understood as a political act with significant liberating or destructive consequences (Davidson 2008; Valentine, Skelton, and Butler 2003).

Identitarian Issues: Being *Autistic or* Having *Autism?*

This brings us back to the fundamental question of identity. Parent and professional associations that support the search for a cure usually refuse to acknowledge the very existence of an identity issue. For them, autism is simply a disease. Children *are not* autistic, they *have* autism. As Kit Weintraub (2005), the mother of two autistic children and a board member of Families for Early Autism Treatment, wrote in response to the autistic self-advocate Michelle Dawson's (2004) critique of the "autism-ABA industry,"

> I love my children, but *I do not love autism*. My children are not part of a select group of superior beings named "autistics." They have *autism*, a neurological impairment devastating in its implications for their lives, if left untreated. . . . it is no more normal to be autistic than it is to have spina bifida. (Weintraub 2005)

Although online discussion groups demonstrate that some autistics do not see their condition as a positive part of their selves (Brownlow 2007), others do consider it as essentially constitutive of who they are. Autism, they argue, is "pervasive, it colors every experience, every sensation, perception, thought, emotion, and encounter, every aspect of existence" (Sinclair 1993). This is also the reason why many activists adopt self-descriptions such as *autistic* or *aspie*, which present autism as an integral part of their identity (Silverman 2008b). For the autism rights activist Jim Sinclair (1999), "person with autism" suggests that autism "is something *bad*—so bad that it isn't even consistent with being a person." Dawson thinks that using that expression would be as bizarre as using "person with femaleness" to designate a woman (quoted in Harmon 2004c). Attitudes toward cure and therapies are consistent with these various positions.

As mentioned, autistic identity is sometimes experienced as a source of pride, even as a "gift" (Antonetta 2005). The emergence of this feeling may begin with a sense of reassurance. High-functioning autistics have reported the "comfort" they felt upon being diagnosed. "Finally an explanation, finally a sense of why and how," wrote a man diagnosed with Asperger syndrome at age thirty-six, shortly after his four-year-old son was diagnosed with the same disorder (Shapiro 2006). Ian Hacking (2006) has noted that "many misfit adults now recognize themselves as autistics, or so they say. It really helps to be able to put a label to your oddities. It brings a kind of peace: so that is what I am." Judy Singer (1999, 62) expounds on the "benefits of a clear identity," and Jane Meyerding (2003) speaks of the "aha! moment" when she discovered autism as an explanation. That led to her finding a community whose thought patterns and modes of expression she identified with: "All my life, I have been forced to translate, translate, translate. Now, suddenly, I have people who speak my own language." Autistics may use the diagnostic label positively; the autism idiom generates "signposts" and "shorthands," as Meyerding says, that help them position themselves with respect to the surrounding culture. Labeling thus metamorphoses from signal of stigma to instrument of liberation.

Asserting identity is often associated with rejecting psychological explanations and psychotherapies. The latter can be seen not only as a waste of time but also as downright dangerous. For example, a woman diagnosed with autism said that after spending her teens "in a state of suicidal clinical depression as a result of bullying and feeling that I must be a failure or insane for being different," she found this opinion "only reinforced by the psychotherapist I got sent to, who decided that all my problems must be the result of 'sexual repression.'" Proud to have "walked out after six sessions," she welcomed the autism diagnosis as "the best thing" that ever happened to her (quoted in Blume 1997a).

As highlighted by the very notion of neurodiversity, autistics' claim to a specific identity is linked to the cerebralization of their condition. As the anthropologist Tanya Luhrmann shows in her ethnographic account of American psychiatry, the biologization and neurologization of mental illnesses tend to bracket off subjective and experiential dimensions and to convey the positive message that "the body is always morally innocent" (Luhrmann 2000, 8). Talking about her own experience of manic depression, the anthropologist Emily Martin (2007, 13) recounts, "I often heard from my psychiatrist that my problem was related to my neurotransmitters, and I always found this comforting." In contrast, "if something is in the mind, it can be controlled and mastered, and a person who fails to do so is morally at fault" (8). When a biologically oriented psychiatrist speaks of depression as a cardiologist speaks about cardiopathies, a distance is introduced between the patient and the disease.

We have seen that the destigmatizing effects of the biological interpretation of mental illness have been overestimated. Yet, for all the criticism that "blaming the brain" (Valenstein 1998) may deserve, it sometimes has freed both patients and families from blame for manic depression, eating disorders, anorexia, autism, or schizophrenia. Thus, in the case of families who financially support neuroimaging research, their adherence to a neurobiological approach is consistent with the widespread rejection of the notorious late-1940s theory of the schizophrenogenic mother. At the same time, it reflects the conviction that psychological ailments can be cured, should be covered by health insurance, and should benefit from other forms of compensation (Martin 2007). As it turns out, it is easier for patients and their relatives to accept a diagnosis of *bipolar disorder*, which has become associated with brain states, than one of *manic depression* or *manic-depressive disorder* (other labels for the same condition), which tend to be perceived as psychological (Montanini and Banzato 2012). In the latter case, "mental illness is in your mind and in your emotional reactions to people. It is your 'you'" (Luhrmann 2000, 6). In contrast, a cerebral disorder is only connected to the body, in the same sense that a heart attack may affect your mind but "is" in your body.

Biological explanations contribute to bring together patients, families, and scientists to spread information about a condition, combat stigma, support patients, and drive the search for treatments (Gibbon and Novas 2008a, Rose 2007). The claim to neurodiversity is connected to a "naturalized" identity, according to which I am who and what I am because my brain is "wired" in a certain way. In his discussion of the "looping effects" of diagnostic labeling, Ian Hacking (1995, 2002) has distinguished labeling from above and from below. Originating as it does with the patients rather than the doctors, neurodiversity illustrates labeling from below, even if it necessarily feeds on information "from above." For autistic self-advocates, neurologizing their condition helps redefine it in terms of an organically localized *difference*. There is, however, no consensus on the neurobiological etiology of autism.

Contemporary research uses several approaches to define biological markers: one may search for the characteristics of the "autistic brain," look for autism genotype(s), or investigate comorbidity and environmental influences (Nadesan 2005). Viewing autism as a brain dysfunction (Fombonne 2003, Freeman and Cronin 2002, Wing 1997), psychiatrists and neuroscientists have tried to discover the disorder's "brain address" (Wickelgren 2005, 1856) and have even suggested that the autistic brain is an extreme form of the "male brain" (Baron-Cohen 2002). In view of such heterogeneity, it has been proposed that autism is best understood as a "multi-system disorder" (Charman 2006). As far as neuroimaging approaches are concerned, they have generated great expectations; actually, however, they "seldom provide data on an individual level, do not yet have well-accepted standards or replicability across time or site . . . and have rarely addressed questions of specificity of findings" (Lord and Jones 2012, 491). In other words, and despite considerable amounts of research, there is still no convincing, well-replicated brain-based autism biomarker (Walsh et al. 2011). In spite of that, "autism has retained its identity as a genetic disorder of the brain" (Silverman 2012, 155).

Autistic advocates place less emphasis on particular biomarkers than on the more general fact that, as Temple Grandin put it, autism "is a neurological disorder. A child is born with it. It's caused by immature development of the brain . . . and not by bad parenting or the environment" (quoted in Blume 1997a). Similarly, for the Dutch self-advocate Martijn Dekker (2006), autism "is neither a physical (bodily) disability, nor a mental illness: it is a neurological disability." His making the brain something different from the body illustrates that organ's special ontological and functional status. All sides of the autism community share his position. Thus, self-advocates' bête noire, Cure Autism Now, has sponsored the creation of an Autism Genetic Resource Exchange (Silverman 2008a), the "world's first collaborative gene bank for autism."[20] (Gene expression patterns in the brain, rather than genomes per se, are increasingly understood as crucial; hence the rising importance of *neurogenetics*. See for example Jones 2012).

Even though some groups have supported alternative causes for autism (for example, mercury poisoning; Bumiller 2008), there is a generalized preference for brain-based explanations, which have emerged as part of the spread of neuroscientific claims beyond the laboratory. It is therefore not a concern for mental disorders that contributes to the "elevation of neurology" but the other way around: this "elevation" provides a major reason for the increasing attention paid to those disorders, including autism (Blume 1997a).

Loving and Hating One's Brain

Whether superficial or well informed, wacky or serious, neurodiversity advocates' engagement with the neurosciences has become a major vehicle for fashioning personal identities. The process began in the late 1990s. We have already mentioned Jane Meyerding's remembering how "surprised" she was to find herself "moving into the realm of neurology." What does her recollection tell us about subjectivation processes within autistic culture? Can we say that some self-advocates become cerebral subjects by way of their engagement with the neurosciences and their claims to *neuro*diversity? Does defining oneself as *neurodiverse* illustrate what Joseph Dumit (2004) calls "objective self-fashioning," that is, the incorporation into one's self-definition of scientific or expert ideas, terms and metaphors? Do all self-advocates mobilize brain vocabularies in the same way? Are there distinctive versions of the "brain story" (Martin 2009)? And how are they shaped differently in blogs, discussion groups, autobiographies, conferences? What sort of information is being used? Is it drawn from scientific articles, popular accounts in magazines, movies, or novels? Who is addressing whom, and in which arenas? How are discourses adapted to different contexts and audiences? Answering these questions is not easy.

There is a substantial amount of social science research on these matters, conducted especially on online materials or by way of ethnographic approaches, focusing on the social construction of disability in the new media (Coleman 2010; Goggin and Newell 2003; Hallett and Barber 2014; Jaeger 2012; Keim-Malpass, Steeves, and Kennedy 2014; Kozinets 2010; Snodgrass 2014). The blogosphere being a major setting for the development and consolidation of disabled identities, the use of the Internet, especially by blind, deaf, and autistic people, has gained considerable attention (Goggin and Noonan 2006). The web has become an essential space of debate and identity development for autistic persons (Biever 2007, Blume 1997a, Dekker 2006, Kenway 2009). Many empirical studies deal with autism in cyberspace.[21] Qualitative research is also being conducted on autistic people's writings, particularly autobiographies and memoirs (Chamak et al. 2008; Davidson 2007, 2008; Hacking 2009; Osteen 2008). Books such as *Voices from the Spectrum* (Ariel and Naseef 2006) collect first-hand stories by parents, siblings, people diagnosed with autism, and mental health professionals, and ethnographic accounts examine identity construction by autistic individuals (Bagatell 2007, 2010; Bertilsdotter Rosqvistab, Brownlow, and O'Dell 2013; Jurecic 2007; Ochs and Solomon 2010; Prince 2010).

The range of subjects is very broad, encompassing differences in parents' and patients' understanding of autism, social interaction and alienation, perceptual differences and sensory distortion, the expression and management of emotion, comprehension and communication difficulties, desire and relationships, the role of the Internet and support communities, diagnosis, self-diagnosis, and the role of "expert" knowledge. But it is self-advocates' already mentioned "neurological self-awareness" and "preference for neurology" (Singer 1999) that most immediately captures the dynamics of the cerebral subject.

Muskie, the creator of a satirical online Institute for the Study of the Neurologically Typical, declares: "My brain is a jewel." "I am," he writes, "in awe of the mind that I have. I and my experience of life is not inferior, and may be *superior*, to the NT experience of life." Though also branded a "Curebie," aspie Michael John Carley (whose son has likewise been diagnosed) rejoices:

> *I love the way my brain works*, I always have and it's one of the things I can now admit to myself. I like the way I think in terms of numbers. I like the way I visualize things. I like the way most especially that I can bury myself in work that I love to a degree that makes everybody else in the world look at me and go, "God! I wish I could do that." No, I am not changing anything. (Quoted in Shapiro 2006, our emphasis)

Meyerding (1998) illustrates a similar reification of the brain when she notes that her employer and friends "think they have conveyed what it is they expect [her] to do, but they have been speaking in a language [her] brain doesn't understand."

Note how these testimonies slide without warning from "my brain" to "I": I love the way my brain works / I like the way I think; those people speak *to me* in ways *my brain* doesn't understand. Beyond metaphorically personifying the brain, such language conflates persons and brains.[22] The creator of the audio post "Asperger's Conversations"[23] says that "we are a world of *funny brains*" and claims that "neuroscience will help us to understand and appreciate the new mix" (our emphasis). Instead of curing autism, some activists propose curing "Neuro-bigotry,"[24] while others dream of Aspergia, a utopian, autism-friendly "neuro city."[25] "Danni's Blog," by an English self-advocate who defines herself as a "Christian Socialist Computer Addict," is filled with references to the way her brain works:

> *I am hating my brain.* . . . I can't deal with the scary thoughts and brain misfiring that makes me too scared to sleep. . . . I need a *brain transplant*, or for River Tam to kill me with her *brain* I don't want to let people down, and I'm even less reliable now than I was before my *brain* went all bad-funky. . . . I had

an appointment with the learning support officer. . . . Was weird, as my *brain* wasn't working right. . . . By this point my *brain* was making weird associations. . . . My *brain* feels all sluggish and blocked. . . . My anxiety is pretty bad and I have other *brain weirdness* things that mean that normal coping methods and stuff don't help. . . . I can't do the homework, partially because . . . my *brain weirdness* is getting worse. If I fail it, I can retake it (most likely when my *brain* is working better). It can be hard when my *brain is hating me* and I'm struggling to keep calm.[26]

Such language is common. In a post on "Identity Politics and the Language Controversy," Dora Raymaker, co-director of the Academic Autistic Spectrum Partnership in Research and Education and a member of the Autistic Self Advocacy Network's Board of Directors, confides: "My *brain* has been terribly 'sticky' on a proposal I'm writing for a conference presentation, and tearing my *brain* away to even read a news story let alone write about it has failed some uncountable number of times. And not only has my *brain* been sticky on the topic of the proposal but my *brain*'s been sticky on identity politics and language."[27]

We could give many other examples. The question is: What do these individuals mean when they say their brains are jewels or that they hate their brains, or when they refer to their brains as being in a particular state or doing this or that? Are they using merely figurative language, or do they mean to say that they are essentially their brains, that their identity and subjectivity can be somehow reduced to brain neurochemistry and processes? Actually, like the many other protagonists of the *neuro*, from reputed neuroscientists to phony brain trainers, who also speak of brains that think, feel, decide, believe, know, desire, and do various other things only persons can do, they do not seem to believe that they are their brains alone. Yet they rely on brain language to talk about themselves. Why? One clear reason is that they live in an environment where *neuro* talk (if not always neuroscience itself) has become a major source and sign of legitimacy. My brain is more authorized "to do it" than myself.

Sometimes "brain" and "mind" are interchangeable: Muskie simply juxtaposes "My *brain* is a jewel" with "I am in awe of the *mind* that I have." But he is obviously talking about the same thing, and the same applies to Carley's enthusiasm about how his brain works. The brain sometimes stands metonymically for the person or "I," as when self-advocates write "my brain doesn't understand," "whatever phrase that non-voluntary portion of my brain happens to be using," or "we are a world of funny brains." On other occasions, the state of the brain—by which a state of being is designated—inspires self-reproach: "sometimes I hate my brain or my brain hates me." That, however, seems to mean that the brain hates itself, since the brain, identified to "I," is also said to feel "all sluggish and blocked," or depicted as not "working right," or making "weird associations."

Neuroscientific metaphors and vocabularies contribute to give the differences between neurotypicals and people on the autistic spectrum a "real" and "natural" character (Brownlow 2007, Brownlow and O'Dell 2006). Neuroscience helps justify those differences, as when a self-advocate declares, "I know they are all individuals, and that we shouldn't blame every NT [neurotypical] for the action of every other NT . . . but there is a common thread that ties them together, and it is at the core of their being. It is more than cultural; it is *how they are hardwired from the factory*" (quoted in Brownlow and O'Dell 2006, 319, our emphasis). Meyerding too neurologizes difference:

> Here came neurology and the possibility that *my brain really was different*. . . . If I could understand my life for the first time only by understanding *how my brain was different* from the majority of brains, how much did I really have in common with all those neuro-typicals (NTs) out there, compared to whom I'd been judged inadequate so many times? . . . Imagine my surprise, then, when I realized I was able to feel "aligned" with this disparate group of individuals joined together by *neurological differences*. . . . *My brain works somewhat differently* from most brains (from *"normal"* brains). . . . Most of the ways *I'm different from the neural norm* can be disguised as eccentricities. (Meyerding 1998, our emphasis)

Thus, a neuroscientific idiom is exploited to discard mere eccentricity and place autism in a positive light. Yet both autistics and neurotypicals believe in the "neurological origins of [their] exclusiveness" (Brownlow 2007, 138; Brownlow and O'Dell 2006, 319). Neurobiology thus functions as an instrument to erect identity frontiers, yet it does so on the basis of an underlying commonality: we all are our brains.

By a largely rhetorical reversal of the normalcy discourse, autistics may stress neurotypicals' strange behavior and satirically pathologize neurotypicality. Once he realized "how bizarre and illogical the NTs really are," the self-advocate Archie found "that their comments and insults" had a greatly reduced effect; he could not "blame the people that are afflicted with neurotypicality," but he added: "that does not mean that I am obligated to change my views to see values in traits I dislike" (quoted in Brownlow 2007, 140–141). Neuroscientific claims are mobilized in the construction of NT and autistic experiences so as to highlight their natural difference, yet at the same time even extreme self-advocates know how inextricably they are linked to the neurotypical world. It would be, for example, unfeasible to keep the utopian island Aspergia free from NTs. Indeed, "if an aspergian man and woman get married and have an NT child would we have to kick it out of the country?"[28]

The counterpart of the construction of differences as ontologically real because neurobiologically based is the belief in a certain ontological homogeneity across the autism spectrum. Some activists consider that "low-" and "high-functioning autism" are variations of degree without fundamental "underlying neurological differences" (Nadesan 2005, 208–209). In 2002 Jane Meyerding explained that, since publishing her 1988 essay "Thoughts on Finding Myself Differently Brained," she had realized that classifying people under different categories within the autism spectrum was "seriously misleading" and declared her preference for seeing herself "as autistic, period."[29] Identity politics here implies both essentializing neurological uniqueness and typologizing brain difference. Of course, criticism has also been addressed to the goal of homogenizing the autism spectrum. We have already mentioned that the *DSM-5* eliminated several forms of autism (the most familiar one being Asperger syndrome), integrating them within "Autism Spectrum Disorder." While this shift corresponds to the aspiration of autistic self-advocates such as Meyerding, an online petition by the Global and Regional Asperger's Syndrome Partnership, carrying over eight thousand signatures, regards it as return to a past "when so many of us grew up thinking of ourselves as bad, broken and damaged, not unique and differently-wired."[30] The same petition points out that many children and adults, particularly those with Asperger syndrome, will lose the diagnosis and, with it, "crucial supports, services and legal protections" (see also Lutz 2013). The situation doesn't lack irony since, as the psychotherapist and cultural commentator Gary Greenberg (2013, 182) remarked: "Four decades after homosexuals demanded to be released from their diagnostic chains, groups of patients were pleading with the APA [American Psychiatric Association, which publishes the *DSM*] *not* to set them free."

Ontological homogeneity is to a large extent a linguistic effect. The world of autistic self-advocacy offers the same phenomenon that Emily Martin observed during her fieldwork on bipolar disorder: Remarks about the brain seemed to be "like clones: endlessly replicating but not generating new connections" (Martin 2009, 7). The brain works like a "confining metaphor" that cuts off links among domains and groups of people. The brain-centered lingo is "folk neurology" (Vrecko 2006) or "folk neurosychology" (Rodriguez 2006), that is, the kind of parlance with which eliminative materialists such as Patricia Churchland (1981) would like to replace folk mentalistic idioms. It has not, however, superseded psychological descriptions of subjective experiences. No amount of neuroscientific progress can suffice to make the mind go away. Indeed:

> If a more reductionistic and brain-based picture of human action displaced our current everyday mental concepts, it would not be because (or solely because) the neural net theory had won in the court of scientific opinion. It would be because the environment we live in (and that scientific theories are produced in) had shifted so that a brain-centered view of a person began to make cultural sense. (Martin 2000, 575)

The neurodiversity universe thus exhibits a cohabitation of everyday ontologies (a phenomenon we discussed in Chapter 1). When acting, thinking, or speaking about themselves and their relations to others, individuals shift ontological registers, and *my brain* may designate *my mind* or, perhaps more precisely, just *I* or *me*. Presumably this does not mean that people are unaware of what they are talking about and say "brain" when they mean something else. Rather, metaphors and metonymies express a more or less harmonious cohabitation of everyday conceptions of the self while at the same time contributing to give a bodily organ—the brain—the kind of psychological depth usually, or formerly, attributed to the mind. The pervasive presence of the *neuro* idiom is thereby legitimized and gives expression to the supposed "neuroscience revolution" in the making of identities.

Identity Politics and the "Neuroscience Revolution"

The combination of "neuro" and "diversity" is by no means self-evident. The term locates difference and singularity so as to naturalize or, rather, physicalize human identity thoroughly. Of course, as we saw in previous chapters, research on neuroplasticity demonstrates that experience shapes the brain in hitherto unimagined ways and to an unexpected extent. Networks of neurons are formed and changed by habits, conscious decisions, acts of the will or attention, physical exercises, food intake, or meditation practices. This supports the shift of diversity, singularity, and creativity to the brain. Neuroscientific research, however, is also characterized by the search for regularities and neuroanatomical and neurophysiological constants that would make it possible to distinguish (ideally on the basis of neuroimaging) between autistic, depressed, schizoid, and normal brains. As we documented in some detail for depression, much neuroscience aims at identifying the brain circuits responsible for normal and pathological mental states. This brings about a paradoxical situation: While neuroplasticity helps account for neurodiversity, neurodiversity advocates tend to minimize the differences among brains within the autism spectrum so as to support their claims for the existence of a brain-based autistic identity. Thus, the "autistic brain" is displayed as ontologically homogeneous and radically different from the comparably homogeneous "neurotypical brain." (The move is analogous to that performed by cultural neuroscience when it implicitly turns cultural difference into an instance of neurodiversity and assumes a fundamental neurobiological homogeneity within each of the groups it studies, "East Asian" and "Western.")

Neuroimaging here plays an extremely powerful role by visually confirming the diagnosis and deepening (whether to celebrate it or pathologize it) a person's sense of autistic identity:

> Joe Powell was diagnosed with Asperger's syndrome, a form of autism, 14 years ago. Before his diagnosis, he didn't speak at all.
> Since then, he says he's made big progress in managing his condition.
> His brain scan confirms his ASD [Autism Spectrum Disorder]. He says seeing his diagnosis charted in black and white made a big difference to him.

"You need to physically see it," he says. "I know the autism is still there. The progress I've made in managing my condition is real, but it's still there." (Hughes 2010)

Powell had participated in a study identifying morphometric features and structural patterns of gray matter anatomy in adults with ASD (Ecker et al. 2010; see Deshpande et al. 2013 for a different fMRI approach to identifying "neural connectivity signatures of autism").

The cerebralization of autism may contribute to reify and naturalize differences between autistic and so-called neurotypical brains; the celebration of disability may open the way to an emphasis on difference via comparison and may even sustain hostility toward nondisabled people (Swain and Cameron 1999). Self-criticism, however, has gained ground within the neurodiversity movement. Sinclair (2005) has condemned antineurotypical prejudice, and some Aspergers consider Aspergia as an "Aspie 'Warsaw ghetto.'"[31] Judy Singer (2007) herself warned that the movement is walking on the "dark side" of identity politics, through "its eternal victimhood, its infantilism, its demand for unconditional love and acceptance without concomitant adult self-reflection, self-criticism, a measure of stoicism, and a willingness to see light and dark in oneself as well as in 'the Other.'"

Singer's criticism implies that self-advocates' use of brain-related terms has contributed to the concealment of individual and institutional dimensions that deserve to be openly discussed. In her description of how Ben, a college student she met at a group called Autistic Adults Coming Together, constructed a positive autistic identity, Nancy Bagatell (2007, 423) observes that having to orchestrate the different discourses around him produced "a lot of discomfort—depression, anxiety and sensory overload—and he desperately wanted relief." One of the bipolars interviewed by Emily Martin (2009, 16) remarked that his "brain contains both health and illness, strength and weakness, darkness and light." Self-advocates tend to neglect such tensions and the fact that, as Singer (2007) points out, "not all is for the best in this brave new world that the 'neuroscience revolution' delineates." Some antipsychiatry advocates "fear that the neurodiversity movement too readily embraces a neurological and medical model for all human behavior."[32] On the one hand, seeing oneself as a cerebral subject bolsters one's sense of identity and may help erase the social stigma often associated with mental pathology. On the other hand, however, it can solipsistically narrow the notion of what it is to be a person. Such, then, are the dilemmas and controversies of the neurodiversity movement. Its members' search for community and relation is in tension with its own reductionistic identity politics, in which selves and difference result from the mechanics of the brain.

Finally, the case of the neurodiversity movement contradicts those who believe that the neurosciences fail to provide "a common ground for shared moral intuitions and values" and therefore lead to an "anthropological and ethical vacuum" (Metzinger 2009, 213). In fact, persons' considering themselves essentially as cerebral subjects has sustained identity formation processes at the individual and the social and community levels. The process is far from straightforward since, as we have seen, autistic self-advocates must negotiate their neurocentric identity politics, a desire for significant forms of sociality, the brain as "confining metaphor" and as liberating condition, and various ways of relating to medicine and to the "neurotypical" world.[33] Nevertheless, across the spectrum of positions, neurodiversity has operated both as an empirical fact and as a shared fundamental value—and one whose status and legitimacy rest largely on its being taken as a validated scientific fact.

Depression and autism, two nosological entities, highlight the ambiguities of cerebralizing processes and the versatility of the *neuro*. But don't those features result from insufficient scientific knowledge? Some day in the future, different biomarkers will delineate with certainty various forms of depression, autism, and other forms of mental distress. Nomenclatures, classifications, and etiologies will finally "carve Nature at its joints"; they will follow from and point to causal mechanisms and neurobiological foundations. Such are the hopes. The cases just sketched, however, suggest that the uncomfortably protean nature of the *neuro* does not represent a problem to be solved or temporary flaws to be superseded but instead highlights its fundamental attribute. There are different ways of being a cerebral subject, ways that do not depend directly on scientific results and idioms but on choices of a different nature (psychological, moral, political, social, even rhetorical) that use those idioms and results as resources.

The last contexts we shall deal with here, literature and film, display such distinctive features and functions as through a magnifying glass. Precisely because they are intrinsically independent from issues of factuality and validity, they can do so in radical form, putting at center stage the dilemmas of the cerebral subject, using them as their most substantive raw material, and performing them without trying to settle them.

FOUR

Brains on Screen and Paper

The previous chapters outlined the history of the cerebral subject and explored some of its key forms in contemporary society. This one deals with it in "fiction," a term we place in quotation marks because, in this domain as in others, fiction is far from fictive. It is "real" not only for the simple reason that it exists but also because it has real effects and contributes to shape ways of being and forms of living. On the one hand, fictions of the sort we shall examine here are not merely literary or cinematographic renderings of brainhood-related philosophical problems and thought experiments but specific ways of addressing them (on this point see for example Lardreau 1988). On the other hand, novels and movies can of course be described as reflecting in many ways issues and questions that are independent of them; however, since they are themselves part of the contexts where those issues and questions take root and are mobilized, they must also be considered in themselves as social agents that structure attitudes and beliefs and play an active role in bringing them about. The same can be said of the plastic arts, as well as music and dance, any of which could have been the subject of an entire chapter.

The visual nature of many of the outputs of contemporary neuroscience, including anatomical preparations, PET and fMRI scans, and most recently the dazzling pictures generated by diffusion imaging, has inspired artists (often in collaboration with neuroscientists) to incorporate brain images into their creative work and has given rise to festivals, exhibitions, and at least two annual brain art competitions.[1] In her book on art in the age of techno-science, Ingeborg Reichle (2009, 35) remarks that, although artists may work with brain scans or other scientific materials, the artistic approach tends to deal with "the diversity of human experience, which rarely lends itself to portrayal by standardized scientific procedures." Some artists' testimonies and life stories, such as those of Susan Aldworth (2011) or Katherine Sherwood (2007), who have respectively worked with brain scans and angiograms, corroborate Reichle's observations at an intensely personal level.

An emphasis on the historical and experiential nature of art implies a certain point of view concerning its relationship with science. It has been customary to discuss science fiction, and more generally science-related cinema and literature, as if the main purpose were to assess how accurately the forms and contents of science are "represented" in the various products of the "entertainment industry." Regardless of whether the assessment is positive or negative, the spirit of such analysis is close to the so-called deficit model. This model, which is rather a set of approaches to the relationship between science and the public, often tries to detect and redress unreliability or inaccuracy, underlines the fact that people respond to information in ways that experts find inadequate, and looks for ways to communicate "correct" information efficiently (for an overview see Bucchi and Neresini 2007). In line with many authors who have dealt with both literature and cinema, we stand at the opposite end of such an approach insofar as we consider it more productive to search for the meanings works of fiction inspire or convey, rather than to examine them for informational consistency, exactness, or completeness.

Of course there always are areas of convergence and divergence between "fact" and "fiction," but emphasizing them implies at least two problematic assumptions. First, the deficit model suggests that scientific knowledge is produced in isolation from nonscientific contexts, especially those of its public dissemination. Second, it forgets that entertainment appeal, storytelling, and narrative and filmic codes necessarily take precedence over accuracy. Although scientific expertise may contribute to verisimilitude, it does not dictate a movie's factual and scientific content (Kirby 2003), and (in film as in literature) accuracy must be considered in the context of an entire "rhetorical apparatus" (Mellor 2009). When works are examined as totalities, then it turns out that their deficits (relative to official expert knowledge) and their internal inconsistencies are not informational shortcomings but sources of surplus of meaning. As we shall illustrate here, visual and narrative resources may convey ambivalence and contradiction or simultaneously display apparently incompatible claims. Rather than being an aesthetic or intellectual defect, this feature reveals fiction's ability to unfold the complexity of the questions it explores and suggests that univocal answers are not available.

It is in such a perspective that we shall here approach fictional worlds. We shall explore them as spaces in the topography of the *neuro*, connected to the others we have already discussed and involving processes that give substance to the cerebral subject and shape it as a resource for thinking about the human. We shall first examine two phenomena in the literary field since the 1990s, the appearance of the "neuronovel" and the invention of "neuro literary criticism," and then deal with brain movies. Because memory plays in the Western tradition such a fundamental role for understanding personal identity, we shall pay special attention to its treatment on screen and paper. Our main overall observation is that both literature and film illustrate a general feature of art, namely that it "is allowed to be a locus of contradictions where opposites may coincide" (Zwijnenberg 2011, 303). Far from weakening them, this feature makes them particularly powerful for shaping subjectivities.

On Paper: Neuronarratives and Neuro Lit Crit

The defining feature of the literary subgenres christened "neuronarratives" (G. Johnson 2008) and "neuronovels" (Roth 2009) is that they place neuroscientific discourse center stage and use it as a means of expression consubstantial to the stories they tell. In parallel to the creation of those labels, the "next big thing in English" (Cohen 2010) seemed to be "neuro lit crit," or the use of the cognitive neurosciences to gain "unexpected insights" into individual texts as well as to answer questions such as "Why do we read fiction?" "Why do we care so passionately about nonexistent characters?" and "What underlying mental processes are activated when we read?" The *Guardian* hailed the emerging field as "the cutting edge of literary studies" (Harris and Flood 2010), public relations officers at universities were quick to publicize brain imaging research claiming to prove the value of reading literature (e.g., Goldman 2012), and the *New York Times* went as far as asking whether neuro lit crit can "save the humanities."[2]

Neuro lit crit could be considered as a branch of neuroaesthetics. Its more distant roots, however, predate the neural turn and lie in the work of authors such as Robert de Beaugrande (1987), Norman Holland (1988), and Reuven Tsur (1992), who, starting in the 1980s, approached literary theory and criticism with psychological models from the cognitive sciences (for overviews, see Crane and Richardson 1999, Richardson 2004, Zunshine 2010; for an early debate, see Herbert Simon's 1994 article and the responses it drew). Cognitive literary studies do not necessarily include considerations about the brain. However, by the early 2000s the emphasis of cognitive approaches on mind embodiment or the embodied mind had made them partly dependent on neuroscientific studies (for the epistemological consequences of this situation, see Hart 2001).

From the standpoint of a cultural history of the cerebral subject, the most significant feature of neuro lit crit is that it emerged at a time when some noted writers chose to depict their protagonists and construct their plots using neuroscientific vocabulary. Although only a minority of authors in literature and literary studies has recourse to the *neuro*, the convergence of viewpoints has configured a perspective characterized by its placing brains in literature and literature in the brain. Some writers introduce neuroscientific issues, processes, and terminology into their narratives, and some scholars hope that a neuroscientific approach will furnish the key to literary creation and reception. Although neuronovelists and literary scholars seem to share the belief that the neuroscientific approach can revitalize their fields, their approaches are not strictly symmetrical: For the scholars, neuroscientific realism is or functions as an ontological belief; for the writers, it appears to be mainly a literary tool.

Like neuroaesthetics, neuro lit crit approaches literature as if it embodied neuroscientific knowledge, considers writers as more or less penetrating but generally accurate spontaneous neuroscientists, and therefore analyzes their texts in a neurorealistic key. Among writers, A. S. Byatt gave a silly expression to that frame of reference in her speculations about the role of mirror neurons in her experience of reading John Donne: "The pleasure Donne offers our bodies is the pleasure of extreme activity of the brain" (Byatt 2006a; see also Byatt 2006b and the critique by Tallis 2008a). Apparently better informed, the pop-science blogger and writer Jonah Lehrer claimed that Gertrude Stein, Walt Whitman, George Eliot, Virginia Woolf, Paul Cézanne, Igor Stravinsky, Marcel Proust, and the fin-de-siècle French chef Auguste Escoffier were all "artists who anticipated the discoveries of neuroscience" and "discovered truths about the human mind—real, tangible truths—that science is only now *re*discovering"; their art "proved to be the most accurate, because they most explicitly anticipated our science" (Lehrer 2007, ix, xi; on Lehrer's fall from grace, see for example Kachka 2012). More serious scholars share such an outlook on artistic creation, claiming, for example, that when Virginia Woolf, in *To the Lighthouse* (1927), connects the lighthouse to the mother figure and thereby evokes "the long-term memory of her mother," she "was unwittingly acknowledging the limbic underpinnings of her art" (Nalbantian 2008, 363).

Considered in the longue durée, the realism of neuroliterary readings is a recent avatar of a tradition that goes back at least to the ancient depiction of Homer as an accurate describer of landscapes, customs, peoples, and events and the equally ancient interpretation of mythical beings and episodes as allegories of human or natural events (Ford 1999, Lamberton 1986). And from Antiquity onward, the value of such a hermeneutic stance has been questioned. Already before neuro lit crit was given a name, some considered it doubtful that "neurological findings can actually be translated into terms relevant to cultural artifacts and achievements and specifically to literary analysis" (Adler and Gross 2002, 210; see responses in *Poetics Today*, Summer 2003). The debate has continued (e.g., Lauer 2009) but has not weakened calls for literary studies to open themselves to neuroscience (e.g., Starr 2012). The appearance of the neologism "neuro lit crit," apparently never used in print before articles published in 2010 in the *New York Times* and the *Guardian* (Cohen 2010, Harris and Flood 2010), gives a certain unity to this trend and associates a label with a set of beliefs, assertions, and methods.

However, beyond commonalities, does the neurologizing gaze fulfill the same function in literature and in neuro lit crit? The answer, as we shall see, is no, and the differences throw light on distinct ways of articulating the cerebral subject. We shall show this by exploring three central features of neuronovels.

First, neuronovels do not limit themselves to incorporating brain-related vocabulary from a third-person perspective, as Victorian authors sometimes did when depicting a character in phrenological terms. Rather, they integrate a neuroscientific perspective and language into first-person narratives. Nevertheless (second feature), such a writing strategy does not imply a sort of cerebral solipsism that would confine literary characters inside their own brains. It has been argued that it does and that by turning inward, "to an almost cellular level," neuronovels bypass the self, society, and history and construe a neurological "privacy without individuality" in which "mere biological contingency" replaces meaning (Roth 2009). Against such a reading, we shall argue that what may look like solipsism serves mainly to problematize the notion of the human as cerebral subject.

Finally, we illustrate neuronovels' ambivalence vis-à-vis the brainhood ideology by examining how they dramatize the phenomenological, affective, and embodied dimension of memory, a psychological function considered as fundamental to selfhood in the Western philosophical tradition as well as in the contemporary sciences of mind and brain. We discuss some of the English-language works that directly inspired the invention of the notion of *neuronovel*; we will therefore leave aside recent "brain memoirs" (Tougaw 2012), which are autobiographical disease narratives, and will refer to older novels that incorporate neurological idioms and phenomena only to give historical perspective. Neuronovels exist in languages other than in English, but we have found no critical work about them other than Stephan Besser's (2013, 2016) on the "poetics of the neuromolecular gaze" in contemporary Dutch fiction.

Varieties of Neurological Fiction

In a 2009 article entitled "The Rise of the Neuronovel," the critic Marco Roth claimed that, in Anglo-American fiction, the workings of the mind had been replaced by the mechanics of the brain, and he dated the shift to the English writer Ian McEwan's 1997 *Enduring Love*. In this novel, a young man with De Clérambault syndrome, a condition also known as erotomania, in which patients believe someone is in love with them, stalks a science journalist. The novel's appendix consists of a clinical case history and a scientific bibliography. Prominent examples of the same genre include Mark Haddon's *The Curious Incident of the Dog in the Night-Time* (2003), on autism; McEwan's *Saturday* (2005), on Alzheimer's and Huntington's disease; Jonathan Lethem's *Motherless Brooklyn* (1999), on Tourette syndrome; John Wray's *Lowboy* (2009), on paranoid schizophrenia; and finally Richard Powers's *The Echomaker* (2006) and Rivka Galchen's *Atmospheric Disturbances* (2008), both on Capgras syndrome. We could add Jonathan Franzen's *The Corrections* (2001) and novels and stories by Tom Wolfe, A. S. Byatt, David Foster Wallace, Umberto Eco, and Jeffrey Eugenides. Medical thrillers, science fiction, and cyberpunk novels and short stories by less well-known authors have also explored neurological conditions as well as the consequences of transplanting brains or transferring their contents (Cavallaro 2004; Dinello 2006; Geraci 2010; Hahn 2005; Pethes 2005; Tofts, Jonson, and Cavallaro 2004).

Roth attributed the rise of the neuronovel to the exhaustion of the linguistic turn in the humanities, the fall of psychoanalysis in the Anglo-American world, and the popularization of brain research and the accompanying neural turn. Nevertheless, just as neuro lit crit is only the latest development of cognitive literary studies, the neuronovel is the newest way of using brain-related issues in fiction. Victorian literature, to begin with, gave a considerable role to phrenology (Boshears and Whitaker 2013), which appears in the works of Anne Brontë (*The Tenant of Wildfell Hall*, 1848), Charlotte Brontë (*Jane Eyre*, 1847; *Shirley*, 1849; *Villette*, 1853; *The Professor*, 1857), George Eliot (*The Lifted Veil*, 1859), Wilkie Collins (*The Legacy of Cain*, 1889), and Charles Dickens (*David Copperfield*, 1850; *Great Expectations*, 1861; *Sketches by Boz*, 1836; *Bleak House*, 1853). Phrenology also appears in fiction from Argentina, Brazil, France, Germany, Italy, Russia, Spain, the United States, and most likely from all other countries where literature was informed by romanticism, realism, and naturalism (Bernucci 2008, Bottoni 2012, Cooter 1984, De Giustino 1975, Goscilo 1981, Krow-Lucal 1983, Oehler-Klein 1990, Van Wyhe 2002, Wright 1982).

Regardless of whether phrenology was part of a writer's personal beliefs, it could be used as a literary device. In *Shirley*, for example, Charlotte Brontë described the character of Mr. Yorke as lacking "the organ of Veneration—a great want, and which throws a man wrong on every point where veneration is required," as well the organ "of Comparison—a deficiency which strips a man of sympathy," and as having "too little of the organs of Benevolence and Ideality, which took the glory and softness from his nature, and for him diminished those divine qualities throughout the universe" (Brontë 1985 [1849], 76). Brontë believed in phrenology (Shuttleworth 1996), but on this occasion exploited it primarily as an easily recognizable resource for ironically depicting a character.

Late Victorian neurology is also present in Bram Stoker, Robert Louis Stevenson, Silas Weir Mitchell, and H. G. Wells (Stiles 2007). *The Strange Case of Dr. Jekyll and Mr. Hyde* (1886) was probably informed by documented cases of multiple personality from the 1870s (Stiles 2006a, Harrington 1987). *Dracula* (1897) drew on debates about the localization of brain function (Stiles 2006b), and works by Charlotte Brontë (*Jane Eyre*, 1847; *Villette*, 1853), Wilkie Collins (*Heart and Science*, 1883), and Thomas Hardy (*Tess of the d'Urbervilles*, 1891) incorporated discussions about the gendered brain (Malane 2005). Inversely, a century later, the novel *Albricks Gold* fictionalized its neuroscientist author's theory that certain regions of the hypothalamus are smaller in gay men than in heterosexual men (Levay 1997).

Brain transplantation featured in Michael Bulgakov's 1925 *Heart of a Dog*, a short novel in which a murderer's brain is transplanted into a dog's body, resulting in a dog with the criminal's personality. Also in 1925, Bulgakov's fellow countryman Alexander Beliaev published *Professor Dowell's Head*, a novel in which a young doctor must take care of the living head an evil scientist exploits for its ideas and knowledge. Both novels allude to research that was being carried out at the time in the Soviet Union (Krementsov 2009). Half a century later, at the other end of the world, the Argentine writer Adolfo Bioy Casares took up the fiction of transplanting a human soul into a dog. Although transplantation here concerns a soul rather than a brain, the procedure involves trepanning the skull and gives rise to a discussion about cerebral localization between the surgeon and the man whose wife's soul is being transplanted:

—Remember what Descartes said? How are you going to remember if you never read him. Descartes thought that the soul was in a gland of the brain.

He said a name which sounded like "pineral" or "mineral."

The missus' soul? I asked.

He was so annoyed when he answered that he confused me.

—Anybody's souls, my good man. Yours, mine.

—What's the gland called?

—Forget it, because it doesn't matter and it doesn't even have the function they attributed to it.

—Then why do you mention it?

—Descartes was not wrong in principle. The soul is in the brain and we can isolate it. (Bioy Casares 2004 [1973], 160)

As other brain fictions, the transplantation theme explores the belief that "the soul" (or, in more secular terms, that which makes us the persons we are), is in the brain and is, in fact, consubstantial with it.

The brain in a vat (mentioned in Chapter 1) is the most recognizable twentieth-century variation on this theme, and in the 1960s it became one of the professional philosophers' favorite thought experiments about personal identity. It might have become especially familiar after the philosopher Hilary Putnam used it in the first chapter of *Reason, Truth, and History* (1981), but it had by then made a good number of literary (and filmic) appearances, from J. D. Bernal's futuristic *The World, the Flesh, and the Devil* of 1929 to Curt Siodmak's 1942 *Donovan's Brain* and the 1964 story "Whither Go the Cephalomos? [¿A dónde van los cefalomos?]" by the Cuban science fiction pioneer Angel Arango. In *A Scanner Darkly* (1977), Philip K. Dick incorporated research on the split brain and hemispheric lateralization and quoted articles of the late 1960s by the neuroscientists Joseph E. Bogen and Michael Gazzaniga. Dick (1977, 144) also referred to Arthur L. Wigan's 1844 *The Duality of Mind*, one of the first works to present the idea of the "double brain" (also discussed in Chapter 1). The 1970s witnessed the rise of medical thrillers, which sometimes rehearse brain-related theories and practices (Hahn 2005, Pethes 2005); a famous example is Michael Crichton's 1972 *The Terminal Man*, in which a microcomputer is inserted into a computer scientist's brain.

Science fiction and, later, cyberpunk authors such as William Gibson (*Neuromancer*, 1985), Pat Cadigan (*Mindplayers*, 1987; *Fools*, 1992), Harry

Harrison and Marvin Minsky (*The Turing Option*, 1992), John Darnton (*Mind Catcher*, 2002), Bruce Sterling (*Schismatrix*, 1985), George Alec Effinger (*When Gravity Fails*, 1987; *A Fire in the Sun*, 1990; *The Exile Kiss*, 1991), and William Hjortsberg (*Odd Corners*, 1971) have dealt more or less critically with brain death and brain transplantation, the cerebral localization of memory and personal identity, and the technological means to preserve brain contents and, thereby, the person (see also Cadigan 2002; Dinello 2006; Geraci 2010; Guidotti 2003; Sterling 1986; Tofts, Jonson, and Cavallaro 2004).

The neuronovel of the 1990s continues in some respects this line of neurological fiction. Nevertheless, while the earlier works usually juxtaposed *neuro* themes to the treatment of characters or were merely instrumental (as for example when someone's personality is described in phrenological terms or modified by a head injury, or even when brain contents are uploaded to a computer), neuronovels provide a neurochemical depiction of interiority. Rather than using neuroscientific vocabulary as an explanation of psychology and behavior, they turn brain mechanisms into a constitutive element of their characters.

Neurological Narrative

Both David Lodge's *Thinks . . .* (2001) and Richard Powers's *Galatea 2.2* (1995) involve writers who are guests at prestigious cognitive neuroscience centers, and both deal with the clash between the scientific and humanist worldviews. Helen Reed, the protagonist of *Thinks . . .* , is a novelist visiting at the Centre for Cognitive Science in the imaginary British Gloucester University. Her relationship with the Centre's director Ralph Messenger gives Lodge the occasion to explore the nature of the human self and consciousness. In one of their first encounters, Helen skeptically remarks that consciousness is "the thing at the moment" among cognitive scientists, who consider it a "'problem' which has to be 'solved'" (Lodge 2001, 61). For the humanist, she notes, the scientific approach to consciousness is not only unfamiliar but also threatening:

> I've always assumed, I suppose, that consciousness was the province of the arts, especially literature, and most specifically the novel. . . . Consciousness is simply

the medium in which one lives and has a sense of personal identity. The
problem is how to represent it, especially in different selves from one's own (61).

The challenge for Helen is how to assimilate the neurocognitive perspective while keeping consciousness as that which enables a sense of personal
identity and, therefore, as appropriate material for literary creation.

In contrast to Helen, Richard Powers, the protagonist of *Galatea 2.2* and
namesake of the novel's author, becomes a convert to the neurological vocabulary and is exhilarated to realize that reading has changed the "physical
structure" of his brain and thus "deformed the cell map of the mind" (Powers 1995, 56). After some hesitation, Richard joins an artificial intelligence
project whose goal is to assemble a device capable of passing the final exam
for a master's degree in English literature. The goal of the project is to
emulate the workings of the human brain through "Implementations"
based on neural networks. Richard then spends months reading literary classics to Helen, as the computer is called, until he believes it can mimic self-
consciousness and has become not only "operationally equivalent" to a
human mind but actually "indistinguishable" from it (52).

At the same time, however, the novelist-protagonist asserts that "knowledge is physical" and that embodiment is crucial for personal identity
(Adams 2008). Words and their neural representations do not suffice for
personhood: the crucial element is not "what your mother reads you" but
"the weight of her arm around you" as she reads (Powers 1995, 147). Even
the computer Helen comes to realize this and to desire full human embodiment. Neuronovels thus problematize the belief that humans are essentially
their brains; films, as we shall see below, do very much the same. And while
it may seem that neuronovelists try to bridge the gap between the "two cultures" (G. Johnson 2008, Max 2007), the tensions they dramatize suggest
insuperable differences.

The relationship between the novelist Helen and Ralph the cognitive scientist in *Thinks* . . . as well as the writer Richard Powers's conversations
with the scientist Philip Lentz in *Galatea 2.2* function as vehicles for conveying knowledge on cognitive neuroscience, artificial intelligence, and
connectionism. That is why some commentators feel that Lodge and Powers are closer to science writing than to fiction (Deresiewicz 2006). Others,
such as the philosopher Daniel Dennett (2008, 160), consider that to ask if

Powers is "doing science in a new, informal, 'artistic' way, or . . . 'just' writing fiction" is to ask a "bad question." But the dilemma remains, and many neuronovels explore the difficulties of reconciling science popularization with basic research and scientific writing. In *Enduring Love*, the popular science writer Joe claims that "someone has to go between the researcher and the general public, giving the higher order explanations that the average laboratory worker is too busy, or too cautious, to indulge" (McEwan 1997, 75). Joe, however, had been a basic scientist and remains torn between his two vocations.

Richard Powers's *The Echo Maker* (2006) depicts an analogous situation. The novel narrates the story of Mark Schluchter, a twenty-seven-year-old meatpacker who, as a consequence of severe head trauma in an almost fatal car accident, develops Capgras syndrome, a rare neurological disorder that makes him believe his sister Karin is an impostor who looks exactly like her. Dr. Hayes and Dr. Weber take up his case. The former sees Capgras exclusively as a brain disorder. For Weber, who also writes popular science, the delusion involves the whole person, his life, history, and personality. Inspired by such scientist-writers as Oliver Sacks, V. S. Ramachandran, Daniel Dennett, and Gerald Edelman, Weber, like Joe in *Enduring Love*, hesitates between basic neurological research and a science popularization style focused on the existential-narrative aspects of illness (Draaisma 2009, Herman and Vervaeck 2009, Tabbi 2008). But a review in *Harper's Magazine* dismissing one of his books as "slightly cartoonish," "entirely predictable," and based on "unacknowledged research" wounds his narcissism and makes him want to return to pure science (Powers 2006, 221). In rehearsing such difficulties at the borderline of science and nonscience, Lodge, Powers, and McEwan seem to remain confident that narratives exploring neurological research and conditions have the "potential to refresh and redeem the field of literature" (G. Johnson 2008, 184). Neuronovels would thus be a creative response to the exhaustion of the traditional consciousness novel, and instead of a deficient instance of psychological realism, a piece like *The Echo Maker* emerges as the "first fully realized novel of *neurological realism*," that is, a novel in which neuroscientific perspectives are "fully enfolded" into the narrative strategies (Harris 2008, 243, 258).

By fictionalizing neurological conditions, neuronovels also offer a mirror image of the kind of clinical narrative so brilliantly practiced by Oliver Sacks

(1985, 1995) or Paul Broks (2003). The Russian neurophysiologist Alexander Luria (1979), who pioneered the genre, referred to it as "romantic neurology," an approach that attempts to recover the "I" or "who" of the patient's subjectivity from the "it" or "what" of the physical illness (Couser 2004, 75–76). We have already mentioned the confrontation of perspectives through the characters of the doctors Haynes and Weber in *The Echo Maker*. Whereas the former reduces Capgras syndrome to neurology, the latter emphasizes the patient's individual story and inner experience. Looking at the patient's brain scans, Haynes perceives only "structure. . . . Something that looks like possible discrete injury near the anterior right fusiform gyrus, as well as the anterior middle and inferior temporal gyri." Weber, in contrast, sees "the rarest of butterflies, fluttering mind, its paired wings pinned to the film in obscene detail." He asks for "something more than neurons," since "Capgras may not be caused so much by the lesion per se as by large-scale psychological reactions to the disorientation" characteristic of the condition (Powers 2006, 131, 132). Dr. Weber's demanding "something more than neurons" points to a second central feature of neuronovels, namely that they need not imply cerebral solipsism.

Neuronovels and Solipsism

The Echo Maker's epigraph is the last sentence from a passage where Alexander Luria explains that, in order to find the sources of human consciousness and freedom, "it is necessary to go outside the limits of the organism . . . into the objective forms of social life" and into "the social history of humanity." And then comes the sentence used as the novel's epigraph: "To find the soul it is necessary to lose it" (Luria 1966, 96–97). The choice of such a statement makes sense because *The Echo Maker* exhibits neurological realism at least as much as it challenges the belief that personhood and selfhood reside essentially in the brain.

Neuronovels, of course, often make characters speak in a *neuro* idiom supposed to account for their psychological situation. For example, in Jonathan Franzen's *The Corrections*, the protagonist Gary finds it hard to believe that his depression "wasn't neurochemical but personal" (Franzen 2001, 198). Franzen writes:

Various chemicals that molecular floodgates had been holding back all afternoon burst loose and flooded Gary's neural pathways. A cascade of reactions initiated by Factor 6 relaxed his tear valves and sent a wave of nausea down his vagus, a "sense" that he survived from day to day by distracting himself from underground truths that day by day grew more compelling and decisive. The truth that he was going to die. (156–157)

The writer here attenuates his own neurobiological depiction by combining it with references to the felt experiences of nausea and of the anticipation of death. Similarly, in *Saturday*, Ian McEwan portrays the main characters through their relation to the brain but lyrically extrapolates the brain-based vocabulary, as in the following description of two nurses crossing a square:

In the lifeless cold, they pass through the night, hot little biological engines with bipedal skills suited to any terrain, endowed with innumerable branching neural networks sunk deep in a knob of bone casing, buried fibres, warm filaments with their invisible glow of consciousness—these engines devise their own tracks. (McEwan 2005, 12)

It has been argued that neuronovels transform the self "into an object whose intricacies can only be described by future science" (Roth 2009). The examples from Franzen and McEwan nevertheless demonstrate that the neurobiological vocabulary may coexist with ambivalence regarding the biological determinism it seems to convey.

Beyond the literary realm, neuronovels illustrate the psychosocial phenomenon we reported in the chapter on cerebralizing distress: *Neuro* idioms are used pragmatically, and neurological solipsism does not necessarily follow from conceiving oneself and others as cerebral subjects. *The Echo Maker*, for instance, transforms contemporary neuroscientific insights about the nature of the self and neurological disorders into a fictional device to convey the fragility of selfhood as well as of collective and individual experience. The feeling of estrangement that characterizes Capgras syndrome becomes the "baseline condition for life in terrorized America" after September 11 (Powers 2007). America, said Powers in an interview, has become a place that "looks like my country, sounds like my country, acts like my country, but it's no place that I can recognize. It must be an impostor" (quoted in Gennero 2008, 96). Thus, starting with a pathological condition,

The Echo Maker explores the social and political predicament of an entire nation. It illustrates how, instead of isolating personhood inside the individual brain, neuronovels use neuropsychiatric disorders as an existential-phenomenological mirror for the undiagnosed.

Cerebralizing Memory?

Neuronovels' treatment of memory provides a major example of the tension between the cerebralizing of selfhood and the desire to maintain what Jonathan Franzen (2002, 19), in a memoir on his father's decline with Alzheimer's, characterized as "the more soul-like aspects of the self." In *Galatea 2.2*, after a few weeks living among cognitive scientists, the humanist protagonist, Powers, questions his basic certainties. He asks,

> What was memory? Where, if anywhere, did it reside? How did an idea look? Why was comprehension bred, or aesthetic taste, or temperament? Predicates threaded my neural maze. After great inference, I came to the conclusion that I hadn't the foggiest idea what cognition was. (Powers 1995, 28)

Powers's quest for an answer leads him away from neuroscientific reduction and toward the phenomenological, embodied, and affective dimension of human interaction. He notes, for example, that although a child who suffers from Down syndrome is described as less able to understand literature than the computer Helen, the "incredible bodily empathy" that makes him capable of interpersonal connection is a dimension of humanity that Helen will never have (Powers 1995, 134; Bould and Vint 2007). Helen possesses a huge memory storage and associative capacity, but that alone does not make her human.

Similarly, in the case of Capgras syndrome as depicted in Powers's *The Echo Maker* and Galchen's *Atmospheric Disturbances*, the derangement of memory is not treated primarily in a neuroscientific perspective. Since at least Antonio Damasio's popular *Descartes' Error* was published in 1995, it has become commonplace to emphasize the links between "emotion" and "reason." Neuroscientists have turned their attention to the integration of emotion, memory, and cognition, and studies have also been devoted to how the brain organizes experience narratively and thereby produces a crucial dimension of "self" (Young and Saver 2001). Writers such as Powers and Galchen are aware of these trends yet focus on how emotion, memory, and self are apprehended in phenomenological terms. In Capgras, emotion is severed from memory: patients recognize the beloved others but, feeling no emotion toward them, perceive them as impostors. In order to sustain our relationship to the world, memory must involve emotion (Harris 2008); the narrative deals with the breakdown of this connection.

Again in *The Echo Maker*, another perspective on memory emerges from a narrative about geographic location as phenomenological and symbolic space. As Powers (2007) explains, "the book is about memory and recognition, but those mental skills are themselves deeply linked to the brain's spatial abilities." Hence his decision to focus on a particular location, a town in Nebraska. At the same time, his treatment of place highlights its significance for the American collective psyche. As Powers (2007) put it in an interview:

> If space is the field for memory, and if memory is the basis of our narrative
> self-invention, then we must live in some seam between inside and outside, some
> corridor between the place we make and the place that makes us. That's why I
> went to this crossroads, the empty, remote center of the Great American Desert.

Instead of situating psychological reality in the brain circuits responsible for memory, the author turns to a reflection on the imaginary geographies and the lived experiences of post-9/11 Americans. Such narrative treatment of memory is consistent with neuronovels' implicit critique of the cerebralizing reduction of self. Speaking about the brain is mainly a trope to deal with sociality and experience, a way for making fiction in a context where *neuro* discourses have emerged as a prominent way of understanding the human.

Neuronovels seem to adhere to a neuroscientific view of the self and memory mainly when read through the lens of cognitive literary criticism and neuro lit crit. For example, in several novels by Ian McEwan, details of certain memories are magnified, and narrative closeups signify the emotions attached to those details: the protagonist's narration of an encounter by a fountain in *Atonement*, the recollection of the hot-air balloon crash in *Enduring Love*, the circumstantial account of the attack on the main character's house and family in *Saturday*. A neuro lit crit analysis of those episodes reads like the following:

When an individual encounters a stressful situation an instinctive response is prompted. . . . This response suggests that in situations where we undergo stress, our neurons process more information than they usually would do, due to the production of adrenaline. Therefore, a great number of synaptic connections are formed in the brain, allowing us to build up a stronger memory of the event just experienced, which then can be strengthened farther through recall. (Ash 2012)

The commentary assumes that the novel's narrative choices are designed to depict neurobiological processes. That, however, is patently false, and the interpretation, illustrative of the neural turn in literary studies, does not throw any light on the novel or on the reader's experience.

Since McEwan is known for his "empirical temperament" and "hostility to irrational thinking," and since *Saturday* is supposed to be "a direct assault on the modern novel's skepticism toward science" (Zalewski 2009, 2, 17, 21), readings of his work in a neuroscientific key may seem appropriate. Such a key, however, does not open any significant hermeneutic door. Even the more elaborate *neuro* interpretations merely juxtapose the writer and the neuroscientist (as Thrailkill 2011 does with McEwan, Gerald Edelman, and Antonio Damasio) and note the convergence of their treatments of the mind/brain. Such readings do little more than redescribe in superficially neuroscientific terms events and characters depicted in the novels. They assume, as we noted, that writers are intuitive neuroscientists or (as Thraikill 2011, 197, claims of McEwan) that they "import" into their narratives neuroscience's "wisdom about the human mind." In contrast, we have argued that neuronovels' treatment of memory is better understood as a distancing device, as a critical statement about neurorealistic exegesis.

David Lodge's *Thinks . . .* develops a fictional framework to dramatize the confrontation between humanist and scientific theories about human consciousness and selfhood. However, if the novel, and to a certain extent also Powers's *Galatea 2.2*, stages the conflict of the Two Cultures as a way of aesthetically conveying opposing worldviews, it also proposes ways of transcending it. At a certain moment the protagonist Helen Reed, a novelist, attends a lecture by Robin Penrose, a feminist postmodern literary scholar who had appeared in Lodge's earlier novel *Nice Work* (1988). The title of Penrose's lecture is "Interrogating the Subject." Helen is dismayed because, instead of a plea for the humanist subject endowed with a cohesive identity, she must listen to a defense of the view that the subject does not exist. The general argument of Penrose's lecture, she explains,

> was that the Subject in all these senses is a Bad Thing, that there is some kind of equivalence between the privileging of the ego in classical psychoanalysis, the fetishization of formal correctness in traditional grammar, the exploitation and oppression of subject races by colonialism, and the idea of a literary canon, they are all repressive and tyrannical and phallocentric and have to be deconstructed. (Lodge 2001, 225)

Helen realizes that Penrose's discourse parallels the neurological deconstruction of selfhood endorsed by her lover, the cognitive scientist Ralph Messenger. She is astonished to see that both Ralph and Penrose "deny that the self has any fixed identity, any 'centre.'" As she puts it: "He says it's a fiction that we make up; she says it is made up for us by culture. It's alarming that there should be so much agreement on this point between the most advanced thinking in the sciences and the humanities" (Lodge 2001, 224, 225–226; see also Gennero 2011). Similarly, in Powers's *The Echo Maker* neuroscience undermines the belief in an integrated and self-conscious self. Both novels depict a neurobiological program deemed coherent with both a biologizing perspective and deconstructionist theories of the subject. On this point, fiction (or, rather, some central fictional characters' view) is consistent with scholarly opinions.

For example, taking for granted an adaptationist view of Darwinism that has been intensely debated since at least Stephen J. Gould and Richard Lewontin's (1979) famous article "The Spandrels of San Marco and the Panglossian Paradigm," Ellen Spolsky (2002, 56) writes that "nothing could be more adaptationist, more Darwinian than deconstruction and poststructuralism, since both understand structuration . . . as an activity that happens within and in response to a specific environment." Her assertion is consistent with the assumption of neuroaesthetics (mentioned here in chapter 2) that human aesthetic preferences reflect evolutionary adaptations. Such an evolutionary perspective gives those preferences an apparently solid foundation but also makes them contingent upon environmental demands. Confronted with positions akin to Spolsky's, Helen, in *Thinks . . .*, notes that the cognitive sciences move in the same direction as psycho-analysis, feminist theory, and deconstruction—all arguing for the ultimately illusory character of the humanist "subject."

On Screen: Brain Transplants and Memory Losses

The novels we just examined variously dramatize the relationship between neuroscientific knowledge (and its supposed larger consequences) and the notion of a "subject" whose embodiment and substantiality is potentially reducible to the networks of neurons out of whose interactions it "emerges." When film scrutinizes similar issues, it obviously does so with its own means. This implies a predominance of visual materials over verbal ones. Dialogues of course remain essential, but embodiment is given a central role as that which displays the issues at hand. The physical criteria of personal identity take center stage, and in that case the main character often is an individual made up of one person's brain and another person's noncerebral body. But when the psychological criteria of personal identity are emphasized, then the dramatic focus shifts to memory, to amnesia, and to situations where characters are given memories that are not originally their own. Within the larger universe of movies that deal with personal identity and its conditions of embodiment, the ones we have chosen here explore specifically the promises and difficulties of the brainhood ideology.

Cerebralizing Frankenstein

Early on in James Whale's classic film *Frankenstein* (1931), Dr. Henry Frankenstein and his hunchbacked assistant Fritz plan to steal a hanged man's body. However, when the body falls to the ground, the neck breaks and makes the brain "useless." The new brain that Fritz steals from the laboratory of Henry's former professor, Dr. Waldman, is in a jar labeled "abnormal." Waldman had explained that the "scarcity of convolutions on the frontal lobe" and the "distinct degeneration of the middle frontal lobe" corresponded exactly with its owner's life "of brutality, violence, and murder." Without noticing its anatomical flaws, Henry uses the brain for his creature and urges a bewildered Fritz to "think of it: the brain of a dead man, waiting to live again in a body I made with my own hands!"

Frankenstein highlights the ontological function of the brain. Unless the brain lives, the person doesn't. In theory then (for the audience is rather captivated by the creature's outer appearance), the "monster" is essentially a brain. But does the film make the theory hold? Although Waldman anticipates the consequences of having given him a "criminal brain," the creature actually becomes aggressive only in reaction to human violence. Yet most commentators have followed Waldman and found in the brain the obvious cause for the creature's alleged killing urges. In 1931, a *New York Times* critic correctly noted that the brain was the reason *given in the film* for the creature's "murderous onslaughts."[3] Sixty years later, a Frankenstein historian suggested that the brain plot attenuated the doctor's guilt (Forry 1990, 92), and the organizers of a 1997 exhibition at the National Library of Medicine in Washington, D.C., wrote that the makers of the 1931 *Frankenstein* exploited a widespread belief in biological determinism "when they created a monster whose evil results from the lobes of his brain rather than his experiences or character" (Lederer 2002, 46).

The film certainly illustrates how natural it was in the 1920s to think of a person's inclinations as determined by the brain and to believe that a diagnosis could be established by observing cortical morphology with the naked eye. At the same time, by emphasizing that the creature's violence is a reaction to human brutality, it questions Waldman's determinism. Historically, even more significant than the existence of a brain subplot is the insistence with which it has been singled out. The exhibition curators who in the 1990s juxtaposed Whale's *Frankenstein* with a 1921 display of "criminal brains" not only highlighted a common historical context of belief about criminality (and, more generally, about racial and individual traits), but they also helped perpetuate the "cerebralizing" interpretation of Whale's classic.[4] In that too, they went along with the later Frankenstein industry.

Indeed, most Frankenstein pictures after Whale's *Bride of Frankenstein* (1935) replace the original theme of the creation of life with a scenario about brain transplantation and its consequences. The shift begins with *Ghost of Frankenstein* (1942) and persists across genres, from gore to mild porn, to Mel Brooks's satirical *Young Frankenstein* (1974) and beyond (Vidal 2016). Everything hinges on whose brain goes into whose body. That is why most films do not include a creature at all in the sense of Whale's "monster" but instead a freak composed of A's body and B's brain. "My brain," says a desolate character in *Frankenstein Must Be Destroyed* (1969), "is in someone else's body . . ." He may as well have said, "*I* am in someone else's body." The brain makes the person. At the same time, as we shall see, precisely because the action results from the person's "being in" an alien body, most films necessarily problematize the reduction of the person to the brain.

The relevant productions in this respect are either "brain movies" or memory movies. We shall not limit the former category to productions involving evil brains or brainlike entities (Senn and Johnson 1992, 99–109) but enlarge it to films that display brains and deal with brain transplantation. The latter category includes films where personal identity is essentially defined by personal memories; although these memories are located in the brain, the organ is not itself a major protagonist. Films produced since the 1970s avoid surgery in favor of manipulating brain contents via microelectronic procedures for inserting, copying, transferring, selling, buying, controlling, implanting, or deleting "embrained" information. In spite of considerable differences—in their scripts, representations of technology, special effects, characters and narrative types, filmic strategies and ambitions—both memory and brain movies explore the relationships between having a body and being a brain and assume that personhood is essentially defined by memories. Memory transfers are functional equivalents of brain transplantations: Saying that A has B's memories is the same as saying that A has B's brain, and giving A's memories to B generally implies relocating some kind of brain substance. The cerebral self is mostly memorial, and the memorial self is necessarily cerebral.

Body Parts and Living Heads

Have other body parts ever been depicted as fulfilling the same function as the brain? For centuries since Antiquity, mind and the physical basis of personality were also located in the heart. The heart played a crucial role in Christian mysticism, and objects in the shape of the cross and other instruments of the Passion were found in holy persons' hearts. It did not, however, play for personal identity a role equivalent to the brain's, and it was predominantly seen as the locus and symbol of emotions (Bound Alberti 2010). The same applies to heart transplantation movies. In the comedy *Heart Condition* (1990), a racist policeman receives the heart of a murdered black lawyer, who returns as a ghost to urge the cop to catch his killers. The drama *Heart* (1999) revolves around a mother's obsession with her son's heart, now located in another man's body. Yet B's heart in A's body never turns A into B. An exception might be the 1969 Mexican cheapie *La horripilante bestia humana* (The horrifying human beast, aka *Night of the Bloody Apes*), in which a mad doctor tries to cure his son's leukemia by giving him a gorilla's heart, which accidentally transforms him into a violent monster.

More relevant at first sight are the adaptations (four between 1924 and 1960) of the French writer Maurice Renard's 1921 novel *Orlac's Hands* (*Les mains d'Orlac*), which, as far as the hands are concerned, all roughly follow the original story. The pianist Stephen Orlac's hands are mangled in an accident, and he receives those of a man executed for murder. He becomes obsessed with them, and crimes are committed that make him look like the culprit. In the novel and most of the movies, someone else is actually framing him. In all cases, the hands affect Orlac not because they carry another person's identity but because of his own psychological instability.

With the exception of Thing, the friendly forearm of the TV series *The Addams Family* (1962–1964), disembodied hands like those in *The Beast with Five Fingers* (1946), *Dr. Terror's House of Horrors* (1965), or *The Hand* (1981) roam about murderously but do not incarnate personhood. *Body Parts* (1991), based on Boileau-Narcejac's novel *Choice Cuts* (. . . *Et mon tout est un homme*, 1965), seems to realize Orlac's fears. A female surgeon grafts onto different bodies the head, legs, and arms of an executed murderer called Fletcher. The legs and arms turn out to be autonomous. Their present owners display various asocial behaviors, and a painter of corny landscapes becomes successful creating violent scenes; they also have memory flashes of Fletcher's crimes. The character with the murderer's head manages to recover one arm and the legs before having his neck broken by the recipient of the other arm. Given that after the head dies, the memory flashes cease and the surviving patient regains control of the grafted arm, the scenario depends on Fletcher's brain, not on body parts that would autonomously function as loci of personhood. In short, neither the heart nor the hands ever reach the status of the brain in the making of personal identity.

Living severed heads are special because they both display a person's face and enclose his or her brain. In *The Brain That Wouldn't Die* (1959), for example, a surgeon saves the head of his girlfriend Jan out of a car wreck and wants to graft it onto another female body so as to make Jan "complete again." In *Professor Dowell's Testament* (1984), based on Alexander Beliaev's 1925 *Professor Dowell's Head*, the title character is kept alive as a detached head. Similarly, in the British TV miniseries *Cold Lazarus* (1996), scientists activate memories from Daniel Feeld's cryogenically preserved head and project them as scenes onto a screen. As visible bodies with recognizable features, Dowell, Jan, and Feeld are their heads; as persons, they are their brains. Yet they do not consider themselves persons and refuse to survive attached to perfusion machines. More dramatically than brain movies, living-head films stage the tragedy of bodily loss, asking how much of our body we could give up and still wish to live; more obviously than brain movies, they probe limits that are fundamentally existential and only secondarily physiological or anatomical.

Brain movies, in contrast, deal with the more disembodied situation of naked brains kept alive in isolation from their natural bodily environment, generally waiting in vats to be transplanted. These "ectobrains," which were popular in B-movies from the 1940s to the 1970s (Vidal 2009b), embody the person to whom they originally "belonged." The most interesting films of this sort involve individuals of the same species. Whether we say that A is the brain donor or the body recipient and B the brain recipient or the body donor, the outcome of the operation is a person with the external appearance and (except for the brain) the internal anatomy of B but with the life history and psychological features of A—essentially, therefore, according to the movies, A *in* or *with* B's body. The plots then focus on the consequences of the surgery, which range from the hybrid's desire for revenge to the choice of a new life in the guise of B; they may also touch on the resurgence in A of personality features originally belonging to B.

Brain Transplants: Staying the Same or Becoming Someone Else?

Three movies stand out in the transplant category: *The Man Who Changed His Mind* (1936), *Change of Mind* (1969), and *L'homme au cerveau greffé* (*The*

Man with the Transplanted Brain, 1971). In contrast to most others, they take the assumption that humans are cerebral subjects as the core of their scenario rather than as a mere trigger for (usually violent) action, and they deal soberly with the transformations that result from a person's survival as that person's brain in someone else's extracerebral body.

At the beginning of Robert Stevenson's *The Man Who Changed His Mind*, we learn that the young Dr. Clare Wyatt is going to work with the aged Dr. Laurience (Boris Karloff in one of his usual roles), a once-respectable scientist turned "mad brain specialist." Laurience has invented a machine for relocating mental contents between brains. "Until now," he explains to an increasingly worried Clare, "it's never been possible to, as it were, extract the thought content from a living brain, and leave it alive but empty. I can do it; I can take the thought contents of the mind of a living animal, and store it, as you would store electricity." As he demonstrates with two chimps, he can then transfer these contents to another living animal's brain. When Clare realizes that Laurience wants to try this out with humans, she refuses to help him any further. The doctor seeks to convince her: "I can take a young body, and keep my own brain." And he can do the same for her: "Think of it: I offer you eternal youth, eternal loveliness!"

In the meantime, Clare has met the journalist Dick Haslewood, the son of the press magnate Lord Haslewood. Although Laurience is discredited in the scientific community, the lord offers him a laboratory in his London "Haslewood Institute of Modern Science." However, after the doctor is ridiculed at a conference, his patron fires him. But Laurience manages to tie Haslewood to the cerebral transfer chair and replace the lord's mind with that of his crippled assistant Clayton. Haslewood (that is, Haslewood's mind in Clayton's body) dies immediately after the transfer; Clayton's mind survives in the nobleman's body. In the meantime, Clare has become engaged to Dick. Laurience, who is in love with her, plans to occupy Dick's body. For that, he needs Clayton's help. However, since Haslewood's body turns out to have a fatal heart disease, Clayton claims Dick's body for himself. Laurience kills him and forces the mind transfer with Dick. Clare later succeeds in reversing the exchange. With his mind back in his original brain, Laurience dies repentant after asking Clare to destroy all his equipment.

The Man Who Changed His Mind stages a metonymy where the brain stands for the mind it "contains." Although Laurience claims he can take a young body and keep his own brain, he does not transplant brains but transfers without surgery the "thought contents" that define personhood. When Laurience tells Clare that he can keep his own brain in a new young body, he makes *brain* stand for *self* or *personality*. Thus, even though the film enacts psychological criteria of personal identity, it is not without good reasons that it is also known as *The Brainsnatcher*. Both the conflict and the empirical or logical dependence between the cerebral and psychological criteria of personhood emerge here in the context of a scientist's quest for brain-based immortality. The film both assumes and relativizes the view of humans as cerebral subjects; it alternately tells us that we are our minds and that we are our brains.

Two later productions explore the brain transplantation drama in the social realm. Robert Stevens' *Change of Mind* (1969) uses it to deal with race relations in 1960s America. The movie begins with closeups of an open-skull brain operation. District Attorney David Rowe, a white man, has terminal cancer; Dr. Bornear transplants his brain into the body of Ralph Dickson, a black man run over by a car. It is therefore as a white lawyer in a black man's body that Rowe will prosecute the white racist sheriff Gene Webb for the murder of a young black woman. The scenario is thoroughly structured by the racial divide.

The Washington, D.C., mayor finds the looks of the postoperative Rowe "a little shocking"; Rowe's visitors and colleagues are embarrassed; his wife, Margaret, won't let him touch her; his mother feels she cannot accept him as her son ("It's me, mother.—How can you look so different without being different?"); he and Dickson's widow Elizabeth sleep together, but she interrupts their lovemaking with the words "Ralph is dead"; in the nightclub where Elizabeth sings, an attractive black woman recognizes Rowe but treats him as black, while a man comments, "Oh. . . It's the brother with the new brain. . . . You know what's stitched inside this freak's head? A white politician's brain." At first, Rowe's political party no longer wants him to run for district attorney: "To a white man," they tell him, "you're black. To a black man, you're a freak." Predictably, however, Rowe's pursuit of the case against the white sheriff attracts black voters, and his party lets him run. But after he discovers that the murder was actually committed by a black man and asks the case against Webb to be dismissed, his black supporters turn against him.

Such situations define the social drama of the movie. How does it relate to brainhood? When a journalist asks Dr. Bornear, "What is he now, doctor: a white man with a black body, or a black man with a white brain?" the surgeon does not reply, but the film's answer seems unambiguous. Several times Rowe asserts the continuity of his personal identity under a different appearance. To his wife, he explains, "the brain is a wonderful thing. It's really everything: David in me, Margaret in you." And the country's top authority on forensic medicine concludes "that since the brain of David Rowe survives, the brain that reasons, has compassion, personality, loyalties and love, which has memory, instinct and sensation, then David Rowe, district attorney of Dorene County, survives medically *and* legally."

Yet Rowe's medical and legal permanence conflicts with the historical conditions of his social survival. Rowe grows increasingly disappointed and estranged from Margaret and becomes cynical about politics. In a country that had only recently passed civil rights laws, his reembodiment transforms not only his mind and body but also his position in society, attitudes toward him, and his sense of self. His preoperative defense of racial desegregation takes on an existential significance that was completely absent when he was still a white district attorney. Hence the movie title's "change of mind."

The plot offers no resolution and finishes with Rowe's flying toward an unknown destination to "think things over." Such an open ending corresponds to the film's nuanced notion of personhood and results from the ways it problematizes brainhood. The movie declares that we are cerebral subjects—where Rowe's brain goes, there goes Rowe. Yet the protagonist's story is entirely mediated by the historical circumstances and interpersonal relations of his newly embodied self. In the end, a new person materializes from the composite of a "white brain," a "black body," and an explosive political context.

Jacques Doniol-Valcroze's *L'homme au cerveau greffé* (*The Man with the Transplanted Brain*) also focuses on the emergence of a new postoperative person. The middle-aged, fatally ill surgeon Jean Marcilly and his collaborator Robert Desagnac have been experimenting with brain transplantation in animals. When the body of the young Franz Eckermann, fatally injured in a car crash, arrives at Marcilly's clinic, the surgeon asks Desagnac to graft his brain into the victim's body. In the end, the resulting individual (let's call him Franz2) voluntarily assumes Franz's persona.

In *Change of Mind*, the grafted brain completely dominated the rest of the body, and the transformations of the self derived from the hybrid's encounter with his social environment. In contrast, *The Man with the Transplanted Brain* stages the reemergence of Franz's personality, accompanied by certain antagonisms of brain and body, as crucial elements for the metamorphosis of Franz2 into a genuinely new individual. Right after the operation, Franz2 introduces himself as Professor Marcilly, and Desagnac calls him "Jean." He speaks with a slight accent, but his demeanor is Marcilly's. Franz2, however, meets Elena, the beautiful Italian who was about to divorce Franz at the time of the accident, and moves into Franz's apartment. Desagnac intuits the coming difficulties and insists that Franz2 is the surgeon living "in someone else's body."

Franz2 and Elena revive their relationship. In the meantime, some of Franz's usual behaviors reappear. Franz was a racecar driver; Franz2 drives recklessly. Franz2 handles a cat affectionately and then suddenly drops it with violence; a rapid alternation of Marcilly's and Franz2's faces highlights the conflict of the two personalities. Franz2 is puzzled; Elena tells him he loved dogs but hated cats. Franz had a penchant for alcohol; Franz2 drinks without realizing it a bottle of rum, and when Elena scolds him, he denies having drank. As Franz2 explains in his medical persona, "the subject seems to have abruptly become prisoner of the body's drives, as if the cerebral element had suddenly ceased to govern."

Scenes demonstrating the coexistence of two persons in Franz2 follow the remark about the power of Franz's body and the identification of the brain with "reason." For example, Franz2 visits Marcilly's widow Elisabeth in her vast bourgeois apartment. Desagnac arrives and is surprised to find Franz2. Franz2 explains that he wished to see his family and environment, even though he does not miss them. In the following scene, Franz2 returns home late, quarrels with Elena, fills a glass with alcohol; as he is about to drink, he sees himself in a mirror, and throws the glass on the floor. Back at Marcilly's, Elisabeth finds a piece of paper with a doodle her husband was in the habit of making and which was not there before Franz2's visit. In the end, the automatisms and conflicts are superseded by Franz2's choice of a new life. When Desagnac objects, "You are not Franz Eckermann," Franz2 responds, "I know, Robert, I know Eckermann is dead, but the truth is that I, Jean Marcilly, love this woman [Elena]." While the dialogues reproduce traditional associations of the brain with reason and of the body with passions and emotions, Franz2 is trying to bring them together.

Marcilly's decision to impersonate Eckermann is catalyzed by the surgeon's attractive daughter Marianne, who falls in love with Franz2. After narrowly escaping from being seduced, he phones Desagnac, introduces himself as Franz Eckermann, calls him "doctor," addresses him as *vous*, and asks him to tell Marianne no longer to follow Franz. While phoning, however, he automatically draws the surgeon's idiosyncratic scribble. When Marcilly deliberately takes on Eckermann's civil identity, which had until then been a mere cover for the surgeon's persistence as a newly embodied brain, he turns the doctor's death and the patient's survival into public and irreversible social realities.

In spite of the surgery and the connections the movie makes between the brain and the noncerebral body, the brain appears as different from the body rather than as an integral part of it. The film thus rehearses the ancient question of the union of two essentially different substances and gives form to speculations derived, as we saw in Chapter 1, from the Lockean redefinition of personal identity. The brain takes on the functions of the soul as substantial foundation of self and even as that which assures its persistence beyond bodily decay. Marcilly is supposed to survive in Franz's body; successive transplants of A's brain into younger bodies could insure A's immortality.

Like *Change of Mind*, *The Man with the Transplanted Brain* lacks resolution: We do not know what ultimately happens to the main protagonists, nor do we find out if there is a limit to the number of successive brain transplants, a threshold beyond which they would definitely not be "the same." The question these films choose to leave open is, at bottom, the ancient one of Theseus's ship: The vessel the hero used to return from Crete after slaying the Minotaur was kept in the Athens harbor, with its parts being replaced as they decayed. Thus, as Plutarch reports in his *Life of Theseus* (here in John Dryden's version), "this ship became a standing example among the philosophers, for the logical question of things that grow; one side holding that the ship remained the same, and the other contending that it was not the same."

Philosophers never stopped pondering the paradox of Theseus's ship, but the puzzle was of course not their preserve. By the early 1960s, when Anglo-American professional philosophy began discussing personal identity with the help of brain transplantation and other cerebral fictions (duplication, grafting of brain halves onto different bodies or of a full brain onto a body cloned from the original), cinema and literature had been covering much of the same ground for several decades. It took thirty more years before a high-profile neuroscientist (Michael Gazzaniga, as quoted in Chapter 1) reified them as a "simple fact" that proves "you are your brain." That such reification could be so naturally asserted illustrates the power of the cerebral subject as an anthropological figure of modern times. In a period when brain death had just been defined (Beecher et al. 1968), movies such as *Change of Mind* and *The Man with the Transplanted Brain* dramatized an extreme version of the predicament that drove the definition: the entire extracerebral body of a brain-dead person becomes a vessel for the brain of a person whose extracerebral body is fatally ill. Yet the films we discussed transmit conflicting messages and support competing views, dramatizing above all the tension between the possibility that someone might be an isolated brain and the fact that we are more fully embodied social beings. By virtue of its capacity to display incompatible doctrines within one production, cinema both reinforces and subverts; it represents and spreads certain beliefs while simultaneously questioning them and suggesting the existence of alternatives.

Memory Movies

In spite of the nuances and ambiguities we just sketched, the result of transplanting A's brain into B's body is A's persistence in B's body. Precisely because they entail the replacement of the entire outer and visible body, brain transplantations lend themselves particularly well to exploring experiences of reembodiment in relational contexts, and since transplanting the entire brain transfers the totality of mental contents, there is no need to single out any of the psychological features that, in the Western tradition, define personal identity. Among these features, memory has been the most closely associated with visual media. This, no doubt, is partly due to the visuality of memory itself, to the fact that memories often arise in visual form. Cinema has a longstanding relationship with memory and has enacted it in many of its different forms (Greenberg and Gabbard 1999, Radstone 2010). Some films have explored heritage, nostalgia, and trauma at the individual and collective levels; others have taken memory itself as their subject; yet others have modeled themselves on the workings of memory or have sought visually and narratively to mirror its mechanisms and evoke the experience of remembrance.

Yet relatively few films have dealt directly with memory and brain, and most of these few concern an amnesic protagonist. Forgetting has served dramatic and comical purposes from cinema's earliest days, but it was mainly in the 1980s that movies began connecting memory explicitly to the brain and to various neurotechnologies. A central feature of those films is that although they often involve amnesia, the condition makes sense only in the light of protagonists' personal histories, experiences, and existential quests. Both before and after venturing a cerebral diagnosis, films (like short stories, novels, and the sciences of memory themselves) must turn toward the mind. Once again, while many filmic elements convey the ideology of the cerebral subject, others resist it, dispute it, contradict it, or give it subtler forms. Both ways, movies incorporate and reinforce elements of widespread views about the primordial role of memories in defining the individual self and of the brain in constituting human personhood.

The "Cerebrality" of Self and Memory

Given the role attributed to memory in the making of personal identity, the importance of amnesia as a dramatic motif comes as no surprise. Yet it seems that "most amnesic conditions in films bear little relation to reality" (Baxendale 2004). For example, although anterograde amnesia (the inability to recall events that take place after the onset of the disease) is more common and incapacitating than retrograde amnesia (the inability to recall events that preceded the onset of the disease), cinema has overwhelmingly focused on the loss of memories of the past. This attests to the vitality of the Lockean view of personal identity.

Indeed, in film, memory has become so much "a shorthand for identity" that, without it, "we cease to exist as who we are and become only receptors of current data" (Bowman 2004, 85, 88. Outside the cinema, this view prevails, but is not the only one; see Tougaw (2016) for examples in patients' memoirs and literary fiction). Moreover, as we shall see, reality and authenticity are crucial: In most movies, "memory of a 'real' past remains a defining criterion of being a 'real' person" (Marsen 2004). The explicit connection to the brain, however, is accessory, and it arose late, after the wave of brain movies examined in the previous section. For example, in the original version of *The Manchurian Candidate* (1962), hypnosis and behavioristic conditioning are used to brainwash soldiers into having false memories of their commander's behavior. In the 2004 remake, implanting a microchip in the brain augmented the initial psychological means, thus highlighting the extent to which the memory theory of personal identity came to emphasize memory-in-the-brain while not displacing first-person psychological experience as the core event.

In *Total Recall* (1990), set in the year 2084, Douglas Quaid has recurrent dreams about Mars, where, as far as he knows, he has never been. At a holiday agency that implants artificial memories of visits to exotic places, Quaid signs up for a vacation as a secret agent to the Red Planet. It turns out, however, that he actually has been a secret agent on Mars. Another personality surfaces during implantation, and, for most of the movie, we cannot tell if the events we see are "reality" or programmed memories. A video of someone identical to Quaid tells him, "you are not you, you are me." This "me" is Hauser, an agent for the dictator of Mars. They had used Quaid to lead them to a rebel leader, and after the leader is killed, Hauser wants his body back. Quaid, however, manages to escape, and in the end, we see him contemplating the fertile landscape of a free Mars in the company of a lovely woman who had also appeared in his dreams. Throughout the movie, tormented by doubts about the reality of his memories (which, a character reminds him, lie in "that black hole you call a brain"), Quaid keeps asking, "If I'm not me, who the hell am I?" Yet he never finds his "real" self, and the rebel leader's dictum summarizes the lesson of the whole situation: "A man is defined by his actions, not his memories."

In *Magdalena's Brain* (2006), the protagonist's husband, Arthur, is a paralyzed Stephen Hawking–type genius. Magdalena helps him continue his work in artificial intelligence, with the new goal of developing an artificial brain "smart enough to repair [his] mind and get [him] out of this chair." The procedure involves a "memory transfer" in which Arthur's mind's contents are downloaded into a bluish liquid. A small amount of the fluid is to be injected into the brain of a good-looking young man who suffers from a brain tumor; once the material is "integrated," his tumor will be removed, and Magdalena will benefit from having her husband's mind in a fitter body. (None of this happens, and somehow Arthur gets back on his feet.) We are not told how the fluid encodes information, but the point is that memory must appear as a physical cerebral substance. Simple visuals (a glass container with a colored liquid) suffice to demonstrate that manipulating memory amounts to handling neural matter. No scientifically sophisticated detail is necessary to communicate the idea that persons are their memories and that memories are a brain substance or involve neurobiological processes.

Dark City (1998) takes a similar approach. A race of Strangers has invaded the Earth. Once a day, they carry out "tuning," an operation in which they stop everything, make everybody lose consciousness, and refashion a Metropolis-like artificial city; they then change people's identities by "imprinting them," that is, injecting other individuals' memories into their brains. The Strangers, who use dead human bodies as "vessels," are a vanishing race of jellyfish-like intelligent creatures that can survive only by becoming like humans. They do not know what defines humanity, and manipulate memories in order to find out. The manipulations are carried out by a human doctor, Daniel P. Schreber, the namesake of the German judge on whose memoirs Sigmund Freud based his 1911 monograph on paranoia. With a syringe, the doctor first extracts memory fluids through people's foreheads. He mixes them—the painful memory of a great love, a dose of unhappy childhood, youthful rebellion, a death in the family—and injects the preparation into other individuals. The resulting people look just as they did before imprinting but have different memories. So who are they?

Personal Identity and the Authenticity of Memory

Dark City illustrates another filmic leitmotif: the authenticity of memories. The humans whose identities the Strangers transform ignore that their recollections are false. The only exceptions are Schreber, who knows the entire truth, and a certain John Murdoch, who is partly immune to the Strangers' manipulation and is able to "tune." Murdoch does not know if his remembered past actually happened or if his beloved wife, "Emma," ever existed. Yet, unlike other humans, he is aware of his predicament, knowing for example that he is not a serial prostitute killer (the imprint attempted on him) and faintly longing for a place called Shell Beach. Together with Schreber, he embarks on the quest to find that town, only to reach a brilliantly colored billboard behind which there is nothing but deep space.

Schreber eventually helps Murdoch defeat the Strangers, reshape the city, and bring back the sun and natural landscapes. At a recreated Shell Beach, Murdoch meets Emma; the fact that she has been imprinted as Anna and does not remember him is no obstacle to a happy ending. Except for Schreber and Murdoch, nobody's personal identity in the new world is made up of personal memories of experienced events. All memory has become "prosthetic," to use with some latitude a term coined to designate memories that "are adopted as the result of a person's experience with a mass cultural technology of memory that dramatizes or recreates a history that he or she did not live" (Landsberg 2004, 28). Yet that does not make identities fake.

Similarly, in *Blade Runner* (1982), Rachel, a genetically manufactured "replicant," is fitted with long-term implanted memories going back to childhood. That those memories are someone else's does not make them any less significant for her sense of self. On the contrary, the psychological continuity between the alien memories and her own coalesce into one personality. To sum up: In film, amnesia disrupts personal identity more than false memories. The integrity of the self largely depends on the integrity of memory; memory lost is more problematic than memory received, and recovering memory is a cerebral affair.

In *Johnny Mnemonic* (1995), the hero is a "mnemonic courier" who gives up all his childhood memories to make room for data he carries in a brain implant. After accomplishing his mission, Johnny wants his memory back. That requires expensive surgery. In order to pay for it, he agrees to transport a dangerously large amount of information about the treatment of a pandemic "nerve attenuation syndrome." Criminal gangs want the information, and a rebel movement wishes to make it freely available. After it turns out that the memory-retrieval surgery cannot take place, Johnny must "hack" his own brain. When he enters the virtual reality of his implant, its contents are broadcast worldwide, thus universally releasing the information about the treatment to the tune of appropriately stirring music. Thanks to the brain space made available, Johnny relives his childhood memories, and the restoration of his full identity opens the way for a happy ending.

Johnny Mnemonic assumes that memory is a collection of fixed data items that can be stored and recovered. Such a view has been called into question by cognitive neuroscience, with its emphasis on plasticity and its demonstration that long-term memories are not embodied in singular locations but in neural connections as well as in different types of memory processed by different brain systems (Eichenbaum 2012). At the same time, like most movies of its kind, *Johnny Mnemonic* highlights connections between memory and emotion that have been corroborated by the mind and brain sciences, especially the role of emotion in encoding and retrieval (see Dunsmoor et al. 2015 for a recent example).

Movies tend to imply that personal identity requires a repertoire of long-term eidetic or "photographic" episodic memories of real autobiographical events stored in discrete brain locations. For instance, the protagonist of *Strange Days* (1995) deals in illegal recordings made directly from the cerebral cortex, allowing future viewers to live others' experiences as if they were their own. This provides a good example of how discredited notions with a high representational potential (memory as a repository of fixed memories) may be conveyed in ways that make them compatible with the reconstructive, malleable, and manipulable nature of memory, which has been the object of outstanding research since Frederick Bartlett's 1932 *Remembering* and of heated discussions during the "false memories" controversy of the 1990s (e.g., Loftus and Ketcham 1994).

Despite the significance of their brain motifs, movies offer psychological, not neurological dramas. Johnny retrieves his childhood memory by hacking his brain, the *Strange Days* devices record from one brain and deliver to another, and Schreber engineers memories by manipulating a cerebral fluid. Like the neurosciences, movies proclaim that memories are cerebral entities and that experiences have neural correlates, yet, also like the neurosciences, they must place psychology center stage if those mechanisms and correlates are to have any meaning.

The same applies to scenarios about erasing memories. In *Eternal Sunshine of the Spotless Mind* (2004), Joel discovers that his ex-girlfriend Clementine has had him and their failed relationship erased from her memory. Enraged, he decides to undergo the same procedure at a clinic that carries out the "focused erasure" of troubling memories. Such an approach to eliminating memories (by destroying localized clusters of neurons) differs from contemporary neuroscience. A 2014 review noted that "enhancing recall, deleting knowledge of the past and implanting fictitious memories—once the preserve of Hollywood blockbusters—are now becoming a reality" thanks to an expanding repertoire of methods (Spiers and Bendor 2014, 2). These methods don't include *Eternal Sunshine*'s, but otherwise the movie has been widely considered as a nuanced representation of how the brain forms memories of intense emotional experiences and how traces of those experiences may resurface in amnesia.

However, commentators' emphasis on memory erasure overlooks one of the film's crucial features. Erasure is of course essential to the film—but chiefly because both the plot and the shooting are driven by its *failure*. At one point during the procedure, Joel, who is unconscious, wants to call it off and preserve certain targeted memories. The movie focuses on resistance to memory deletion rather than on the consequences of its success.

Moreover, some films convey the idea that natural memories are never totally lost. In *Johnny Mnemonic*, where the deletion is voluntary, the memory of the protagonist's childhood is entirely retrievable. Sometimes, traces remain operational. *Eternal Sunshine* opens with Joel impulsively taking the train to the forgotten place where he first met Clementine. In *Paycheck* (2003), which only briefly shows that the procedure involves intervening in the brain, an engineer agrees to have erased from his memory the three years during which he cracked the design of a machine to see into the future. But when he attempts to claim the millions owed him, he discovers that he has signed away the money, and in exchange he receives an envelope with miscellaneous everyday objects. Although he does not recognize them, he uses them appropriately to escape from the killers who chase him; after memory erasure, the unconscious traces of such knowledge (what psychologists call "implicit memory") serve him well.

For most of the movies that have explored memory, identity, and the brain since the 1980s, humans are essentially defined by the relationship of their present to their remembered past: totally forgetting our past would turn us into different persons. Moreover, most films tend to assume a storehouse model of memory, perhaps because it is familiar and facilitates representation. The notion of discrete brain locations can be more readily incorporated into filmic action than neural networks and chemical events happening at the synaptic gap. Concurrently, though, filmic representation goes beyond that basic theory, and the same relativizing effect concerns the belief that memories are indestructible. Movies give it a central role, yet they also stage discordant characterizations of memorative experiences and of memories themselves. Although characters associate their recollections to events that actually happened and when in doubt look for empirical confirmation, film also portrays memories as (re)constructed, whether by their subjects' unconscious or by neurotechnological means.

Thus, a movie may insist on antithetical properties: on the one hand, the authenticity of memories and historical truth as criteria for a genuine self and, on the other, the primacy of narrative truth and a psychological reality that eludes the dichotomies of objective and subjective, natural and artificial, true and false (for *historical* and *narrative truth*, see Spence 1984). Even alien or fabricated memories are frequently displayed as "flashbulb memories" (a term coined to convey the vividness with which highly emotional memories are recalled; Brown and Kulik 1977) and are given the eidetic and visual qualities cinematography often associates with veracity and authenticity. Finally, how memories are felt, perceived, and integrated into webs of social relations is more important than their source. As psychology and the neurosciences themselves demonstrate, memory "is dialogic and arises not only from direct experience but from the intercourse of many minds" (Sacks 2013).

As for personal identity, in spite of their great diversity, memory movies, including brain-and-memory productions, adhere to the classical Lockean theory. Only when apparently accurate visual-emotional memories of an experienced past reach consciousness does an amnesic protagonist recover his or her original and ipso facto authentic identity and becomes again himself or herself. This contributes to turn protagonists into cerebral subjects. At the same time, when cinema displays the phenomenological complexity of memory and embeds it in social relations, it downplays the brainhood ideology. The same observation applies to the movies discussed earlier. They typically begin by proclaiming that humans are essentially their brains, and that assumption sustains brain transplantation with the purpose of perpetuating someone's personal identity "in" a different body. But the plots actually depend on the failure of those initial expectations, and the surgical hybrids voluntarily redefine themselves as new whole persons—bodily, cognitively, emotionally, socially. With specifically cinematographic means, these films thus question the dichotomy brain-body and other central tropes of the neurocultural outlook that began to emerge a few decades after they were themselves produced.

In spite of obvious differences, the "neuroliterary" field points to social and experiential features of the cerebral subject's mode of existence that coincide with those manifest in film. As we have seen, the novelistic genre that neurologizes consciousness and analyzes characters in *neuro* terms parallels the neurologization of literary criticism. Although they seem to share an attitude of neuroscientific realism, critics and writers (as far as can be judged from their writings) differ significantly. Neuro lit crit involves a double reduction: a hermeneutic one, whereby authors are regarded as intuitive neuroscientists and their texts as expressions or vehicles of implicit neuroscientific knowledge, and an ontological one, whereby the acts involved in literary creation require a brain and barely anything else. In line with the other neurodisciplines, neuro lit crit sees culture as accessory insofar as literature is the product of a brain whose functioning is merely modulated by contextual and historical factors. Such modulation may be considered necessary but is nevertheless subsidiary to the brain and functions at the most as a secondary cause.

Neuronovels too may seem committed to an ideology of the cerebral subject. However, in contrast to neuro lit crit, they introduce brain-based views chiefly as narrative tools that generate ambivalence vis-à-vis the neurosolipsism of some of their situations and characters. They thus proclaim that fictionally bringing brains into literature is likely to be more productive for understanding the function and workings of literary creation and reception than the scholarly hermeneutics that places literature in the brain. Like brain and memory movies, their choice *not* to deliver a verdict on the situations they depict and the thought experiments they dramatize implies that the situations have no theoretical, only existential or phenomenological resolution and that those experiments either are not useful tools or stage badly formulated questions. For the late Oxford philosopher Kathleen Wilkes (1988), the theoretical impossibility of thought experiments rendered them irrelevant. But the ways literature and cinema have rehearsed them suggest otherwise, since they underline the profound differences between being a brain and not being able to be without one.

We have just observed that literature and cinema perform the cerebral subject and the ideology of brainhood in ways that both assert them and challenge them and that they enact the radical difference between the claim that "we are our brains" and the fact that we cannot be without one. This raises the question of who has the authority to examine whether and how we "are our brains" and points to the human sciences in their interpretive, contextualizing, and historicizing dimensions.[1] Yet in most recent and contemporary contexts, it is chiefly the neurosciences that have claimed and generally obtained that authority.

Of course, it is up to these sciences to inform us about the brain and the nervous system and to document how sociocultural processes are neurobiologically "implemented." Beyond the widespread mirages of inter- and transdisciplinarity, collaborations between them and other scientific endeavors increasingly bring to light the complexity of the brain's interactions with internal and external milieus, both organic and social. But this is no reason to accept the ideology of brainhood. Presumably no one believes we could be brains in vats and at the same time fully human. However, as we have illustrated here, there are well-funded research programs and successful commercial enterprises based not on simply acknowledging that we need a brain to be what we are and do what we do but on claiming that we are essentially our brains and that there can be no valid knowledge of human phenomena unless one shows what happens in the brain when those phenomena take place.

In a longue durée perspective, such views have been ultimately made possible by the historical transformations of debates about personhood within the Latin Christian tradition, and they followed from the early modern psychologization and partial disincarnation of the self. "Partial" because it was understood that humans, though no longer obliged to possess a complete body to be persons, could not be fully incorporeal, and insofar as personhood was redefined in purely psychological terms, human persons became essentially their brains. "Essentially" first qualified a material entity, the minimal body needed for personhood and personal identity. With time, however, it also came to signal the conviction that the ultimate level of explanation for the vast range of human behavior is neurobiological. Those who sincerely refuse such reductionism but make more or less angelic calls for critical friendship between the human and the natural sciences in this domain tend to go along with it. This is not what they want, since (on the contrary) they emphasize full contextualized embodiments, but it is a structural effect of their position, for they imply that the lack of collaborative goodwill proceeds from human sciences that are afraid of losing their prerogatives instead of willingly embracing the neural turn.

The opposite, however, is the case. While claiming to leave behind the supposed speculations of the human sciences, the *neuro* is moved by their basic agendas, even if it does not share their purposes and barely takes into account their concepts and empirical work. It is from the cultural and historical import of these agendas that its own attempts at cerebralizing the human may gain some luster. Obviously, all sciences are "human" insofar as they are our creation, and the adjective thus applies in a twofold manner to those that concern human beings. However, beyond objects and methods, what differentiates the natural from the human sciences (and here we include the humanities and social sciences) may still be captured by the distinction between a causal, ontologically or methodologically reductionist *Erklären* and an interpretive and historicizing *Verstehen* (on these matters see for example Smith 2007). Both are necessary, their divergences are not absolute, and conversations between them are possible and may sometimes be both desirable and rewarding. But so are disciplinary autonomy and the maintenance of boundaries when they contribute to a division of labor that helps make sense of the human world and experience. These considerations apply to the *neuro* itself, which the life sciences seem to perceive as the natural and in any case intrinsically justified consequence of neuroscientific progress but which the human sciences recognize as a historically rooted and contextually dependent cultural phenomenon.

Is it, however, the same phenomenon throughout? We know today much more about the brain than was known in the late seventeenth century. But we don't deal here with the brain—only with an ideology whose origins can be traced to that period and with its subsequent expansion and materializations. The *neuro*, as we explained, is not a single entity but the sum of those materializations. Neither does it relate in a straightforward manner to how the histories of the self and the body have intertwined since the mid–twentieth century. On the one hand, the neural turns and the late forms of the cerebral subject have been interpreted as the extension to the brain of a broader transformation in which bodies and selves have grown increasingly close and interdependent by way of genetics, molecular biology, and biomedical technologies. Although the neurobiological dimension is said to be essential for the resulting "somatic individuality," the self and personhood do not emerge as essentially "neuro." On the other hand, the body is seen as increasingly obsolete, as an archaic vessel to be technologically enhanced and then replaced as we move from a transhuman to a posthuman condition (Ortega 2014). Ultimately personhood would be synthetically replicated and the cerebral subject realized as an *in silico* programmable and networked model of and for human properties (Stollfuß 2014).

The dialectic between the organic embodiment of the somatic individual and the relative dematerialization of the posthuman one (relative because even virtual bodies must run on some sort of matter) is taking place as we write, and much of it lies in futures envisioned as utopian or dystopian. Yet, for all its novelty, when looked at from the appropriate distance, it has the odd look of a familiar scene from the times when Charles Bonnet imagined that if a Huron's soul could inherit Montesquieu's brain, Montesquieu would keep on thinking.

As we saw, in connection precisely with Bonnet's post-Lockean thought experiment, the cerebral subject is a product of history, not an organism identified in nature thanks to the advancement of science. That reason for refusing its naturalness is reinforced by how it functions as an anthropological figure. In contrast to the unequivocal quality it irradiates in the framework of the *neuro*, its real impact is characterized by what we have termed "ambivalence." On the one hand, we do not need to engage in systematic Foucauldianism to recognize in the *neuro* a universe that, like the "discourse" of *The Archaeology of Knowledge*, "is characterized not by privileged objects, but by the way in which it forms objects that are in fact highly dispersed," as well as by a capacity to generate "mutually exclusive objects, without having to modify itself" (Foucault 1969, 49). Discourse in such a framework is not mere language but a "formation" involving objects, concepts, and practices as well as subjective positions and power relations. The *neuro* possesses the resilience of such a formation, its aptitude both to generate and hold contradiction, to embrace incongruities without going to pieces. The ambivalences we have examined with regard to the ideology of brainhood are at the core of an eclectic, fragmented, yet robust system.

On the other hand (and that is part of the same phenomenon), when people can, they resist or adopt the *neuro* depending on local and fluctuating interests. In the neurodiversity movement, a fitting exemplar of its complex modus operandi, neuroscientific information is used to rethink disease as difference, but the redefinition of mental illness as brain disorder justifies both resisting and advocating therapy for the diagnosed individuals. In the neurodisciplines of culture, the "neuro" is largely a matter of opportunity. Goals are defined, ranging from the theoretical to the pragmatic (from, say, naturalizing art to getting funds); neuroscientific idioms and research are an advantageous way to pursue them. The protagonists of the *neuro* nonetheless explain that their choice has inherent necessity by virtue of what humans fundamentally are. The same applies to the global challenges of mental illness and the promises of neuroscience to uncover their ultimate causes. The fulfillment of the hopes raised in this domain would certainly be beneficial for humanity, but the promises conveying those hopes rest on a desire for cerebral causality that tends to exclude elements of context and relationality recognized as important for understanding and treating psychological distress.

All these fields share what we characterized as a "modern creed." It is *modern* both because of its chronology (not thinkable before the late seventeenth century) and because it is an element of the psychological, philosophical, political, and scientific cosmologies usually labeled "modern." It is a *creed* because it states basic beliefs that guide action. "Belief," in our usage, neither stands as a term of abuse nor serves to place the *neuro* in a closed world of faith, opinion, and subjectivity; knowledge, after all, is a species of belief (according to a widespread view, justified true belief acquired by a method that is reliable in the relevant context). Rather, our emphasis is on the fact that, if it were formulated as a statement opening with *Credo*, with the words *I believe*, the set of basic shared beliefs of the *neuro* community would begin with the assertion that we are essentially our brains or (if the formulation were extreme) that "everything, absolutely everything, is in the brain."[2]

We have here argued that such a creed neither historically derives from nor currently depends on neuroscientific knowledge (though it may feel that way for certain individuals and communities) but that it must be traced to early modern scientific and philosophical developments that transformed the notions of personhood and personal identity. Later research into the brain, up to the present day, has buttressed the "cerebralization" of personhood but (contrary to what is often claimed) cannot substantiate it either conceptually or empirically; neuroscience sustains and embodies various "neural turns," but it cannot turn the view of humans as cerebral subjects into a piece of "natural knowledge" based on empirical evidence.

Brainhood, in short, is best understood as a historically contingent resource, born to uphold and make plausible a redefinition of personhood. The arts, by simultaneously asserting it and denying it, infuse it with ambiguity and inconclusiveness and display it for what it is: an ideology, a complex set of notions, beliefs, and ideals in whose making empirical knowledge about the brain plays a role at best comparable to that of a supporting actor. And this makes sense: As Louis Menand (2002) put it, "every aspect of life has a biological foundation in exactly the same sense, which is that unless it was biologically possible it wouldn't exist. After that, it's up for grabs."

ACKNOWLEDGMENTS

Being Brains is based on some of the research we carried out, together and separately, over a number of years. From the beginning we had a book in mind, and some of the material we published as articles or chapters has been reworked here. It is also both together and separately that we have incurred many personal and intellectual debts. We can here acknowledge only a few.

Various institutions have supported us generously: the Brazilian National Council for Scientific and Technological Development (CNPq), the Carlos Chagas Filho Research Support Foundation of the State of Rio de Janeiro (FAPERJ), the Brazilian Coordination for the Improvement of Higher-Level Personnel (CAPES), the German Academic Exchange Service (DAAD), the Institute for Social Medicine of the State University of Rio de Janeiro, the Max Planck Institute for the History of Science (Berlin), ICREA (Catalan Institution for Research and Advanced Studies), and CEHIC (Center for the History of Science, Autonomous University of Barcelona).

We are grateful to our friend Aurore Millet for graciously letting us use *Reminiscence I* no. 4 (2005) for the cover of the book. Her red pencil drawing comes from a series reflecting her experiences on both sides of Israel's "security fence." In it, "a sense of spirituality and silent minimalism" (as Tina Sherwell puts it in Millet 2009) powerfully combines the intimate psychological sphere with the political and social world—a combination that, in a different way, is also at the heart of *Being Brains*.

From Tokyo to Mexico City and from Montreal to Buenos Aires, we have exchanged ideas and opinions with very many individuals and presented our work on numerous occasions to extremely varied audiences in a large number of institutions. We can here do no more than to mention a few.

We owe special gratitude to Lorraine Daston, the director of Department II at the Max Planck Institute for the History of Science, for her insightful feedback as well as for facilitating many opportunities for us to carry out our joint work.

Joelle Abi-Rached, Maurizio Meloni, and Nikolas Rose have been cherished critical interlocutors. Antonio Battro, a dear friend, invited us to the Mind, Brain, and Education Summer School at the Ettore Majorana Foundation and Centre for Scientific Culture in Erice (Sicily), where we benefited from valuable discussions in a unique setting.

Francisco Ortega thanks in particular the Department of Social Science, Health, and Medicine of King's College, London, where he was Senior Visiting Research Fellow in 2012–2013. He is also grateful to Dominique Behague, Benilton Bezerra, Jurandir Freire Costa, and Rafaela Zorzanelli.

Fernando Vidal is especially grateful for the hospitality of the École des Hautes Études en Sciences Sociales, Paris, and the Pontifical Catholic University of Rio de Janeiro. He also thanks Nicole Becker, Suparna Choudhury, Nicolas Langlitz, and Claudia Swan.

1. On the use and abuse of the prefix *neuro-*, see for example Muzur and Rinčić (2013).

2. Moreover, unlike a critic such as the German philosopher Markus Gabriel (2015), we have taken into account and entered into dialogue with the very large body of research that deals with the phenomenon of "neurocentrism" (a term Gabriel gives the impression of having coined himself).

3. BRAIN stands for "Brain Research through Advancing Innovative Neurotechnologies."

4. The extremes of this ribbon touch. In 2014, tensions related to the Human Brain Project (HBP) became public after several hundred scientists published a protest letter. While the core problems concerned governance and transparency, as well as the place of cognitive and systems neuroscience within HBP, numerous neuroscientists had from the beginning questioned its basic scientific logic, and even the *Mediation Report*'s subdued language echoes the many who believe that HBP was sold on unrealistic claims and overstated promises (Marquardt 2015). Bartlett (2015) and Theil (2015) give a good idea of the controversy.

5. See for example EyeWire, "a citizen science project aimed at mapping the neural connections of the retina," launched by a professor of computational neuroscience at MIT: http://scistarter.com/project/566-Eyewire.

6. The singular "neuroscience" is practical, but it masks the heterogeneity of approaches, methods, and concepts used to deal empirically with the brain (on this point see for example Abi-Rached 2008).

7. "Cuando entendamos el cerebro, la humanidad se entenderá a sí misma por dentro por primera vez. No me extrañaría que esto revolucione la cultura y cambie muchísimas cosas como la educación, el sistema legal o la economía. Será un nuevo humanismo" (Yuste 2015).

8. The "Pepsi paradox" refers to the fact that people show a reliable preference for Coke (vs. Pepsi) when they have brand information (as in supermarkets) but not in the absence of such information (as in blind taste tests). Neuroimaging studies of this phenomenon demonstrate a consistent neural response in the ventromedial prefrontal cortex correlated with subjects' behavioral preferences (McClure et al. 2004) and that damage to that brain area abolishes the paradox (Koenigs and Tranel 2008). When it comes to knowing that branding makes a difference in consumer preference, neuroimaging results are superfluous. Moreover, it has been easy to go beyond the neural correlates uncovered by these fascinating studies to claim, improperly, that they show "why" people choose Coke over Pepsi (e.g., Mlodinow 2012).

9. We have drawn inspiration from Nicolai Krementsov's discussion of biology as a cultural resource in his book on the quest for immortality in Bolshevik science (Krementsov 2014, 187–193).

1. GENEALOGY OF THE CEREBRAL SUBJECT

1. The first brain thought-experiment in professional philosophy seems to appear in Sidney Shoemaker's *Self-Knowledge and Self-Identity* (1963), in the form of Brownson, a character with Robinson's body and Brown's brain. The best-known version is Hilary Putnam's "brains in a vat" in the first chapter of *Reason, Truth, and History* (1981). But there are many others. *The Mind's I*, a popular 1981 book edited by Douglas Hofstadter and Daniel Dennett, gathered some particularly extravagant varieties; in *Philosophical Explanations* (1981), Robert Nozick tested his own theory of personal identity against eight brain-fictional situations. Philosophers do not refer to J. D. Bernal's 1929 depiction of a future world where humanity is made up of interconnected brains in vats (*The World, the Flesh, and the Devil. An Enquiry Into the Future of the Three Enemies of the Rational Soul*, chap. 3; see Gere 2004). See Wilkes (1988) for a critical discussion of thought experiments as tools for the philosophy of personal identity.

2. Calls for an "embodied cognitive science" that would give up the "brain-centered view of cognitive function" and would "no longer be able to claim that the brain is the organ of the mind" (Kiverstein and Miller 2015, 9) remain programmatic and represent a minority.

3. Ferret (1993) here summarizes a position widespread in Anglo-American philosophy of personal identity.

4. The "cerebral subject" is thus to be distinguished from the "cerebral self," defined as a prepsychological bodily reality made up of the cortex and its sensorimotor connections (Arminjon, Ansermet, and Magistretti 2011). Beyond that distinction, *self, person,* and *subject* tend to have different connota-

tions. We have chosen not to differentiate these terms rigorously but will use them according to the nuances they convey: While *self* evokes interiority and reflexive consciousness, *person* and *personhood* are connected to attributes more directly relevant to legal and moral contexts, and *subject* may be more associated with the making of subjectivity in particular environments.

5. Such an interpretation and the work on which it is based may be limited, but it is hard to see how, as Cooter (2014, 148) affirms about the work of one of us (FV), historicizing "only serve[s] further to naturalize and sustain neuro representations" and thus becomes one of the "technologies of power for the new regime of truth."

6. http://www.neurosurgery.org/cybermuseum/pre2oth/epapyrus.html, www.ibro1.info/Pub/Pub_Main_Display.asp?LC_Docs_ID=3199.

7. Burton 1651, part. I, memb. III, subsect. I, "Passions and Perturbations of the Mind, how they cause Melancholy."

8. Hippocrates, *On the Sacred Disease*, trans. Francis Adams [*The Genuine Works of Hippocrates*, 1849], http://classics.mit.edu/Hippocrates/sacred.html.

9. Lega (2006) argues that Willis's work on the brain helped Locke go beyond Descartes "and develop a unique and radical philosophy that ultimately served as the foundation for modern neuroscience" (573). The thesis is interesting but is not substantiated in detail. On Willis in connection with the present discussion, see in particular Frank (1990).

10. See also the special section "20 Years of fMRI—What Has It Done for Understanding Cognition?" *Perspectives on Psychological Science* 8 (1), 2013.

11. See also Paterniti (2000), an account of the author's trip across the United States with Dr. Harvey, to return the extant pieces of Einstein's brain to the physicist's granddaughter.

12. See http://www.skepdic.com/braingym.html and http://www .senseaboutscience.org/resources.php/55/sense-about-brain-gym. Paul and Gail Dennison, the authors of the incriminated *Brain Gym®: Teachers Edition*, respond in http://www.braingym.org.nz/articles2.asp. In 2016, a "brain training" company agreed to pay $2 million to settle United States Federal Trade Commission charges of deceptive advertising (https://www.ftc.gov /news-events/press-releases/2016/01/lumosity-pay-2-million-settle-ftc -deceptive-advertising-charges).

13. http://www.positscience.com/about.

2. DISCIPLINES OF THE *NEURO*

1. The Canadian philosopher Ian Hacking is not the only scholar who, following Foucault's lead, has investigated such reflexive action. His depiction of "looping effects" in the processes of "making up people" as well as his notions about "historical ontology" have been perhaps the most influential;

we refer to them later. In a similar spirit, the sociologist Nikolas Rose, often cited in this book, has been investigating subjectivation since his early work on the "psy disciplines" (Rose 1990, 1996). Hacking, Rose, and many others have sought to document subjectivation processes concretely; other authors (e.g., Richards 2002) have remained more purely programmatic. In his *History of the Human Sciences*, Roger Smith (1997, 22) characterizes well the phenomenon in question when he writes that "since ordinary people provided these sciences with their subject matter, the human sciences existed in a circle of interactions between science and ordinary life, a circle in which they influenced and were influenced by popular culture."

2. *Neurobabble* and *neuromythology* are common in blogs and articles of various sorts; the latter also as a book title (Tallis 2004). For the other terms, see, respectively, Tallis (2008a, 2009), Legrenzi and Umiltà (2009), Hasler (2009).

3. Neuroimaging overhype is documented throughout this chapter and other places of this book, but see in addition Rusconi and Mitchener-Nissen (2014).

4. We have used "organic" in the sense that the Italian Marxist thinker Antonio Gramsci applied to the intellectuals who would self-consciously articulate from within the experiences and interests of the working class. Gramsci contrasted organic intellectuals to the "traditional" ones who, though claiming to be disinterested, were tied to the dominant class and culture.

5. http://www.neuroethics.ox.ac.uk.

6. http://www.jsmf.org/programs/uhc.

7. Less simplistic attempts at articulating the human and the biological sciences by way of a "nuanced epigenetic and neurobiological attention" (in the case we mention here, to the links between health and urban life) seem to be emerging (Fitzgerald, Rose, and Singh 2016a). However, even profound thinkers with an original and valuable project may get carried away by the grandiosity of their views, and, in euphuistic language, stick to *neuro*- prefixes, offer the customary prosopopoeia of the brain, and rapturously convey (perhaps against their own viewpoint) the usual ontological illusions about the empirical, theoretical, and political significance of showing that a phenomenon (in this case, "urban citizenship") can be "instantiated neurobiologically" (Fitzgerald, Rose, and Singh 2016b, 234).

8. Rosen presented his richly documented assessment in 2011 as a lecture on the twentieth anniversary of the first public presentation of fMRI (video at http://www.nmr.mgh.harvard.edu/history-fMRI). As of 2016, nothing indicates that it should be revised.

9. It is only fair to underline that the "questionable research practices" discussed in these and other articles are by far not unique to neuroscience. For an example from psychology, see Open Science Collaboration (2015).

10. https://s4sn.org/.

11. http://neuroanthropology.net/ 2008/08/14/cultural-neuroscience/.

12. http://neuroanthropology.net/2008/08/24/the-first-phd-in
-neuroanthropology.

13. Several collective works give a good idea of the range of neuroaesthetic inquiry: Dresler (2009), Lauring (2015), Martín-Aragúz et al. (2010), Martindale, Locher, and Petrov (2007), Skov and Vartanian (2009a). See also the collection of articles devoted to "Perspectives in Neuroaesthetics" in *Rendiconti Lincei di Scienze Fisiche e Naturali* 23(3), 2012. The best examination of neuroaesthetics as a whole remains Cappelletto (2009); like other discussions from outside the field, it has been totally ignored by its practitioners.

14. http://www.aesthetics.mpg.de.

15. http://www.uea.ac.uk/about/media-room/press-release-archive/-/asset
_publisher/a2jEGMiFHPhv/content/cracking-the-real-da-vinci-code

16. For more on Redies's model, see Redies et al. (2007) and Redies, Hasenstein, and Denzler (2007). The model derives largely from the purported existence of scale-invariant, fractal-like properties shared by artworks and complex natural scenes. In yet another attempt to reduce art to basic perceptual processes, these properties were said to characterize Jackson Pollock's drip paintings and thus to provide the clue to their "fundamental content" (Taylor, Micolich, and Jonas 1999; see also Taylor 2002); for a critical discussion (with further references) of the supposed fractality of Pollock's art and its controversial use in authentication, see Schreyach (2007).

17. We draw the distinction between the *aesthetic* and the *artistic* relation from the French literary critic Gérard Genette (1999). As far as we can tell, neither empirical aesthetics in general nor neuroaesthetics in particular use these concepts. However, we shall employ them because those disciplines do convey (often implicitly) a notion of "aesthetic(s)" that overlaps with Genette's usage. They all aim at taking into account the sensory and cognitive dimensions of experience in understanding the human "aesthetic" response to artworks and non-art objects. See also Schaeffer (1997, 2009).

18. We say "functional equivalents" because, for example, it is possible to enjoy colors though suffering from color blindness, as is the case of the artist Neil Harbisson: born with achromatopsia, he carries a permanently implanted "eyeborg" that allows him to hear colors, and he advocates, through his Cyborg Foundation, the development of similar devices for all the senses (http://cyborgism.wix.com/cyborg).

19. Canonical neurons, also in the F5 area of the ventral premotor cortex of monkeys, fire when one sees an object that can be grasped by the prehensile movement of the hand whose movements they encode.

3. CEREBRALIZING DISTRESS

1. *DSM* is now in its fifth edition (*DSM-5* 2013, http://www.dsm5.org); *ICD* is now in its tenth edition (*ICD-10*, http://www.who.int/classifications/icd /en/). The first version of *ICD-10* was released in 1990, and *ICD-11* is scheduled for 2018.

2. E.g., http://www.who.int/mediacentre/factsheets/fs369/en/ (October 2015).

3. The same applies to a much shorter handbook, Anderson and Camm (2014).

4. http://www.nimh.nih.gov/research-priorities/rdoc/nimh-research -domain-criteria-rdoc.shtml.

5. The literature on the history of melancholy is vast. *Saturn and Melancholy* (Klibansky, Panofsky, and Saxl 1964) remains the iconic classic; its scope is matched only by Jackson (1986), which emphasizes the continuity between melancholy and depression and attributes it to their representing the same empirical reality. For a recent and perceptive discussion of the historiography and the historical and conceptual issues involved, see Bell (2014), which focuses on the centuries up to 1800.

6. Summerfield argues against GMH, Miller tries to show its agenda is not "a form of cultural imperialism," and White balances between the two. See Cooper (2016) for a reflection on how to move beyond the current impasse.

7. Amanda Baggs, who speaks through a voice synthesizer, became one of the best-known autism self-advocates after posting, in January 2007, her video *In My Language* (http://www.youtube.com/watch?v=JnylM1hI2jc). There is controversy over whether Baggs is really a person with autism and if she really made the video herself. See among others http://amandabaggscontroversy .blogspot.com/.

8. See http://www.neurodiversity.com and http://www.aspiesforfreedom .com for the sites of two of the most vocal groups.

9. Recently published research demonstrating that the "mentally ill" label prompts significantly lower levels of tolerance and acceptance than person-first language (Granello and Gibbs 2016) will perhaps have an impact among advocates of neurodiversity.

10. http://www.autism-society.org.

11. http://www.naar.org/naar.asp; http://web.archive.org/web /20040927074818/http://www.cureautismnow.org/. See also Silverman (2012).

12. http://www.autismspeaks.org/index.php.

13. To understand better the debate and the positions at play, see Chamak (2008), Clarke and van Amerom (2007, 2008), Silverman (2008a, 2008b, 2012).

14. http://archive.is/3yGy8.

15. See for instance the post "Nevertheless, There Are Differences Between Autistics Who Approve of and Disapprove of Neurodiversity"

(March 3, 2009), http://autismnaturalvariation.blogspot.com.es/2009/03
/nevertheless-there-are-differences.html.

16. See also http://www.jonathans-stories.com.

17. See http://www.wrongplanet.net and http://web.archive.org/web
/20070202083611/http://autisticculture.com/ for particularly interesting and
informative websites in this respect.

18. See http://en.wikipedia.org/wiki/Autistic_Pride_Day.

19. http://cafepress.com/proudlyautistic.

20. http://www.agre.org.

21. See Brownlow (2007), Brownlow and O'Dell (2006), Cascio (2014),
Clarke and Amerom (2007, 2008), Davidson (2008), Goupil (2014), Jones and
Meldal (2001), Jones et al. (2001), Ortega et al. (2013), Waltz (2005).

22. A critique of this "mereological fallacy" (ascribing to the part properties of the whole) is at the heart of M. R. Bennett and P. M. S. Hacker's
Philosophical Foundations of Neuroscience (2003, part I, chap. 3).

23. http://welkowitz.typepad.com.

24. http://wrongplanet.net/forums/viewtopic.php?t=184314.

25. http://www.aspiesforfreedom.com/archive/index.php/thread-11062.
html.

26. http://dannilion.com/category/wow/page/2/, our emphasis.

27. "Identity politics and the language controversy," http://doraraymaker
.com/wp/change/?p=3853 (emphasis added)

28. http://www.aspiesforfreedom.com/showthread.php?tid=11062.

29. "Snippet" from Jane Meyerding's website, http://web.archive.org/web
/20050205082305/http://mjane.zolaweb.com/snipframe.html.

30. http://www.causes.com/actions/1593909-dsm-5-committee-dont
-reduce-the-criteria-for-an-autism-spectrum-disorder-in-the-dsm-5.

31. http://www.aspiesforfreedom.com/showthread.php?tid=11062.

32. http://everything2.com/title/neurotypical.

33. Whereas the radicalism of some self-advocates in the United States,
Canada, and Australia has pushed the movement toward identity politics by
essentializing neurological uniqueness and typologizing brain difference, this
does not seem to have happened in countries like France or Brazil (Block and
Cavalcante 2014, Chamak 2008, Ortega et al. 2013, Rios and Andrada 2015).

4. BRAINS ON SCREEN AND PAPER

1. One competition is organized by the Neuro Bureau, an "open neuroscience" initiative; the other one, called Art of Neuroscience, is run by the
Netherlands Institute for Neuroscience, with sponsorship from the publisher
Springer (http://www.neurobureau.org/galleries/brain-art-competition-2015;
http://aon.nin.knaw.nl/). On brain/art as well as exhibition catalogs, see

Albano, Arnold, and Wallace (2002); Aldworth et al. (2008); Anker and Frazzetto (2006); Frazzetto and Anker (2009); Gilmore (2006); Kwint and Wingate (2012); Landi (2009); Pepperell (2011).

2. "Can 'Neuro Lit Crit' Save the Humanities?" *New York Times* (5 April 2010), http://roomfordebate.blogs.nytimes.com/2010/04/05/can-neuro-lit-crit-save-the-humanities/.

3. *New York Times* review of Frankenstein (5 December 1931); in *Tabula Rasa* 3 (1994), http://www.tabula-rasa.info/Horror/FrankensteinFiles.html.

4. The image, "Massachusetts Department of Mental Diseases Exhibits Pictures of 50 Criminal Brains," announces a display at the American Museum of Natural History during the Second International Congress of Eugenics (1921); reproduced in Lederer (2002, 45) and available at http://www.eugenicsarchive.org/html/eugenics/index2.html?tag=567.

"UP FOR GRABS"

1. Similarly, in connection with climate change, the historian Julia Thomas (2015, 255) concludes: "Engaging with biology reveals a multiplicity of human figures and delimits the possible answers to humanistic questions of value—but cannot decide them. Ultimately, defining what is most endangered by climate change is the role of the humanists."

2. Statement by the distinguished Spanish neuroscientist Manuel Martín-Loeches. http://www.tendencias21.net/Martin-Loeches-Todo-absolutamente-todo-esta-en-el-cerebro_a7130.html. We have provided throughout this book many other examples of such a stance.

Abbott, Alison. 2011. "Novartis to Shut Brain Research Facility." *Nature* 480 (8 December): 161–162.

Abi-Rached, Joelle M. 2008. "The New Brain Sciences: Field or Fields?" Brain Self and Society working paper no. 2. London: BIOS/London School of Economics and Political Science.

Abi-Rached, Joelle M., and Nikolas Rose. 2010. "The Birth of the Neuromolecular Gaze." *History of the Human Sciences* 23: 11–36.

Abi-Rached, Joelle M., Nikolas Rose, and Andrei Mogoutov. 2010. "Mapping the Rise of the New Brain Sciences." Brain Self and Society working paper no. 4. BIOS/London School of Economics and Political Science.

Abou-Saleh, Mohammed T. 2006. "Neuroimaging in Psychiatry: An Update." *Journal of Psychosomatic Research* 61: 289–293.

Adams, Jon. 2008. "The Sufficiency of Code, *Galatea 2.2*, and the Necessity of Embodiment." In Burn and Dempsey 2008, 137–150.

Adler, Hans, and Sabine Gross. 2002. "Adjusting the Frame, Comments on Cognitivism and Literature." *Poetics Today* 23 (2): 195–220.

Albano, Caterina, Ken Arnold, and Marina Wallace, eds. 2002. *Head On: Art with the Brain in Mind*. London: Artakt.

Alcaro, Antonio, et al. 2010. "Is Subcortical-Cortical Midline Activity in Depression Mediated by Glutamate and GABA? A Cross-Species Translational Approach." *Neuroscience and Biobehavioral Reviews* 34: 592–605.

Aldworth, Susan. 2011. "The Physical Brain and the Sense of Self: An Artist's Exploration." In Ortega and Vidal 2011, 273–292.

Aldworth, Susan, Paul Broks, Robert Mason, and Gill Saunders. 2008. *Scribing the Soul*. London: Susan Aldworth.

Álvarez-Jiménez, Mario, et al. 2008. "Non-pharmacological Management of Antipsychotic-Induced Weight Gain: Systematic Review and Meta-analysis of Randomised Controlled Trials." *British Journal of Psychiatry* 193: 101–107.

Ambady, Nalini, and Jamshed Bharucha. 2009. "Culture and the Brain." *Current Directions in Psychological Science* 18 (6): 342–345.

Ames, Daniel L., and Susan T. Fiske. 2010. "Cultural Neuroscience." *Asian Journal of Social Psychology* 13: 72–82.

Aminoff, J. Michael. 1993. *Brown-Sequard: A Visionary of Science.* New York: Raven.

Ananthaswamy, Anil. 2015. *The Man Who Wasn't There: Investigations Into the Strange New Science of the Self.* New York: Dutton.

Anderson, Ian, and John Camm, eds. 2014. *Handbook of Depression.* 2nd ed. New York: Springer.

Anderson, Rodney J., et al. 2012. "Deep Brain Stimulation for Treatment-Resistant Depression: Efficacy, Safety, and Mechanisms of Action." *Neuroscience and Biobehavioral Reviews* 36: 1920–1933.

Andreasen, Nancy Coover. 2004. *Brave New Brain: Conquering Mental Illness in the Era of the Genome.* New York: Oxford University Press.

———. 2006. *The Creating Brain: The Neuroscience of Genius.* New York: Plume.

Angell, Marcia. 2004. *The Truth About the Drug Companies: How They Deceive Us and What to Do About It.* New York: Random House.

———. 2011. "The Epidemic of Mental Illness: Why?" *New York Review of Books* (23 June). http://www.nybooks.com/articles/archives/2011/jun/23/epidemic-mental-illness-why/.

Angermeyer, Matthias C., and Herbert Matschinger. 2005. "Causal Beliefs and Attitudes to People with Schizophrenia: Trend Analysis Based on Data from Two Population Surveys in Germany." *British Journal of Psychiatry* 186 (3): 331–334.

Anker, Suzanne, and Giovanni Frazzetto. 2006. *Neuroculture: Visual Art and the Brain.* Westport, Conn.: Westport Arts Center.

Anonymous. 2006. "Retraining the Brain. Doctors Test Drug-Free Methods to Restore Lost Mental Capabilities." *CBS News* (15 January). http://www.cbsnews.com/stories/2006/01/15/sunday/main1209916.shtml.

Antonetta, Susanne. 2005. *A Mind Apart: Travels in a Neurodiverse World.* Tarcher: Penguin.

APA 2007. "Functional Magnetic Resonance Imaging: A New Research Tool." Washigton, D.C.: American Psychological Association. http://www.apa.org/research/tools/fmri-booklets.aspx.

Applbaum, Kalman. 2006. "Educating for Global Mental Health: The Adoption of SSRIs in Japan." In Petryna, Lakoff, and Kleinman 2006, 85–110.

Arango, Ángel. 1964. *¿A dónde van los cefalomos?* Havana: Ediciones R.

Arbabshirani, Mohammad R., et al. 2013. "Classification of Schizophrenia Patients Based on Resting-State Functional Network Connectivity." *Frontiers in Neuroscience* 7, art. 133. doi:10.3389/fnins.2013.00133.

Ariel, Cindy N., and Robert A. Naseef, eds. 2006. *Voices from the Spectrum: Parents, Grandparents, Siblings, People with Autism, and Professionals Share Their Wisdom*. London: Jessica Kingsley.

Arikha, Noga. 2007. *Passions and Tempers: A History of the Humours*. New York: Ecco/HarperCollins.

Arminjon, Mathieu, François Ansermet, and Pierre Magistretti. 2011. "Émergence du moi cérébral de Théodore Meynert à Antonio Damasio." *Psychiatrie Sciences Humaines Neurosciences* 9: 153–161.

Armstrong, Thomas. 2010. *Neurodiversity: Discovering the Extraordinary Gifts of Autism, ADHD, Dyslexia, and Other Brain Differences*. Cambridge, Mass.: Da Capo.

Arpaly, Nomy. 2005. "How It Is Not 'Just Like Diabetes': Mental Disorders and the Moral Psychologist." *Philosophical Issues* 15: 282–298.

Ash, Imogen. 2012. "How Is the Selective Nature of Memory Explored by Ian McEwan and in Biology?" *PsyArt. An Online Journal or the Psychological Study of the Arts*. http://www.psyartjournal.com/article/show/ash-how_is _the_selective_nature_of_memory_ex.

Bagatell, Nancy. 2007. "Orchestrating Voices: Autism, Identity, and the Power of Discourse." *Disability and Society* 22 (4): 413–426.

———. 2010. "From Cure to Community: Transforming Notions of Autism." *Ethos* 38: 33–55.

Baker, Dana L. 2011. *The Politics of Neurodiversity: Why Public Policy Matters?* Boulder, Colo.: Lynne Rienner.

Balt, Steve. 2014. "Assessing and Enhancing the Effectiveness of Antidepressants." *Psychiatric Times*. http://www.psychiatrictimes.com /psychopharmacology/assessing-and-enhancing-effectiveness -antidepressants.

Balter, Michael. 2014. "Talking Back to Madness." *Science* 343 (6176): 1190–1193.

Bareither, Isabelle, Felix Hasler, Anna Strasser. 2015. "9 Ideen für eine bessere Neurowissenschaft." *Spektrum* (9 January). http://www.spektrum.de/news /9-ideen-fuer-eine-bessere-neurowissenschaft/1324147.

Barilan, Yechiel M. 2002. "Head-Counting vs. Heart-Counting: An Examination of the Recent Case of the Conjoined Twins from Malta." *Perspectives in Biology and Medicine* 45 (4): 593–603.

———. 2003. "One or Two: An Examination of the Recent Case of the Conjoined Twins from Malta." *Journal of Medicine and Philosophy* 28 (1): 27–44.

Baron-Cohen, Simon. 2002. "The Extreme Male Brain Theory of Autism." *TRENDS in Cognitive Sciences* 6 (6): 248–254.

Bartlett, Frederick. 1932. *Remembering*. Cambridge: Cambridge University Press.

Bartlett, Tom. 2015. "Can the Human Brain Project Be Saved? Should It Be?" *Chronicle of Higher Education* 61 (23): A6.

Bass, Alison. 2010. "Emory Neurologist Has History of Failing to Disclose Conflicts of Interest." 15 November. http://alison-bass.blogspot.com.es /2010/11/emory-neurologist-has-history-of.html

Battaglia, Fortunato, Sarah H. Lisanby, and David Freedberg. 2011. "Cortico-motor Excitability During Observation and Imagination of a Work of Art." *Frontiers in Human Neuroscience.* doi:10.3389/fnhum.2011.00079.

Baxendale, Sallie. 2004. "Memories Aren't Made of This: Amnesia in the Movies." *British Medical Journal* 329: 1480.

Beaulieu, Anne. 2012. "Fast-Moving Objects and Their Consequences. A Response to the NT in Practice." In Littlefield and Johnson 2012, 152–159.

Becker, Nicole. 2006. *Die Neurowissenschaftliche Herausforderung der Pädagogik.* Bad Heilbrun: Klinkhardt.

Beckham, Eduard E., and William R. Leber, eds. 1985. *Handbook of Depression: Treatment, Assessment, and Research.* Homewood, Ill.: Dorsey.

———. 1995. *Handbook of Depression: Treatment, Assessment, and Research.* 2nd ed. New York: Guilford.

Beecher, Henry K., et al. 1968. "A Definition of Irreversible Coma. Report of the Ad Hoc Committee of the Harvard Medical School to Examine the Definition of Brain Death." *JAMA* 205 (6): 337–340.

Begley, Sharon. 2010. "How different cultures shape the brain." *Newsweek,* 18 February. http://europe.newsweek.com/how-different-cultures -shape-brain-75289?rm=eu.

Béhague, Dominique. 2009. "Psychiatry and the Politicization of Youth in Pelotas, Brazil: The Equivocal Uses of 'Conduct Disorder' and Related Diagnoses." *Medical Anthropology Quarterly* 23 (4): 455–482.

Beliaev, Alexandre. 1980 [1925] *Professor Dowell's Head.* New York: Macmillan.

Bell, Matthew. 2014. *Melancholia: The Western Malady.* New York: Cambridge University Press.

Belluck, Pam. 2006. "As Minds Age, What's Next? Brain Calisthenics." *New York Times* (27 December). http://query.nytimes.com/gst/fullpage.html?sec =health&res=9E0DEED71E31F934A15751C1A9609C8B63.

Bennett, Laura, Kathryn Thirlaway, and Alexandra J. Murray. 2008. "The Stigmatising Implications of Presenting Schizophrenia as a Genetic Disease." *Journal of Genetic Counseling* 17 (6): 550–559.

Bennett, M. R., and P. M. S. Hacker. 2003. *Philosophical Foundations of Neuro-science.* Malden, Mass.: Blackwell.

Bentall, Richard P. 2009. *Doctoring the Mind: Why Psychiatric Treatments Fail.* New York: New York University Press.

Bernal, John Desmond. 1969 [1929]. *The World, the Flesh, and the Devil: An Enquiry Into the Future of the Three Enemies of the Rational Soul.* Bloomington: Indiana University Press.

Bernat, James L. 2005. "The Concept and Practice of Brain Death." *Progress in Brain Research* 150: 369–379.

———. 2009. "Contemporary Controversies in the Definition of Death." *Progress in Brain Research* 77: 21–31.

———. 2013. "Controversies in Defining and Determining Death in Critical Care." *Nature Reviews Neurology* 9: 164–173.

Bernucci, Leopoldo. 2008. "Cientificismo e aporias em Os Sertões." In *Discurso, Ciência e controvérsia em Euclides da Cunha*, edited by Leopoldo Bernucci. São Paulo: Edusp.

Bertilsdotter Rosqvist, Hanna, Charlotte Brownlow, and Lindsay O'Dell. 2013. "Mapping the Social Geographies of Autism—On- and Offline Narratives of Neuro-shared and Neuro-separate Spaces." *Disability and Society* 28 (3): 367–379.

Besser, Stephan. 2013. "From the Neuron to the World and Back: The Poetics of the Neuromolecular Gaze in Bart Koubaa's *Het gebied van Nevski* and James Cameron's *Avatar*." *Journal of Dutch Literature* 4 (2): 43–67.

———. 2015. "Mixing Repertoires: Cerebral Subjects in Contemporary Dutch Neurological Fiction." In *Illness and Literature in the Low Countries from the Middle Ages Until the Twenty-First Century*, edited by Jaap Grave, Rick Honings, and Bettina Noak, 253–272. Göttingen: V&R Unipress.

Bhar, Sunil, and Aaron Beck. 2009. "Treatment Integrity of Studies That Compare Short-Term Psychodynamic Psychotherapy with Cognitive-Behaviour Therapy." *Clinical Psychology: Science and Practice* 16: 370–378.

Bickle, John. 2013. "Multiple Realizability." In *The Stanford Encyclopedia of Philosophy*. http://plato.stanford.edu/archives/spr2013/entries/multiple-realizability.

Biehl, João. 2005. *Vita: Life in a Zone of Social Abandonment*. Berkeley: University of California Press.

———. 2006. "Pharmaceutical Governance." In Petryna, Lakoff, and Kleinman 2006, 206–209.

Biever, Celeste. 2007. "Let's Meet Tomorrow in Second Life." *New Scientist* 2610: 26–27.

Bioy Casares, Adolfo. 2004 [1973]. *Asleep in the Sun*. Translated by Suzanne Jill Levine. New York: NYRB.

Blais, Mark A., et al. 2013. "Treatment as Usual (TAU) for Depression: A Comparison of Psychotherapy, Pharmacotherapy, and Combined Treatment at a Large Academic Medical Center." *Psychotherapy* 50 (1): 110–118.

Blakemore, Sarah-Jayne. 2008. "The Social Brain in Adolescence." *Nature Reviews Neuroscience* 9: 267–277.

Blakeslee, Sandra. 2007. "A Small Part of the Brain, and Its Profound Effects." *New York Times* (6 February). http://www.nytimes.com/2007/02/06/health/psychology/06brain.html.

Blank, Robert H. 1999. *Brain Policy: How the New Neuroscience Will Change Our Lives and Our Politics.* Washington, D.C.: Georgetown University Press.

———. 2013. *Intervention in the Brain: Politics, Policy, and Ethics.* Cambridge, Mass.: MIT Press.

Block, Pamela, and Fatima Cavalcante. 2014. "Historical Perceptions of Autism in Brazil: Professional Treatment, Family Advocacy, and Autistic Pride, 1943–2010." In *Disability Histories,* edited by Susan Burch and Michael Rembis. Champaign: University of Illinois Press, 77–97.

Blume, Harvey. 1997a. "Autism and the Internet, or It's the Wiring, Stupid." http://web.mit.edu/m-i-t/articles/index_blume.html.

———. 1997b. "Autistics, freed from face-to-face encounters, are communicating in cyberspace." *New York Times* (June 30). http://www.nytimes.com /1997/06/30/business/autistics-freed-from-face-to-face-encounters-are -communicating-in-cyberspace.html.

Boekel, Wouter, Eric-Jan Wagenmakers, Luam Belay, Josine Verhagen, Scott Brown, and Birte U. Forstmann. 2015. "A Purely Confirmatory Replication Study of Structural Brain-Behavior Correlations." *Cortex* 66: 115–133.

Boekel, Wouter, Birte U. Forstmann, and Eric-Jan Wagenmakers. 2016. "Challenges in Replicating Brain-Behavior Correlations: Rejoinder to Kanai (2015) and Muhlert and Ridgway (2015)." *Cortex* 74: 348–352.

Bogen, Joseph E. 1969. "The Other Side of the Brain: An Appositional Mind." *Bulletin of the Los Angeles Neurological Society* 34: 135–162.

———. 1971. "Neowiganism." In *Drugs, Development, and Cerebral Function,* edited by W. Lynn Smith. Springfield, Ill.: C. C. Thomas, 358–361.

———. 1985. "Foreword." In *A New View of Insanity: The Duality of the Mind Proved by the Structure, Functions, and Diseases of the Brain,* by Arthur L. Wigan [1844]. Malibu, Calif.: Joseph Simon, IX–XV.

Boileau-Narcejac, Pierre L. 1965. . . . *Et mon tout est un homme.* Paris: Denoël.

Bonnet, Charles. 1760. *Essai analytique sur les faculties de l'âme.* In *Œuvres d'histoire naturelle et de* philosophie. Neuchâtel: Samuel Fauche, 1779–1783, tome 6 (=vol. 8).

Borck, Cornelius. 2005. *Hirnströme. Eine Kulturgeschichte der Elektroenzephalographie.* Göttingen: Wallstein.

Borg, Emma. 2007. "If Mirror Neurons Are the Answer, What Was the Question?" *Journal of Consciousness Studies* 14 (8): 5–19.

Borgelt, Emily L., Daniel Z. Buchman, and Judy Illes. 2012. "Neuroimaging in Mental Health Care: Voices in Translation." *Frontiers in Human Neuroscience* 6, art. 293: 1–5.

Boshears, Rhonda, and Harry Whitaker. 2013. "Phrenology and Physiognomy in Victorian Literature." In *Literature, Neurology, and Neuroscience: Historical and Literary Connections,* edited by Anne Stiles, Stanley Finger, and François Boller, 87–112. Amsterdam: Elsevier.

Bottoni, Patrizia. 2012. "Il romanzo gotico di Francesco Mastriani." PhD diss., University of Toronto.

Bould, Mark, and Sherryl Vint. 2007. "Of Neural Nets and Brains in Vats: Model Subjects in *Galatea 2.2* and *Plus*." *Biography* 30 (1): 84–105.

Bound Alberti, Fay. 2010. *Matters of the Heart: History, Medicine, and Emotion.* New York: Oxford University Press.

Boundy, Kathryn. 2008. "'Are You Sure, Sweetheart, That You Want to Be Well?': An Exploration of the Neurodiversity Movement." *Radical Psychology* 7: 1–20.

Bowers, Jeffrey S. 2016. "The Practical and Principled Problems with Educational Neuroscience." *Psychological Review.* http://dx.doi.org/10.1037/revoooo025.

Bowman, James. 2004. "Memory and the Movies." *New Atlantis: A Journal of Technology and Society* 5: 85–90.

Boyce, Alison C. 2009. "Neuroimaging in Psychiatry: Evaluating the Ethical Consequences for Patient Care." *Bioethics* 23: 349–359.

Boyd, Robynne. 2008. "Do People Use Only 10 Percent of Their Brains?" *Scientific American* (7 February). http://www.scientificamerican.com/article/people-only-use-10-percent-of-brain/.

Braden, Charles S. 1963. *Spirits in Rebellion: The Rise and Development of New Thought.* Dallas: Southern Methodist University Press.

Broca, Paul. 1861. "Remarques sur le siège de la faculté du langage articulé, suivies d'une observation d'aphémie (perte de la parole)." *Bulletin de la Société Anatomique* 6: 330–357.

Broer, Christian, and Marjolijn Heerings. 2013. "Neurobiology in Public and Private Discourse: The Case of Adults with ADHD." *Sociology of Health and Illness* 35 (1): 49–65.

Broks, Paul. 2003. *Into the Silent Land: Travels in Neuropsychology.* New York: Atlantic Monthly Press.

Brontë, Charlotte. 1985 [1849]. *Shirley.* London: Penguin.

Brooks, David. 2013. "Beyond the Brain." *New York Times* (17 June). http://www.nytimes.com/2013/06/18/opinion/brooks-beyond-the-brain.html?_r=0.

Brosnan, Caragh, and Mike Michael. 2014. "Enacting the 'Neuro' in Practice: Translational Research, Adhesion, and the Promise of Porosity." *Social Studies of Science* 44 (5): 680–700.

Brown-Séquard, Charles-Édouard. 1874a. "The Brain Power of Man: Has He Two Brains or Has He One?" *Cincinnati Lancet and Observer* 17: 330–333.

———. 1874b. "Dual Character of the Brain." *Smithsonian Miscellaneous Collections* 15: 1–21.

———. 1890. "Have We Two Brains or One." *The Forum* 9: 627–643.

Brown, Roger, and James Kulik. 1977. "Flashbulb Memories." *Cognition* 5: 73–99.

Brownlee, Christen. 2006a. "Eat Smart. Foods May Affect the Brain as Well as the Body." *Science News* 169 (9): 136–137.

———. 2006b. "Buff and Brainy. Exercising the Body Can Benefit the Mind." *Science News* 169 (8): 122–124.

Brownlow, Charlotte. 2007. "The Construction of the Autistic Individual: Investigations in Online Discussion Groups." PhD diss., University of Brighton.

Brownlow, Charlotte, and Lindsay O'Dell. 2006. "Constructing an Autistic Identity: AS Voices Online." *Mental Retardation* 44 (5): 315–321.

Broyd, Samantha J., Charmaine Demanuele, Stefan Debener, Suzannah K. Helps, Christopher J. James, and Edmund J. S. Sonuga-Barke. 2009. "Default-Mode Brain Dysfunction in Mental Disorders: A Systematic Review." *Neuroscience and Biobehavioral Reviews* 33: 279–296.

Bryson, Norman. 2003. "Introduction: The Neural Interface." In *Blow-Up: Photography, Cinema, and the Brain*, by Warren Neidich. New York: Distributed Art Publishers, 11–19.

Bucchi, Masimiano, and Federico Neresini. 2007. "Science and Public Participation." In *The Handbook of Science and Technology Studies*, 3rd ed., edited by Edward J. Hackett et al. Cambridge, Mass.: MIT Press, 449–472.

Buchman Daniel Z., et al. 2013. "Neurobiological Narratives: Experiences of Mood Disorder Through the Lens of Neuroimaging." *Sociology of Health and Illness* 35 (1): 66–81.

Buckner, Randy L., Jessica R. Andrews-Hanna, and Daniel L. Schacter. 2008. "The Brain's Default Network: Anatomy, Function, and Relevance to Disease." *Annals of the New York Academy of Sciences* 1124: 1–38.

Buford, Chris, and Fritz Allhoff. 2005. "Neuroscience and Metaphysics." *American Journal of Bioethics* 5 (2): 34–36, W33–34.

Bulgakov, Mikhail. 1987 [1925]. *Heart of a Dog*. Translated by Mirra Ginsberg. New York: Grove.

Bülow, Hans-Henrik, et al. 2008. "The World's Major Religions' Points of View on End-of-Life Decisions in the Intensive Care Unit." *Intensive Care Medicine* 34: 423–430.

Bumiller, Kristin. 2008. "Quirky Citizens: Autism, Gender, and Reimagining Disability." *Signs: Journal of Women in Culture and Society* 33 (4): 967–991.

Burkeman, Oliver. 2015. "Therapy Wars: The Revenge of Freud." *Guardian* (7 January). http://www.theguardian.com/science/2016/jan/07/therapy-wars-revenge-of-freud-cognitive-behavioural-therapy.

Burn, Stephen J., and Peter Dempsey, eds. 2008. *Intersections: Essays on Richard Powers*. Champaign, Ill.: Dalkey Archive.

Burton, Robert. 1651 [1621]. *The Anatomy of Melancholy*. http://www.gutenberg.org/files/10800/10800-h/10800-h.htm.

Busso, Daniel S., and Courtney Pollack. 2015. "No Brain Left Behind: Consequences of Neuroscience Discourse for Education." *Learning, Media, and Technology* 40 (2): 168–186.

Button, Katherine S., et al. 2013. "Power Failure: Why Small Sample Size Undermines the Reliability of Neuroscience." *Nature Reviews Neuroscience* 14 (5): 365–376.

Byatt, Antonia S. 2006a. "Observe the Neurones. Between, Above and Below John Donne." *Times Literary Supplement* (September 22). http://www.thetimes.co.uk/tto/others/article1888544.ece.

———. 2006b. "Feeling Thought: Donne and the Embodied Mind." In *The Cambridge Companion to Donne*, edited by Achsah Gibbory, 247–257. New York: Cambridge University Press.

Cabral, Joana, et al. 2013. "Structural Connectivity in Schizophrenia and Its Impact on the Dynamics of Spontaneous Functional Networks." *Chaos: An Interdisciplinary Journal of Nonlinear Science* 23 (4): 046111. doi:10.1063/1.4851117.

Cacioppo, John T., and Gary G. Berntson. 1992. "Social Psychological Contributions to the Decade of the Brain: Doctrine of Multilevel Analysis." *American Psychologist* 47: 1019–1028.

———, eds. 2005. *Social Neuroscience*. New York: Psychology Press.

Cadigan, Pat, ed. 2002. *The Ultimate Cyberpunk*. New York: ibooks.

Callard, Felicity, and Daniel S. Margulies. 2011. "The Subject at Rest: Novel Conceptualizations of Self and Brain from Cognitive Neuroscience's Study of the 'Resting State.'" *Subjectivity* 4: 227–257.

Callard, Felicity, and Des Fitzgerald. 2015. *Rethinking Interdisciplinarity Across the Social Sciences and Neurosciences*. New York: Palgrave Macmillan.

Capacchione, Lucia. 2001. *The Power of Your Other Hand: A Course in Channeling the Inner Wisdom of the Right Brain*. Franklin Lakes, N.J.: New Page.

Cappelletto, Chiara. 2009. *Neuroestetica. L'arte del cervello*. Rome: Laterza.

Caramazza, Alfons, Stefano Anzellotti, Lukas Strnad, and Angelika Lingnau. 2014. "Embodied Cognition and Mirror Neurons: A Critical Assessment." *Annual Review of Neuroscience* 37: 1–15.

Carey, Benedict. 2005. "Can Brain Scans See Depression?" *New York Times* (October 18). http://www.nytimes.com/2005/10/18/health/psychology/can-brain-scans-see-depression.html.

Carey, Nessa. 2012. *The Epigenetics Revolution: How Modern Biology Is Rewriting Our Understanding of Genetics, Disease, and Inheritance*. New York: Columbia University Press.

Carson, Gerald. 1957. *Cornflakes Crusade*. New York: Reinhart.

Carver, Joanna. 2012. "New Look at Einstein's Brain Pictures Show His Genius." *New Scientist* (20 November). http://www.newscientist.com/blogs/shortsharpscience/2012/11/einsteins-brain.html.

Casati, Roberto, and Alessandro Pignocchi. 2007. "Mirror and Canonical Neurons Are Not Constitutive of Aesthetic Response [Comment on Freedberg and Gallese, 2007]." *Trends in Cognitive Science* 11 (10): 410.

Cascio, Ariel. 2014. "New Directions in the Social Study of the Autism Spectrum: A Review Essay." *Culture, Medicine, and Psychiatry* 38: 306–311.

Cavallaro, Dani. 2004. "The Brain in a Vat in Cyberpunk, the Persistence of the Flesh." In Gere 2004, 287–305.

Cavanagh, Patrick. 2005. "The Artist as Neuroscientist." *Nature* 434: 301–307.

CBS. 2006. "Retraining the Brain: Doctors Test Drug-Free Methods to Restore Lost Mental Capabilities." *CBS News* (15 January). http://www.cbsnews.com/news/retraining-the-brain/.

Cela-Conde, Camilo J., et al. 2011. "The Neural Foundations of Aesthetic Appreciation." *Progress in Neurobiology* 94: 39–48.

Chafetz, Michael D. 1992. *Smart for Life: How to Improve Your Brain Power at Any Age.* New York: Penguin.

Chamak, Brigitte. 2008. "Autism and Social Movements: French Parents' Associations and International Autistic Individuals' Organizations." *Sociology of Health and Illness* 30 (1): 76–96.

———. 2014. "Autism as Viewed by French Parents." In *Comprehensive Guide to Autism*, edited by Vinood B. Patel, Victor R. Preedy, and Colin R. Martin. New York: Springer, 2533–2543.

Chamak, Brigitte, and Beatrice Bonniau. 2013. "Changes in the Diagnosis of Autism: How Parents and Professionals Act and React in France." *Culture Medicine and Psychiatry* 37: 405–426.

Chamak, Brigitte, et al. 2008. "What Can We Learn About Autism from Autistic Persons?" *Psychotherapy and Psychosomatics* 77 (5): 271–279.

Charlton, James. 2000. *Nothing About Us Without Us: Disability Oppression and Empowerment.* Berkeley: University of California Press.

Charman, Tony. 2006. "Autism at the Crossroads: Determining the Phenotype Matters for Neuroscience." *Nature Neuroscience* 9 (10): 1197.

Charney, Dennis, et al. 2002. "Neuroscience Research Agenda to Guide Development of a Pathophysiologically Based Classification System." In *A Research Agenda for DSM-V*, edited by David J. Kupfer, Michael B. First, and Darrel A. Regier, 31–84. Washington, D.C.: American Psychiatric Association.

Chatterjee, Anjan. 2010. "Neuroaesthetics: A Coming of Age Story." *Journal of Cognitive Neuroscience* 23: 53–62.

———. 2012. "Neuroaesthetics: Growing Pains of a New Discipline." In *Aesthetic Science: Connecting Minds, Brains, and Experience*, edited by Arthur P. Shimamura and Stephen E. Palmer. New York: Oxford University Press, 299–317.

Chatterjee, Anjan, and Martha J. Farah, eds. 2013. *Neuroethics in Practice: Medicine, Mind, and Society.* New York: Oxford University Press.

Cheek, Joanna. 2012. "Myth: Reframing Mental Illness as a 'Brain Disease' Reduces Stigma." *Canadian Foundation for Healthcare Improvement*. http://www.cfhi-fcass.ca/SearchResultsNews/12-06-04/a078ceca-4a41-4d14-82b5-b60f5a8bb991.aspx.

Cheon, Bobby K., et al. 2011. "Cultural Influences on Neural Basis of Intergroup Empathy." *Neuroimage* 57 (2): 642–650.

———. 2013. "Constraints, Catalysts, and Coevolution in Cultural Neuroscience: Reply to Commentaries." *Psychological Inquiry* 24 (1): 71–79.

Cheu, Johnson. 2004. "De-gene-erates, Replicants, and Other Aliens: (Re) defining Disability in Futuristic Film." In Corker and French 1999, 198–212.

Chiao, Joan Y., ed. 2009a. *Cultural Neuroscience: Cultural Influences on Brain Function*. Progress in Brain Research 178. New York: Elsevier.

———. 2009b. "Cultural Neuroscience: A Once and Future Discipline." In Chiao 2009a, 287–304.

———. 2011. "Cultural Neuroscience: Visualizing Culture-Gene Influences on Brain Function." In *The Oxford Handbook of Social Neuroscience*, edited by Jean Decety and John T. Cacioppo. Oxford: Oxford University Press, 742–762.

Chiao, Joan Y., and Nalini Ambady. 2007. "Cultural Neuroscience: Parsing Universality and Diversity Across Levels of Analysis." In *Handbook of Cultural Psychology*, edited by Shinobu Kitayama and Dov Cohen. New York: Guilford, 237–254.

Chiao, Joan Y., and Katherine D. Blizinsky. 2010. "Culture-Gene Coevolution of Individualism-Collectivism and the Serotonin Transporter Gene." *Proceedings of the Royal Society B* 277: 529–537.

Chiao, Joan Y., and Bobby K. Cheon. 2012. "Cultural Neuroscience as Critical Neuroscience in Practice." In Choudhury and Slaby 2012a, 287–304.

Chiao, Joan Y., et al. 2008. "Cultural Specificity in Amygdale Response to Fear Faces." *Journal of Cognitive Neuroscience* 20 (12): 2167–2174.

———. 2009. "Neural Basis of Individualistic and Collectivistic Views of Self." *Human Brain Mapping* 30 (9): 2813–2820.

———. 2010. "Theory and Methods in Cultural Neuroscience." *Social Cognitive and Affective Neuroscience* 5 (2/3): 356–361.

———. 2013. "Cultural Neuroscience: Progress and Promise." *Psychological Inquiry* 24 (1): 1–19.

Choudhury, Suparna. 2010. "Culturing the Adolescent Brain: What Can Neuroscience Learn from Anthropology?" *Social Cognitive and Affective Neuroscience* 5 (2/3): 159–167.

Choudhury, Suparna, and Laurence J. Kirmayer. 2009. "Cultural Neuroscience and Psychopathology: Prospects for Cultural Psychiatry." In Chiao 2009a, 263–283.

Choudhury, Suparna, and Kelly A. McKinney. 2013. "Digital Media, the Developing Brain, and the Interpretive Plasticity of Neuroplasticity." *Transcultural Psychiatry* 50 (2): 192–215.

Choudhury, Suparna, and Jan Slaby, eds. 2012a. *Critical Neuroscience: A Handbook of the Social and Cultural Contexts of Neuroscience*. Malden, Mass.: Blackwell.

———. 2012b. "Introduction. Critical Neuroscience—Between Lifeworld and Laboratory." In Choudhury and Slaby 2012a, 1–26.

Choudhury, Suparna, Kelly A. McKinney, and Moritz Merten. 2012. "Rebelling Against the Brain: Public Engagement with the 'Neurological Adolescent.'" *Social Science and Medicine* 74: 565–573.

Choudhury, Suparna, Saskia Kathi Nagel, and Jan Slaby. 2009. "Critical Neuroscience: Linking Science and Society Through Critical Practice." *BioSocieties* 4 (1): 61–77.

Churchland, Paul M. 1981. "Eliminative Materialism and the Propositional Attitudes." *Journal of Philosophy* 78 (2): 67–90.

Cikara, Mina, and Jay J. Van Bavel. 2014. "The Neuroscience of Intergroup Relations: An Integrative Review." *Perspectives on Psychological Science* 9 (3): 245–274.

Clarke, Basil. 1987. *Arthur Wigan and the Duality of Mind*. Cambridge: Cambridge University Press.

Clarke, Edwin, and Kenneth Dewhurst. 1996. *An Illustrated History of Brain Function*. Berkeley: University of California Press.

Clarke, Edwin, and L. Stephen Jacyna. 1987. *Nineteenth-Century Origins of Neuroscientific Concepts*. Berkeley: University of California Press.

Clarke, Juanne, and Gudrun van Amerom. 2007. "'Surplus Suffering': Differences Between Organizational Understandings of Asperger's Syndrome and Those People Who Claim the 'Disorder.'" *Disability and Society* 22 (7): 761–776.

———. 2008. "Asperger's Syndrome: Differences Between Parents' Understanding and Those Diagnosed." *Social Work in Health Care* 46 (3): 85–106.

Clausen, Jens, and Neil Levy, eds. 2015. *Handbook of Neuroethics*. Dordrecht: Springer.

Clifford, Jim. 1988. *The Predicament of Culture: Twentieth-Century Ethnography, Literature, and Art*. Cambridge, Mass.: Harvard University Press.

Cohen, Adam B. 2009. "Many Forms of Culture." *American Psychologist* 6: 194–204.

Cohen, Alex, and Oye Gureje. 2007. "Making Sense of Evidence." *International Review of Psychiatry* 19 (5): 583–591.

Cohen, Alex, Vikram Patel, and Harry Minas. 2014. "A Brief History of Global Mental Health." In Patel, Minas, Cohen, and Prince 2014, 3–26.

Cohen, Isabel, and Marcelle Goldsmith. 2002. *Hands On: How to Use Brain Gym in the Classroom*. Ventura, Calif.: Edu Kinesthetics, Inc.

Cohen, Patricia. 2010. "Next Big Thing in English, Knowing They Know That You Know." *New York Times* (1 April). http://www.nytimes.com/2010/04/01/books/01lit.html.

Cohn, Simon. 2010. "Picturing the Brain Inside, Revealing the Illness Outside: A Comparison of the Different Meanings Attributed to Brain Scans by Scientists and Patients." In *Technologized Images, Technologized Bodies*, edited by Jeanette Edwards, Penelope Harvey, and Peter Wade, 65–84. New York: Berghahn.

———. 2012. "Disrupting Images: Neuroscientific Representations in the Lives of Psychiatric Patients." In Choudhury and Slaby 2012a, 179–193.

Coleman, Gabriella E. 2010. "Ethnographic Approaches to Digital Media." *Annual Review of Anthropology* 39: 487–505.

Coles, Romand. 2013. "The Neuropolitical *Habitus* of Resonant Receptive Democracy." In Vander Valk 2012a, 178–197.

Coltheart, Max. 2006. "Perhaps Functional Neuroimaging Has Not Told Us Anything About the Mind (So Far)." *Cortex* 42: 422–427.

———. 2013. "How Can Functional Neuroimaging Inform Cognitive Theories?" *Perspectives on Psychological Science* 8 (1): 98–103.

Combe, Andrew. 1836–1837. "Remarks on the Possibility of Increasing the Development of the Cerebral Organs by Adequate Exercise of the Mental Faculties." *Phrenological Journal* 10: 414–426.

Combe, George. 1828. *The Constitution of Man Considered in Relation to External Objects*. Edinburgh: Mclachlan & Stewardt and John Anderson.

Connolly, William. 2002. *Neuropolitics: Thinking, Culture, Speed*. Minneapolis: University of Minnesota Press.

Conrad, Erin C., and Raymond De Vries. 2011. "Field of Dreams: A Social History of Neuroethics." In Pickersgill and Van Keulen 2011, 299–324.

Constable, Catherine. 2012. "Withdrawal of Artificial Nutrition and Hydration for Patients in a Permanent Vegetative State: Changing Tack." *Bioethics* 26 (3): 157–163.

Conway, Bevil R., and Alexander Rehding. 2013. "Neuroaesthetics and the Trouble with Beauty." *PLOS Biology* 11 (3): e1001504. doi:10.1371/journal.pbio.1001504.

Cook, Richard, Geoffrey Bird, Caroline Catmur, Clare Pressa, and Cecilia Heyes. 2014. "Mirror Neurons: From Origin to Function." *Brain and Behavioral Sciences* 37 (2): 177–192.

Cooper, Rachel. 2015. "Must Disorders Cause Harm? The Changing Stance of the DSM." In *The DSM-5 in Perspective: Philosophical Reflections on the Psychiatric Babel*, edited by Steeves Demazeux and Patrick Singy, 83–96. New York: Springer.

Cooper, Sara. 2016. "Global Mental Health and Its Critics: Moving Beyond the Impasse." *Critical Public Health* 26 (4): 355–358.

Cooter, Roger. 1984. *The Cultural Meaning of Popular Science: Phrenology and the Organisation of Consent in Nineteenth-Century Britain.* Cambridge: Cambridge University Press.

Cooter, Roger. 2014. "Neural Veils and the Will to Historical Critique: Why Historians of Science Need to Take the Neuro-Turn Seriously." *Isis* 105 (1): 145–154.

Corker, Mairian. 1999. "New Disability Discourse, the Principle of Optimization, and Social Change." In Corker and French 1999, 299–324.

Corker, Mairian, and Sally French, eds. 1999. *Disability Discourse.* Philadelphia: Open University Press.

Corker, Mairian, and Tom Shakespeare, eds. 2004. *Disability/Postmodernity: Embodying Disability Theory.* London: Continuum.

Corrigan, W. Patrick, et al. 2002. "Challenging Two Mental Illness Stigmas: Personal Responsibility and Dangerousness." *Schizophrenia Bulletin* 28 (2): 293–309.

Cotman, W. Carl, and Nicole C. Berchtold. 2002. "Exercise: A Behavioral Intervention to Enhance Brain Health and Plasticity." *Trends in Neurosciences* 25 (6): 295–301.

Couser, G. Thomas. 2004. *Vulnerable Subjects: Ethics and Life Writing.* Ithaca, N.Y.: Cornell University Press.

Cowen, Phillip J. 2013. "Classification of Depressive Disorders." In *Behavioral Neurobiology of Depression and Its Treatment*, edited by Philip J. Cowen, Trevor Sharp, and Jennifer Y. F. Lau, 3–13 Berlin: Springer.

Crane, Mary Thomas, and Alan Richardson. 1999. "Literary Studies and Cognitive Science: Toward a New Interdisciplinarity." *Mosaic* 32: 123–140.

Crichton. Michael. 1972. *The Terminal Man.* New York: Ballantine.

Crick, Francis. 1994. *The Astonishing Hypothesis: The Scientific Search for the Soul.* New York: Touchstone.

Cromby, John, and Simon J. Williams. 2011. "Neuroscience and Subjectivity." *Subjectivity* 4: 215–226.

Crossley, Nick. 1998. "R. D. Laing and the British Antipsychiatry Movement." *Social Science and Medicine* 47 (7): 877–889.

———. 2006. *Contesting Psychiatry: Social Movements in Mental Health.* London: Routledge.

Cunningham, John P., and Byron M. Yu. 2014. "Dimensionality Reduction for Large-Scale Neural Recordings." *Nature Neuroscience* 17: 1500–1509.

D'Alembert, Jean. 1986 [1767]. "Eclaircissements sur différents endroits des *Elémens de philosophie*" [Explanations on different places of the *Elements of Philosophy*]. In *Essai sur les elémens de philosophie.* Paris: Fayard.

Damousi, Joy, and Mariano Ben Plotkin, eds. 2009. *The Transnational Unconscious: Essays in the History of Psychoanalysis and Transnationalism.* London: Palgrave Macmillan.

Danto, Arthur C. 1964. "The Artworld." *Journal of Philosophy* 61 (19): 571–584.
———. 1981. *The Transfiguration of the Commonplace*. Cambridge, Mass.:
Harvard University Press.
———. 1993. "Andy Warhol: Brillo Box." *Artforum* 32 (1): 128–129.
———. 1997. *After the End of Art: Contemporary Art and the Pale of History*.
Princeton, N.J.: Princeton University Press.
Daston, Lorraine, and Otto Sibum. 2003. "Introduction: Scientific Personae
and Their Histories." *Science in Context* 16 (1/2): 1–8.
Davidson, Joyce. 2007. "'In a World of Her Own . . .': Re-presenting Alien-
ation and Emotion in the Lives and Writings of Women with Autism."
Gender, Place, and Culture 14 (6): 659–677.
———. 2008. "Autistic Culture Online: Virtual Communication and Cultural
Expression on the Spectrum." *Social and Cultural Geography* 9 (7): 791–806.
Davidson, Richard J., et al. 2002a. "Depression: Perspectives from Affective
Neuroscience." *Annual Review of Psychology* 53: 545–574.
———. 2003. "Alterations in Brain and Immune Function Produced by
Mindfulness Meditation." *Psychosomatic Medicine* 65:564–570.
Davidson, Richard J., Diego Pizzagalli, and Jack Nitschke. 2002b. "The
Representation and Regulation of Emotion in Depression: Perspectives
from Affective Neuroscience." In *Handbook of Depression*, edited by Ian
Gotlib and Constance Hammen, 219–244. New York: Guilford.
———. 2009. "Representation and Regulation of Emotion in Depression:
Perspectives from Affective Neuroscience." In *Handbook of Depression*, edited
by Ian Gotlib and Constance Hammen, 218–248. New York: Guilford.
Davidson, Richard J., and Bruce McEwen. 2012. "Social Influences on
Neuroplasticity: Stress and Interventions to Promote Well-Being." *Nature
Neuroscience* 15 (5): 689–695.
Davies, David. 2014. "'This Is Your Brain on Art.' What Can Philosophy of
Art Learn from Neuroscience?" In *Aesthetics and the Sciences of Mind*, edited
by Gregory Currie, Matthew Kieran, Aaron Meskin, and Jon Robson.
New York: Oxford University Press, 57–74.
Davies, Stephen. 2009. "Evolution, Art, and Aesthetics." In *A Companion to
Aesthetics*, 2nd ed., edited by Stephen Davies, Kathleen Marie Higgins,
Robert Hopkins, Robert Stecker, and David E. Cooper. Oxford: Black-
well, 259–261.
Davis, Lennard J. 1995. *Enforcing Normalcy: Disability, Deafness, and the Body*.
London: Verso.
———. 2002. *Bending Over Backwards: Disability, Dismodernism, and Other
Difficult Positions*. New York: New York University Press.
Dawson, Michelle. 2004. "The Misbehavior of Behaviorists. Ethical Chal-
lenges to the Autism-ABA Industry." http://www.sentex.net/~nexus23/naa
_aba.html.

de Almeida, Jorge C., and Mary Louise Phillips. 2013. "Distinguishing Between Unipolar Depression and Bipolar Depression: Current and Future Clinical and Neuroimaging Perspectives." *Biological Psychiatry* 73: 111–118.

de Beaugrande, Robert. 1987. "Schemas for Literary Communication." In *Literary Discourse: Aspects of Cognitive and Social Psychological Approaches*, edited by Laszlo Halász, 49–99. Berlin: de Gruyter.

De Giustino, David. 1975. *Conquest of Mind: Phrenology and Victorian Social Thought*. London: Croom Helm.

De Grazia, David. 2011. "The Definition of Death." *Stanford Encyclopedia of Philosophy*. http://plato.stanford.edu/entries/death-definition/.

de Vignemont, Frédérique, and Tania Singer. 2006. "The Empathic Brain: How, When, and Why?" *Trends in Cognitive Sciences* 10 (10): 435–441.

De Vos, Jan, and Ed Pluth, eds. 2016. *Neuroscience and Critique: Exploring the Limits of the Neurological Turn*. New York: Routledge.

Dekker, Martijn. 2006. "On Our Own Terms: Emerging Autistic Culture." http://web.archive.org/web/20061111053135/http://trainland.tripod.com/martijn.htm.

Denkhaus, Ruth, and Mathias Bös. 2012. "How Cultural is 'Cultural Neuroscience'? Some Comments on an Emerging Research Paradigm." *BioSocieties* 7 (4): 433–458.

Dennett, Daniel C. 2008. "Astride the Two Cultures: A Letter to Richard Powers, Upadated." In Burn and Dempsey 2008, 151–161.

Dennison, Gail E., Paul E. Dennison, and Jerry V. Teplitz. 1994. *Brain Gym for Business. Instant Brain Boosters for On-The-Job Success*. Ventura, Calif.: Edu-Kinesthetics, Inc.

Deresiewicz, William. 2006. "Science Fiction." *The Nation* (October 9). http://www.thenation.com/article/science-fiction?page=0,0.

Deshpande, Gopikrishna, et al. 2013. "Identification of Neural Connectivity Signatures of Autism Using Machine Learning." *Frontiers in Human Neuroscience* 17 (7), art. 670: 1–15.

Deville, James. 1841. "Account of a Number of Cases in Which a Change Had Been Produced on the Form of the Head by Education and Moral Training." *Phrenological Journal* 14: 32–38.

Dewhurst, Kenneth, trans. 1980. *Thomas Willis's Oxford Lectures*. Oxford: Sandford.

Di Dio, Cinzia, Emiliano Macaluso, and Giacomo Rizzolatti. 2007. "The Golden Beauty: Brain Response to Classical and Renaissance Sculptures." *PLoS ONE* 11 (November): 1–9.

Diamond, Marian C., et al. 1985. "On the Brain of a Scientist: Albert Einstein." *Experimental Neurology* 88: 198–204.

Dick, Philip K. 1991 [1977]. *A Scanner Darkly*. New York: Vintage.

Dinello, Daniel. 2006. *Technophobia! Science Fiction Visions of Posthuman Technology*. Austin: University of Texas Press.

Dobbs, David. 2006. "A Depression Switch?" *New York Times Magazine* (2 April). http://www.nytimes.com/2006/04/02/magazine/02depression.html?pagewanted=all&_r=0.

Doidge, Norman. 2007. *The Brain That Changes Itself: Stories of Personal Triumph from the Frontiers of Brain Science*. New York: Penguin.

———. 2015. *The Brain's Way of Healing: Remarkable Discoveries and Recoveries from the Frontiers of Neuroplasticity*. New York: Viking.

Dolan, Brian. 2007. "Soul Searching: A Brief History of the Mind/Body Debate in the Neurosciences." *Neurosurgical Focus* 23: 1–7.

Domínguez Duque, Juan F. 2012. "Neuroanthropology and the Dialectical Imperative." *Anthropological Theory* 12 (1): 5–27.

———. 2015. "Toward a Neuroanthropology of Ethics: Introduction." In *Handbook of Neuroethics*, edited by Jens Clausen and Neil Levy. Dordrecht: Springer, 289–298.

Domínguez Duque, Juan F., et al. 2009. "The brain in Culture and Culture in the Brain: A Review of Core Issues in Neuroanthropology." In Chiao 2009a, 43–64.

———. 2010. "Neuroanthropology: A Humanistic Science for the Study of the Culture-Brain Nexus." *SCAN* [*Social Cognitive and Affective Neuroscience*] 5 (2/3): 38–147.

Doucet, Hubert. 2005. "Imagining a Neuroethics Which Would Go Further Than Genethics." *American Journal of Bioethics* 5 (2): 29–31, W23–24.

Downey, Greg. 2012a. "Neuroanthropology." In *The SAGE Handbook of Social Anthropology*, vol. 2., edited by Richard Fardon, Oliva Harris, Trevor H. J. Marchand, Cris Shore, Veronica Strang, Richard Wilson, and Mark Nuttall. London: Sage, 243–260.

———. 2012b. "Culture Variation in Rugby Skills: A Preliminary Neuroanthropological Report." *Annals of Anthropological Practice* 36 (1): 26–44.

———. 2012c. "Balancing Across Cultures: Equilibrium in Capoeira." In Lende and Downey 2012a, 169–194.

Downey, Greg, and Daniel H. Lende. 2012. "Neuroanthropology and the Encultured Brain." In Lende and Downey 2012a, 23–65.

Draaisma, Douwe. 2009. "Echos, Doubles, and Delusions: Capgras Syndrome in Science and Literature." *Style* 43 (3): 429–441.

Dresler, Martin, ed. 2009. *Neuroästhetik. Kunst—Gehirn—Wissenschaft*. Leipzig: Seemann.

Drevets, Wayne. 1998. "Functional Neuroimaging Studies of Depression: The Anatomy of Melancholia." *Annual Review of Medicine* 49: 341–361.

Droz, Marion M. 2011. "La plasticité cérébrale de Cajal à Kandel. Cheminement d'une notion constitutive du sujet cérébral." *Revue d'Histoire des Sciences* 63 (2): 331–367.

Dudley, Kevin J., et al. 2011. "Epigenetic Mechanisms Mediating Vulnerability and Resilience to Psychiatric Disorders." *Neuroscience and Biobehavioral Reviews* 35: 1544–1551.

Dumit, Joseph. 2003. "Is It Me or My Brain? Depression and Neuroscientific Facts." *Journal of Medical Humanities* 24 (12): 35–46.

———. 2004. *Picturing Personhood. Brain Scans and Biomedical Identity.* Princeton, N.J.: Princeton University Press.

———. 2012. *Drugs for Life: How Pharmaceutical Companies Define Our Health.* Durham, N.C.: Duke University Press.

Dunsmoor, Joseph E., et al. 2015. "Emotional Learning Selectively and Retroactively Strengthens Memories for Related Events." *Nature* 520: 345–348.

Eagleton, Terry. 2000. *The Idea of Culture.* Malden, Mass.: Blackwell.

Eaton, William R. 2005. *Boyle on Fire: The Mechanical Revolution in Scientific Explanation.* New York: Continuum.

Ecker, Christine, et al. 2010. "Describing the Brain in Autism in Five Dimensions—Magnetic Resonance Imaging–Assisted Diagnosis of Autism Spectrum Disorder Using a Multiparameter Classification Approach." *Journal of Neuroscience* 30 (32): 10612–10623.

Ecks, Stefan. 2013. *Eating Drugs: Psychopharmaceutical Pluralism in India.* New York: NYU Press.

Ecks, Stefan, and Soumita Basu. 2009. "The Unlicensed Lives of Antidepressants in India: Generic Drugs, Unqualified Practitioners, and Floating Prescriptions." *Transcultural Psychiatry* 46: 86–106.

Eco, Umberto. 2007. *On Ugliness.* Translated by Alastair McEwen. New York: Rizzoli.

Edwards, Betty. 1979. *Drawing on the Right Side of the Brain.* Los Angeles: J. P. Tarcher.

Ehrenberg, Alain. 2004. "Le sujet cérébral." *Esprit* 309: 130–155.

Ehrenwald, Jan. 1984. *Anatomy of Genius: Split Brains and Global Minds.* New York: Human Sciences.

Eichenbaum, Howard. 2012. *The Cognitive Neuroscience of Memory: An Introduction.* 2nd ed. New York: Oxford University Press.

Eijkholt, Marleen, James A. Anderson, and Judy Illes. 2012. "Picturing Neuroscience Research Through a Human Rights Lens: Imaging First Episode Schizophrenic Treatment-Naïve Individuals." *International Journal of Law and Psychiatry* 35: 146–152.

Eklund, Anders, Thomas E. Nichols, and Hans Knutsson. 2016. "Cluster Failure: Why fMRI Inferences for Spatial Extent Have Inflated False-Positive Rates." *PNAS* 113 (28): 7900–7905.

Erk, Susanne, Henrik Walter, and Manfred Spitzer. 2002. "Functional Neuroimaging of Depression." *Advances in Biological Psychiatry* 21: 63–69.

Esch, Tobias. 2014. "The Neurobiology of Meditation and Mindfulness." In *Meditation: Neuroscientific Approaches and Philosophical Implications*, edited by Stefan Schmidt and Harald Walach. New York: Springer, 153–173.

Evans, Warren F. 1874. *Mental Medicine: A Theoretical and Practical Treatise on Mental Psychology*. 3rd ed. Boston: Carter & Pettee.

Eyal, Gil, et al. 2010. *The Autism Matrix: The Social Origins of the Autism Epidemic*. Cambridge: Polity.

Falk, Dean, Frederick E. Lepore, and Adrianne Noe. 2012. "The Cerebral Cortex of Albert Einstein: A Description and Preliminary Analysis of Unpublished Photographs." *Brain*. doi:10.1093/brain/aws295.

Falk, John H., and John D. Balling. 2010. "Evolutionary Influence on Human Landscape Preference." *Environment and Behavior* 42: 479–493.

Farah, Martha J., ed. 2010a. *Neuroethics: An Introduction with Readings*. Cambridge, Mass.: MIT Press.

———. 2010b. "Neuroethics: An Overview." In Farah 2010a, 1–10.

———. 2014. "Brain Images, Babies, and Bathwater: Critiquing Critiques of Functional Neuroimaging." In *Interpreting Neuroimages: An Introduction to the Technology and Its Limits, Hastings Center Report* 45 (2): S19–S30. doi:10.1002/hast.295.

Farah, Martha J., and Cayce J. Hook. 2013. "The Seductive Allure of 'Seductive Allure.'" *Perspectives on Psychological Science* 8 (1): 88–90.

Fernandez-Duque, Diego, Jessica Evans, Colton Christian, and Sara D. Hodges. 2015. "Superfluous Neuroscience Information Makes Explanations of Psychological Phenomena More Appealing." *Journal of Cognitive Neuroscience* 27 (5): 926–944.

Ferrari, Alize J., et al. 2013. "Burden of Depressive Disorders by Country, Sex, Age, and Year: Findings from the Global Burden of Disease Study 2010." *PLoS Medicine* 10 (11): e1001547. doi:10.1371/journal.pmed.1001547.

Ferret, Stéphane. 1993. *Le philosophe et son scalpel. Le problème de l'identité personnelle*. Paris: Minuit.

Fimiani, Filippo. 2009. "Simulations incorporées et tropismes empathiques. Notes sur la neuro-esthétique." *Images Re-vues. Histoire, Anthropologie et Théorie de l'Art* 6 (June). http://imagesrevues.revues.org /426.

Finger, Stanley. 2000. *Minds Behind the Brain: A History of the Pioneers and Their Discoveries*. New York: Oxford University Press.

Finger, Stanley, Dahlia W. Zaidel, François Boller, and Julien Bogousslavsky, eds. 2013. *The Fine Arts, Neurology, and Neuroscience. New Discoveries and Changing Landscapes*. Progress in Brain Research 204. Amsterdam: Elsevier.

Finn, Emily S. 2015. "Brain Activity Is as Unique—and Identifying—as a Fingerprint." *The Conversation* (12 October). http://theconversation.com/brain-activity-is-as-unique-and-identifying-as-a-fingerprint-48723.

Finn, Emily S., et al. 2015. "Functional Connectome Fingerprinting: Identifying Individuals Using Patterns of Brain Connectivity." *Nature Neuroscience* 18 (11): 1664–1671.

Fitch, Tecumseh W., Antje von Graevenitz, and Eric Nicolas. 2009. "Bio-Aesthetics and the Aesthetic Trajectory: A Dynamic Cognitive and Cultural Perspective." In Skov and Vartanian 2009a, 59–102.

Fitzgerald, Des, and Felicity Callard. 2014. "Social Science and Neuroscience Beyond Interdisciplinarity: Experimental Entanglements." *Theory, Culture, and Society* 32 (1): 3–32.

Fitzgerald, Des, et al. 2014. "Ambivalence, Equivocation, and the Politics of Experimental Knowledge: A Transdisciplinary Neuroscience Encounter." *Social Studies of Science* 44 (5): 701–721.

Fitzgerald, Des, Nikolas Rose, and Ilina Singh. 2016a. "Revitalizing Sociology: Urban Life and Mental Illness Between History and the Present." *British Journal of Sociology* 67 (1): 138–160.

———. 2016b. "Living Well in the *Neuropolis*." *Sociological Review Monographs* 64: 221–237.

Fitzgerald, Paul B., Angela R. Laird, Jerome Maller, and Zafiris J. Daskalakis. 2008. "A Meta-Analytic Study of Changes in Brain Activation in Depression." *Human Brain Mapping* 29 (6): 683–695.

Fitzpatrick, Susan M. 2012. "Functional Brain Imaging. Neuro-Turn or Wrong Turn?" In Littlefield and Johnson 2012, 180–198.

Foley, Debra L., and Katherine I. Morley. 2011. "Systematic Review of Early Cardiometabolic Outcomes of the First Treated Episode of Psychosis." *Archives of General Psychiatry* 68: 609–616.

Fombonne, Eric. 2003. "Modern Views on Autism." *Canadian Journal of Psychiatry* 48 (8): 503–506.

Fonagy, Peter, et al. 2015. "Pragmatic Randomized Controlled Trial of Long-Term Psychoanalytic Psychotherapy for Treatment-Resistant Depression: The Tavistock Adult Depression Study (TADS)." *World Psychiatry* 14: 312–321.

Ford, Andrew. 1999. "Performing Interpretation: Early Allegorical Exegesis of Homer." In *Epic Traditions in the Contemporary World: The Poetics of Community*, edited by Margaret Beissinger, Jane Tylus, and Susanne Wofford, 33–53. Berkeley: University of California Press.

Forest, Denis. 2014. *Neuroscepticisme. Les sciences du cerveau sous le scalpel de l'épistémologue.* Paris: Ithaque.

Forry, Steven Earl. 1990. *Hideous Progenies: Dramatizations of Frankenstein from Mary Shelley to the Present.* Philadelphia: University of Pennsylvania Press.

Foucault, Michel. 1969. *The Archaeology of Knowledge.* Translated by A. M. Sheridan Smith. London: Routledge.

———. 1983. "Afterword: The Subject and Power." In *Michel Foucault: Beyond Structuralism and Hermeneutics,* edited by Hubert L. Dreyfus and Paul Rabinow. Brighton: Harvester, 206–208.

———. 1986. *The Care of the Self.* Translated by R. Hurley. New York: Pantheon.

———. 1988. "Technologies of the Self." In *Technologies of the Self,* edited by Luther H. Martin, Huck Gutman, and Patrick H. Hutton. Amherst: University of Massachusetts Press, 16–49.

———. 1990. *The Use of Pleasure.* Translated by R. Hurley. New York: Vintage.

Fournier, Jay C., et al. 2010. "Antidepressant Drug Effects and Depression Severity: A Patient-Level Meta-analysis." *JAMA* 303 (1): 47–53.

Fraenkel, Béatrice. 2007. "L'invention de l'art pariétal préhistorique. Histoire d'une expérience visuelle." *Gradhiva. Revue d'Anthropologie et d'Histoire des Arts* 6: 18–31.

Frances, Allen J. 2013. *Saving Normal: An Insider's Revolt Against Out-of-Control Psychiatric Diagnosis, DSM-5, Big Pharma, and the Medicalization of Ordinary Life.* New York: William Morrow.

Franzen, Jonathan. 2001. *The Corrections.* New York: Picador.

———. 2002. "My Father's Brain." In *How to Be Alone,* 7–38. New York: Farrar, Straus & Giroux.

Frazzetto, Giovanni, and Suzanne Anker. 2009. "Neuroculture." *Nature Reviews Neuroscience* 10: 815–821.

Freedberg, David. 1985. *Iconoclasts and Their Motives.* Maarssen: Gary Schwartz.

———. 1989. *The Power of Images: Studies in the History and Theory of Response.* Chicago: University of Chicago Press.

———. 2007. "Empathy, Motion, and Emotion." In *Wie sich Gefühle Ausdruck verschaffen. Emotionen in Nahsicht,* edited by Klaus Herding and Antje Krause-Wahl, 17–51. Berlin: Driesen.

———. 2008. "Antropologia e storia dell'arte: la fine delle discipline?" *Ricerche di Storia dell'arte* 94: 5–18.

———. 2009a. "Immagini e risposta emotiva: la prospettiva neuroscientifica." In *Prospettiva Zeri,* edited by Anna Ottani Cavina. Turin: Umberto Allemandi, 85–105.

———. 2009b. "Choirs of Praise: Some Aspects of Action Understanding in Fifteenth-Century Painting and Sculpture." In *Medieval Renaissance Baroque: A Cat's Cradle for Marilyn Aronberg Lavin,* edited by David A. Levine and Jack Freiberg. New York: Italica, 65–81.

———. 2009c. "Movement, Embodiment, Emotion." In *Histoire de l'art et anthropologie.* Paris, INHA / Musée du quai Branly. http://actesbranly.revues.org/330.

Freedberg, David, and Vittorio Gallese. 2007. "Motion, Emotion, and Empathy in Aesthetic Experience." *Trends in Cognitive Science* 11 (5): 197–203.

Freeman, Jonathan B. 2013. "Within-Cultural Variation and the Scope of Cultural Neuroscience." *Psychological Inquiry* 24: 26–30.

Freeman, Jonathan B., and Pegeen Cronin. 2002. "Diagnosing Autism Spectrum Disorder in Young Children: An Update." *Infants and Young Children*, 14 (3): 1–10.

Freeman, Jonathan B., et al. 2009. "Culture Shapes a Mesolimbic Response to Signals of Dominance and Subordination That Associates with Behavior." *NeuroImage* 47: 353–359.

Frith, Chris. 2007. *Making up the Mind: How the Brain Creates Our Mental World*. Hoboken, N.J.: Wiley.

Fuller, Robert C. 1982. *Mesmerism and the American Cure of Souls*. Philadelphia: University of Pennsylvania Press.

———. 1989. *Alternative Medicine and American Religious Life*. New York: Oxford University Press.

———. 2001. *Spiritual, But Not Religious: Understanding Unchurched America*. New York: Oxford University Press.

Furey, Maura L., et al. 2013. "Potential of Pretreatment Neural Activity in the Visual Cortex During Emotional Processing to Predict Treatment Response to Scopolamine in Major Depressive Disorder." *JAMA Psychiatry* 70 (3): 280–290.

Gabriel, Markus. 2015. *Ist ist nich Gehirn. Philosophie des Geistes für das 21. Jahrhundert*. Hamburg: Ullstein.

Gabrieli, John D., Satrajit S. Ghosh, and Susan Whitfield-Gabrieli. 2015. "Prediction as a Humanitarian and Pragmatic Contribution from Human Cognitive Neuroscience." *Neuron* 85 (1): 11–26.

Gaddy, James. 2007. "Shadow Boxer." *Print* (July/August). http://www.printmag.com/Article/Shadow_Boxer.

Gainer, Ruth S., and Harold Gainer. 1977. "Educating Both Halves of the Brain: Fact or Fancy?" *Art Education* 30 (5): 20–22.

Galchen, Rivka. 2008. *Atmospheric Disturbances*. London: Harper Perennial.

Gall, Franz J. 1835. *On the Functions of the Brain and of Each of Its Parts: With Observations on the Possibility of Determining the Instincts, Propensities, and Talents, or the Moral and Intellectual Dispositions of Men and Animals, by the Configuration of the Brain and Head*. 6 vols. Translated by Winslow Lewis. Boston: Marsh, Capen & Lyon.

Gall, Franz J., and Johann-Caspar Spurzheim. 1809. *Recherches sur le système nerveux en général, et sur celui du cerveau en particulier. Mémoire présenté à l'Institut de France, le 14 mars 1808, suivi d'observations sur le rapport qui en a été fait à cette compagnie par ses commissaires*. Paris: F. Schoelle et H. Nicolle.

Gallagher, Shaun. 2010. "The Body's Architecture." Lecture delivered at the Third International Arakawa and Gins: Architecture and Philosophy Conference. http://web.archive.org/web/20110811053018/http://ag3.griffith.edu.au/node/223?offset=-120

Gallagher, Shaun, and Dan Zahavi. 2008. *The Phenomenological Mind: An Introduction to Philosophy of Mind and Cognitive Science.* London: Routledge.

Gallese, Vittorio. 2007. "Before and Below Theory of Mind: Embodied Simulation and the Neural Correlates of Social Cognition." *Philosophical Transactions of the Royal Society B* 362: 659–669.

———. 2008. "Empathy, Embodied Simulation, and the Brain: Commentary on Aragno and Zepf/Hartmann." *Journal of the American Psychoanalytic Association* 56: 769–781.

———. 2009. "Motor Abstraction: A Neuroscientific Account of How Action Goals and Intentions Are Mapped and Understood." *Psychological Research* 73 (4): 486–498.

———. 2011. "Embodied Simulation Theory: Imagination and Narrative." *Neuropsychoanalysis: An Interdisciplinary Journal for Psychoanalysis and the Neurosciences* 13 (2): 196–200.

Gallese, Vittorio, and David Freedberg. 2007. "Mirror and Canonical Neurons Are Crucial Elements in Aesthetic Response [Reply to Casati and Pignocchi, 2007]." *Trends in Cognitive Science* 11 (10): 411.

Garland, David. 2014. "What Is a 'History of the Present'? On Foucault's Genealogies and Their Critical Preconditions." *Punishment and Society* 16 (4): 365–384.

Gazzaniga, Michael S. 1967. "The Split Brain in Man." *Scientific American* 217 (2): 24–29.

———. 2005. *The Ethical Brain.* New York: Dana.

Genette, Gérard. 1999 [1997]. *The Aesthetic Relation.* Translated by G. M. Goshgarian. Ithaca, N.Y.: Cornell University Press.

Gennero, Valeria. 2008. "Gli inganni del cervello. Intervista a Richard Powers." *Acoma, Rivista Internazionale di Studi Nord-Americani* 37: 91–96.

———. 2011. "Larger Than Our Biologies. Identity and Consciousness in Contemporary Fiction." In Ortega and Vidal 2011, 307–323.

Geraci, Robert M. 2010. *Apocalyptic AI: Visions of Heaven in Robotics, Artificial Intelligence, and Virtual Reality.* New York: Oxford University Press.

Gere, Cathy, ed. 2004. *The Brain in a Vat.* Special issue of *Studies in History and Philosophy of Biology and the Biomedical Sciences* 35.

———. 2011. "'Nature's Experiment:' Epilepsy, Localization of Brain Function, and the Emergence of the Cerebral Subject." In Ortega and Vidal 2011, 235–247.

Gibbon, Sahra, and Carlos Novas. 2008a. "Introduction: Biosocialities, Genetics, and the Social Sciences." In Gibbon and Novas 2008b, 1–18.

————, eds. 2008b. *Biosocialities, Genetics, and the Social Sciences: Making Biologies and Identities*. London: Routledge.

Gibbons, Robert D., et al. 2012. "Benefits from Antidepressants: Synthesis of 6-Week Patient-Level Outcomes from Double-Blind Placebo-Controlled Randomized Trials of Fluoxetine and Venlafaxine." *Archives of General Psychiatry* 69 (6): 572–579.

Gibbons, Robert V., et al. 1998. "A Comparison of Physicians' and Patients' Attitudes Toward Pharmaceutical Industry Gifts." *Journal of General Internal Medicine* 13 (3): 151–154.

Gilbert, Scott F. 1995. "Resurrecting the Body: Has Postmodernism Had Any Effect on Biology?" *Science in Context* 8 (4): 563–577.

Gilmore, Jonathan. 2006. "Brain Trust." *Artforum* (July 1): 121–122.

Giordano, James, and Bert Gordijn, eds. 2010. *Scientific and Philosophical Perspectives in Neuroethics*. New York: Oxford University Press.

Glannon, Walter, ed. 2007. *Defining Right and Wrong in Brain Science: Essential Readings in Neuroethics*. Washington, D.C.: Dana.

Glannon, Walter. 2011. "Brain, Behavior, and Knowledge [Commentary on Pardo and Patterson 2011]." *Neuroethics* 4: 191–194.

Goggin, Gerard, and Christopher Newell. 2003. *Digital Disability: The Social Construction of Disability in New Media*. Lanham, Md.: Rowman & Littlefield.

Goggin, Gerard, and Tim Noonan. 2006. "Blogging Disability: The Interface Between New Cultural Movements and Internet Technology." In *Uses of Blogs*, edited by Axel Bruns and Joanne Jacobs, 161–172. New York: Peter Lang.

Goh, Joshua O., et al. 2010. "Culture Differences in Neural Processing of Faces and Houses in the Ventral Visual Cortex." *Social Cognitive and Affective Neuroscience* 5: 227–235.

Goldacre, Ben. 2013. *Bad Pharma: How Drug Companies Mislead Doctors and Harm Patients*. New York: Faber & Faber.

Goldberg, Elkhonon. 2001. *The Executive Brain: Frontal Lobes and the Civilized Mind*. Oxford: Oxford University Press.

Goldman, Corrie. 2012. "This Is Your Brain on Jane Austen, and Stanford Researchers Are Taking Notes." *Stanford Report* (7 September). http://news.stanford.edu/news/2012/september/austen-reading-fmri-090712.html.

Gombrich, Ernst. 1990. "The Edge of Delusion [Review of Freedberg 1989]." *New York Review of Books* (15 February). http://www.nybooks.com/articles/archives/1990/feb/15/the-edge-of-delusion/.

Gong, Qiyong, and Yong He. 2015. "Depression, Neuroimaging, and Connectomics: A Selective Overview." *Biological Psychiatry* 77: 223–235.

Good, Byron J. 2010. "The Complexities of Psychopharmaceutical Hegemonies in Indonesia." In *Pharmaceutical Self: The Global Shaping of Experience in an Age of Psychopharmacology*, edited by Janis H. Jenkins. Santa Fe, N.M.: School for Advanced Research Press, 117–144.

Goscilo, Helena. 1981. "Lermontov's Debt to Lavater and Gall." *Slavonic and East European Review* 59 (4): 500–515.

Gotlib, Ian, and Paul J. Hamilton. 2008. "Neuroimaging and Depression: Current Status and Unresolved Issues." *Current Directions in Psychological Science* 17: 159–163.

Gotlib, Ian, and Constance Hammen, eds. 2014. *Handbook of Depression*. 3rd ed. New York: Guilford.

Gould, Stephen J., and Richard C. Lewontin. 1979. "The Spandrels of San Marco and the Panglossian Paradigm: A Critique of the Adaptationist Programme." *Proceedings of the Royal Society of London, Series B* 205 (1161): 581–598.

Goupil, Georgette, et al. 2014. "L'utilisation d'internet par les parents d'enfants ayant un trouble du spectre de l'autisme / Internet Use by Parents of Children with Autism Spectrum Disorders." *Canadian Journal of Learning and Technology / La Revue Canadienne de l'Apprentissage et de la Technologie* 40: 1–18.

Graby, Steven. 2015. "Neurodiversity: Bridging the Gap Between the Disabled People's Movement and the Mental Health System Survivors' Movement?" In *Madness, Distress, and the Politics of Disablement*, edited by Helen Spandler, Jill Anderson, and Bob Sapey. Bristol: Policy, 231–243.

Graham, Daniel J., and David J. Field. 2007. "Statistical Regularities of Art Images and Natural Scenes: Spectra, Sparseness, and Nonlinearities." *Spatial Vision* 21 (1/2): 149–164.

Graham, Julia, et al. 2013. "Meta-analytic Evidence for Neuroimaging Models of Depression: State or Trait?" *Journal of Affective Disorders* 151: 423–431.

Grande, David. 2010. "Limiting the Influence of Pharmaceutical Industry Gifts on Physicians: Self-Regulation or Government Intervention?" *Journal of General and Internal Medicine* 25 (1): 79–83.

Grande, David, Judy Shea, and Katrina Armstrong. 2012. "Pharmaceutical Industry Gifts to Physicians: Patient Beliefs and Trust in Physicians and the Health Care System." *Journal of General Internal Medicine* 27: 274–279.

Grandin, Temple. 1995. *Thinking in Pictures and Other Reports from My Life with Autism*. New York: Vintage.

Granello, Darcy Haag, and Todd A. Gibbs. 2016. "The Power of Language and Labels: 'The Mentally Ill' Versus 'People with Mental Illnesses.'" *Journal of Counseling and Development* 94: 31–40.

Gray, Kurt, et al. 2011. "More Dead Than Dead: Perceptions of Persons in the Persistent Vegetative State." *Cognition* 121 (2): 275–280.

Green, Michael J., et al. 2012. "Do Gifts from the Pharmaceutical Industry Affect Trust in Physicians?" *Family Medicine* 44: 325–331.

Greenberg, Gary. 2010. *Manufacturing Depression: The Secret History of a Modern Disease*. New York: Simon & Shuster.

———. 2013. *The Book of Woe: The DSM and the Unmaking of Psychiatry*. New York: Blue Rider.

Greenberg, Harvey, and Krin Gabbard. 1999. "Reel Recollection: Notes on the Cinematic Depiction of Memory." *PsyArt: A Hyperlink Journal for the Psychological Study of the Arts*. http://www.psyartjournal.com/article/show /greenberg-reel_recollection_notes_on_the_cinematic.

Greicius, Michael D., et al. 2007. "Resting-State Functional Connectivity in Major Depression: Abnormally Increased Contributions from Subgenual Cingulate Cortex and Thalamus." *Biological Psychiatry* 62 (5): 429–437.

Grimm, Simone, et al. 2009. "Increased Self-Focus in Major Depressive Disorder Is Related to Neural Abnormalities in Subcortical-Cortical Midline Structures." *Human Brain Mapping* 30: 2617–2627.

Gross, Sky. 2011. "A Stone in a Spaghetti Bowl: The Biological and Metaphorical Brain in Neuro-Oncology." In Pickersgill and Van Keulen 2011, 99–119.

Guidotti, Francesca. 2003. *Cyborg e dintorni. Le formule della fantascienza*. Bergamo: Bergamo University Press.

Gupta, Akhil, and James Ferguson. 1992. "Beyond 'Culture': Space, Identity, and the Politics of Difference." *Cultural Anthropology* 7: 6–23.

Gupta, Mona. 2014. *Is Evidence-Based Psychiatry Ethical?* Oxford: Oxford University Press.

Gusfield, Joseph R. 1992. "Nature's Body and the Metaphors of Food." In *Cultivating Differences: Symbolic Boundaries and the Making of Inequality*, edited by Michèle Lamont and Marcel Fournier. Chicago: University of Chicago Press, 75–103.

Gutchess, Angela H., et al. 2010. "Neural Differences in the Processing of Semantic Relationships Across Cultures." *Social Cognitive and Affective Neuroscience* 5: 254–263.

Gutchess, Angela H., and Joshua O. Goh. 2013. "Refining Concepts and Uncovering Biological Mechanisms for Cultural Neuroscience." *Psychological Inquiry* 24: 31–36.

Hacking, Ian. 1995. "The Looping Effects of Human Kinds." In *Causal Cognition: A Multidisciplinary Approach*, edited by Dan Sperber, David Premack, and Ann J. Premack, 351–383. Oxford: Clarendon.

———. 2002. "Making Up People." In *Historical Ontology*, 99–114. Cambridge, Mass.: Harvard University Press.

———. 2006. "What Is Tom Saying to Maureen?" *London Review of Books* 28 (9). htttp://www.lrb.co.uk/v28/no9/ian-hacking/what-is-tom-saying-to -maureen.

———. 2009. "Autistic Autobiography." *Philosophical Transactions of the Royal Society B* 364: 1467–1473.

Hagner, Michael. 2001. "Cultivating the Cortex in German Neuroanatomy." *Science in Context* 14: 541–564.

———. 2004. *Geniale Gehirne: Zur Geschichte der Elitenhirnforschung*. Berlin: Wallstein.

———. 2009 [2006]. "The Mind at Work: The Visual Representation of Cerebral Processes." Translated by U. Froese. In *Body Within: Art, Medicine, and Visualization*, edited by Renée van de Vall and Robert Zwijnenberg, 67–90. Leiden: Brill.

Hagner, Michael, and Cornelius Borck. 2001. "Mindful Practices: On the Neurosciences in the Twentieth Century." *Science in Context* 14: 507–510.

Hahn, Torsten. 2005. "Risk Communication and Paranoid Hermeneutics, Towards a Distinction Between 'Medical Thrillers' and 'Mind-Control Thrillers' in Narrations on Biocontrol." *New Literary History* 36 (2): 187–204.

Hallett, Ronald E., and Kristen Barber. 2014. "Ethnographic Research in a Cyber Era." *Journal of Contemporary Ethnography* 43 (3): 306–330.

Hamilton, Paul J., et al. 2012. "Functional Neuroimaging of Major Depressive Disorder: A Meta-analysis and New Integration of Baseline Activation and Neural Response Data." *American Journal of Psychiatry* 169: 693–703.

Han, Shihui. 2013. "Culture and Brain: A New Journal." *Culture and Brain* 1 (1): 1–2.

Han, Shihui, and Yina Ma. 2014. "Cultural Differences in Human Brain Activity: A Quantitative Meta-analysis." *NeuroImage* 99: 293–300.

Han, Shihui, and Georg Northoff. 2008. "Culture-Sensitive Neural Substrates of Human Cognition: A Transcultural Neuroimaging Approach." *Nature Reviews Neuroscience* 9: 646–654.

Han, Shihui, and Ernst Pöppel, eds. 2011. *Culture and Neural Frames of Cognition and Communication*. Berlin: Springer.

Han, Shihui, et al. 2013. "A Cultural Neuroscience Approach to the Biosocial Nature of the Human Brain." *Annual Review of Psychology* 64: 335–359.

Han, Ying, et al. 2009. "Gray Matter Density and White Matter Integrity in Pianists' Brains: A Combined Structural and Diffusion Tensor MRI Study." *Neuroscience Letters* 459: 3–6.

Hanakawa, Takashi, et al. 2003. "Neural Correlates Underlying Mental Calculation in Abacus Experts: Functional Magnetic Resonance Imaging Study." *Neuroimage* 19: 296–307.

Hanegraaf, Wouter. 1998. *New Age Religion and Western Culture: Esotericism in the Mirror of Secular Thought*. Albany: SUNY Press.

Hanlon, Charlotte, Abebaw Fekadu, and Vikram Patel. 2014. "Interventions for Mental Disorders." In Patel, Minas, Cohen, and Prince 2014, 252–276.

Hansen, Helena, and Mary Skinner. 2012. "From White Bullets to Black Markets and Greened Medicine: The Neuroeconomics and Neuroracial Politics of Opioid Pharmaceuticals." *Annals of Anthropological Practice* 36 (1): 167–182.

Hanson, Allan F. 1992. *Testing Testing: Social Consequences of the Examined Life.* Berkeley: University of California Press.

Hardcastle, Valerie G., and Matthew C. Stewart. 2002. "What Do Brain Data Really Show?" *Philosophy of Science* 69: S72–S82.

Harmon, Amy. 2004a. "Adults and Autism; An Answer, but Not a Cure, for a Social Disorder." *New York Times* (April 29).

———. 2004b. "Neurodiversity Forever: The Disability Movement Turns to Brains." *New York Times* (May 9).

———. 2004c. "How About Not 'Curing' Us, Some Autistics Are Pleading." *New York Times* (December 20).

Harrington, Anne. 1987. *Mind, Medicine, and the Double Brain: A Study in Nineteenth-Century Thought.* Princeton, N.J.: Princeton University Press.

———. 1991. "Beyond Phrenology: Localization Theory in the Modern Era." In *The Enchanted Loom: Chapters in the History of Neuroscience*, edited by Pietro Corsi. New York: Oxford University Press, 207–239.

———. 2008. *The Cure Within: A History of Mind-Body Medicine.* New York: Norton.

Harrington, Anne, and Godehard Oepen. 1989. "Whole Brain Politics and Brain Laterality Research." *European Archives of Psychiatry and Neurological Science* 239 (3): 141–143.

Harrington, Jean, and Christine Hauskeller. 2014. "Translational Research: An Imperative Shaping the Spaces in Biomedicine." *TECNOSCIENZA: Italian Journal of Science and Technology Studies* 5 (1): 191–201.

Harris, Charles B. 2008. "The Story of the Self, *The Echo Maker*, and Neurological Realism." In Burn and Dempsey 2008, 230–259.

Harris, Lauren J. 1980. "Left-Handedness: Early Theories, Facts, and Fancies." In *Neuropsychology of Left-Handedness*, edited by Jeannine Herron. New York: Academic Press, 3–78.

———. 1985. "Teaching the Right Brain: Historical Perspective on a Contemporary Educational Fad." In *Hemispheric Function and Collaboration in the Child*, edited by Catherine T. Best. New York: Academic Press, 231–274.

Harris, Paul, and Alison Flood. 2010. "Literary Critics Scan the Brain to Find Out Why We Love to Read." *The Observer* (11 April). http://www.guardian.co.uk/science/2010/apr/11/brain-scans-probe-books-imagination.

Hart, F. Elizabeth. 2001. "The Epistemology of Cognitive Literary Studies." *Philosophy and Literature* 25 (2): 314–334.

Hart, Sarah, et al. 2013. "Altered Fronto-limbic Activity in Children and Adolescents with Familial High Risk for Schizophrenia." *Psychiatry Research: Neuroimaging* 212: 19–27.

Harvey, Ruth. 1975. *The Inward Wits: Psychological Theory in the Middle Ages and the Renaissance.* London: Warburg Institute.

Hasler, Felix. 2009. "Stoppt den Neurowahn!" *Das Magazin* (23 October). http://web.archive.org/web/20101202114243/http://dasmagazin.ch/index .php/stoppt-den-neurowahn/.

———. 2013. *Neuromythologie. Eine Streitschrift gegen die Deutungsmacht der Hirnforschung.* Bielefeld: transcript.

Hasler, Gregor, 2010. "Pathophysiology of Depression: Do We Have Any Solid Evidence of Interest to Clinicians?" *World Psychiatry* 9: 155–161.

Healy, David. 1997. *The Antidepressant Era.* Cambridge, Mass.: Harvard University Press.

———. 2002. *The Creation of Psychopharmacology.* Cambridge, Mass.: Harvard University Press.

———. 2004. *Let Them Eat Prozac: The Unhealthy Relationship Between the Pharmaceutical Industry and Depression.* New York: New York University Press.

———. 2008. *Mania: A Short History of Bipolar Disorder.* Baltimore, Md.: Johns Hopkins University Press.

———. 2013. *Pharmageddon.* Berkeley: University of California Press.

Healy, Melissa. 2013. "Einstein's Brain a Wonder of Connectedness." *Los Angeles Times* (10 October). http://articles.latimes.com/2013/oct/10/science /la-sci-einstein-brain-connectedness-20131010.

Hedden, Trey, et al. 2008. "Cultural Influences on Neural Substrates of Attentional Control." *Psychological Science* 19: 12–17.

Heim, Christine, and Elisabeth B. Binder. 2012. "Current Research Trends in Early Life Stress and Depression: Review of Human Studies on Sensitive Periods, Gene-Environment Interactions, and Epigenetics." *Experimental Neurology* 233 (1): 102–111.

Heinz, Andreas, et al. 2014. "The Uncanny Return of the Race Concept." *Frontiers in Human Neuroscience* 8, art. 836. doi:10.3389/ fnhum.2014.00836.

Hendrickx, Sarah. 2010. *The Adolescent and Adult Neuro-Diversity Handbook: Asperger Syndrome, ADHD, Dyslexia, Dyspraxia, and Related Conditions.* London: Jessica Kingsley.

Hendrix, Scott E., and Christopher J. May. 2012. "Neuroscience and the Quest for God." In Littlefield and Johnson 2012, 105–119.

Herman, Luc, and Bart Vervaeck. 2009. "Capturing Capgras, *The Echo Maker* by Richard Powers." *Style* 43 (3): 407–428.

Hickok, Gregory. 2009. "Eight Problems for the Mirror Neuron Theory of Action Understanding in Monkeys and Humans." *Journal of Cognitive Neuroscience* 21 (7): 1229–1243.

Hodges, Brian. 1995. "Interactions with the Pharmaceutical Industry: Experiences and Attitudes of Psychiatry Residents, Interns, and Clerks." *Canadian Medical Association Journal* 153 (5): 553–559.

Hofmann, Bjørn. 2015. "Exit Exceptionalism: Mental Disease Is Like Any Other Medical Disease." *Journal of Psychiatry and Neuroscience* 45 (6): E36.

Hofstadter, Douglas R., and Daniel C. Dennett, eds. 1981. *The Mind's I: Fantasies and Reflections on Self and Soul.* Toronto: Bantam.

Holland, Norman. 1988. *The Brain of Robert Frost.* New York: Routledge.

Holland, Stephen, Celia Kitzinger, and Jenny Kitzinger. 2014. "Death, Treatment Decisions and the Permanent Vegetative State: Evidence from Families and Experts." *Medicine, Health Care, and Philosophy* 17 (3): 413–423.

Holtzheimer, Paul E., and Helen S. Mayberg. 2011. "Stuck in a Rut: Rethinking Depression and Its Treatment." *Trends in Neurosciences* 34 (1): 1–9.

Horgan, John. 2014. "Much-Hyped Brain-Implant Treatment for Depression Suffers Setback." *Scientific American* (11 March). http://blogs .scientificamerican.com/cross-check/2014/03/11/much-hyped-brain -implant-treatment-for-depression-suffers-setback/.

Horwitz, Allan, and Jerome Wakefield. 2007. *The Loss of Sadness: How Psychiatry Transformed Normal Sorrow Into Depressive Disorder.* New York: Oxford University Press.

Hoyer, Armin. 2010. *Neurotechnologie, Philosophie und Hirnforschung: Zur Entstehung und Institutionalisierung der Neuroethik.* MA thesis, Johann Wolfgang Goethe-Universität, Frankfurt am Main.

Hsu, Chung-Ting, et al. 2015. "The Magical Activation of Left Amygdala when Reading Harry Potter: An fMRI Study on How Descriptions of Supra-Natural Events Entertain and Enchant." *PLoS ONE.* doi:10.1371/ journal.pone.0118179.

Huarte de San Juan, Juan. 1698. *The tryal of wits. Discovering the great difference of wits among men, and what sort of learning suits best with each genius.* Translated by "Mr. Bellamy." London: Printed for Richard Sare.

Hubbard, Ruth, and Elijah Wald. 1993. *Exploding the Gene Myth.* Boston: Beacon.

Hughes, Jane. 2010. "New Brain Scan to Diagnose Autism." *BBC News Health* (August 10). http://www.bbc.co.uk/news/health-10929032.

Hultman, Rainbo, Stephen D. Mague, Qiang Li, et al. 2016. "Dysregulation of Prefrontal Cortex-Mediated Slow-Evolving Limbic Dynamics Drives Stress-Induced Emotional Pathology." *Neuron* 91: 439–452.

Hunter, Madeleine. 1976. "Right-Brained Kids in Left-Brained Schools." *Today's Education* 65 (4): 45–48.

Hyde, Luke W., et al. 2015. "Cultural Neuroscience: New Directions as the Field Matures. What Do Cultural Neuroscience Findings Mean?" *Culture and Brain* 3:75–92.

Hyman, John. 2006. "Art and Neuroscience." http://web.archive.org/web /20070309014832/http://www.interdisciplines.org/artcognition/papers/15.

Hyman, Steven E. 2007. "Can Neuroscience Be Integrated Into the DSMV?" *Nature Reviews Neuroscience* 8: 725–732.

———. 2008. "A Glimmer of Light for Neuropsychiatric Disorders." *Nature* 455: 890–893.

———. 2009. "How Adversity Gets Under the Skin." *Nature Neuroscience* 12 (3): 241–243.

Illes, Judy, ed. 2006. *Neuroethics: Defining the Issues in Theory, Practice, and Policy*. New York: Oxford University Press.

Illes, Judy, and Eric Racine. 2005. "Imaging or Imagining? A Neuroethics Challenge Informed by Genetics." *American Journal of Bioethics* 5 (2): 5–18.

Illes, Judy, Eric Racine, and Matthew P. Kirschen. 2006. "A Picture Is Worth a Thousand Words, but Which One Thousand?" In Illes 2006, 149–168.

Illes, Judy, et al. 2008. "In the Mind's Eye: Provider and Patient Attitudes on Functional Brain Imaging." *Journal of Psychiatric Research* 43 (2): 107–114. doi:10.1016/j.jpsychires.2008.02.008.

Illes, Judy, and Barbara J. Sahakian, eds. 2011. *Oxford Handbook of Neuroethics*. New York: Oxford.

Ingram, Rick E., ed. 2009. *International Encyclopedia of Depression*. New York: Springer.

Insel, Thomas R. 2012. [Interview with]. *Psychiatric Annals* 42 (9): 350–351.

———. 2013. "Transforming Diagnosis." http://www.nimh.nih.gov/about /director/2013/transforming-diagnosis.shtml.

Insel, Thomas R., and Remi Quirion. 2005. "Psychiatry as a Clinical Neuro-science Discipline." *JAMA* 294 (17): 2221–2224.

Insel, Thomas, et al. 2010. "Research Domain Criteria (RDoC): Toward a New Classification Framework for Research on Mental Disorders." *American Journal of Psychiatry* 167: 748–751.

Ioannidis, John P. A. 2015. "Translational Research May Be Most Successful When It Fails." *Hastings Center Report* 45 (2): 39–40.

Ione, Amy. 2003. "Examining Semir Zeki's 'Neural Concept Formation and Art: Dante, Michelangelo, Wagner.'" *Journal of Consciousness Studies* 10 (2): 58–66.

Ishizu, Tomohiro, and Semir Zeki. 2011. "Toward a Brain-Based Theory of Beauty." *PloS ONE* 6 (7): 1–10.

Jackson, John. 1905. *Ambidexterity or Two-Handedness and Two-Brainedness: An Argument for Natural Development and Rational Education*. London: Kegan Paul, Trench, Trübner & Co.

Jackson, Stanley W. 1986. *Melancholia and Depression: From Hippocratic Times to Modern Times*. New Haven, Conn.: Yale University Press.

Jacob, K. Stanly, and Vikram Patel. 2014. "Classification of Mental Disorders: A Global Mental Health Perspective." *The Lancet* 383: 1433–1435.

Jacob, Pierre. 2008. "What Do Mirror Neurons Contribute to Human Social Cognition?" *Mind and Language* 23 (2): 190–223.

Jacobsen, Thomas, Ricarda I. Schubotz, Lea Höfel, and D. Yves von Cramon. 2006. "Brain Correlates of Aesthetic Judgment of Beauty." *NeuroImage* 29: 276–285.

Jaeger, Paul T. 2012. *Disability and the Internet: Confronting a Digital Divide*. Boulder, Colo.: Lynne Rienner.

Jaroff, Leon. 1989. "The Gene Hunt." *Time* (20 March). http://www.time.com /time/magazine/article/0,9171,957263,00.html.

Joel, Daphna, et al. 2015. "Sex Beyond the Genitalia: The Human Brain Mosaic." *PNAS*. http://www.pnas.org/cgi/doi/10.1073/pnas.1509654112.

Johnsen, Tom J., and Oddgeir Friborg. 2015. "The Effects of Cognitive Behavioral Therapy as an Anti-Depressive Treatment Is Falling: A Meta-analysis." *Psychological Bulletin* 141 (4): 747–768.

Johnson, Davi. 2008. "'How Do You Know Unless You Look?': Brain Imaging, Biopower, and Practical Neuroscience." *Journal of Medical Humanities* 29: 147–161.

Johnson, Gary. 2008. "Consciousness as Content: Neuronarratives and the Redemption of Fiction." *Mosaic* 41 (1): 169–184.

Jones, Gareth D. 1989. "Brain Birth and Personal Identity." *Journal of Medical Ethics* 15: 173–178.

———. 1998. "The Problematic Symmetry Between Brain Birth and Brain Death." *Journal of Medical Ethics* 24: 237–242.

Jones, Nev, and Timothy Kelly. 2015. "Inconvenient Complications: On the Heterogeneities of Madness and Their Relationship to Disability." In *Madness, Distress, and the Politics of Disablement*, edited by Helen Spandler, Jill Anderson and Bob Sapey. Bristol: Policy, 43–55.

Jones, Rachel. 2012. "What Makes a Human Brain?" *Nature Reviews Neuroscience* 13 (10): 655.

Jones, Robert S. P., and Tor O. Meldal. 2001. "Social Relationships and Asperger's Syndrome. A Qualitative Analysis of First-Hand Accounts." *Journal of Intellectual Disabilities* 5 (1): 35–41.

Jones, Robert S. P., Andrew Zahl, and Haci C. Huws. 2001. "First-Hand Accounts of Emotional Experiences in Autism: A Qualitative Analysis." *Disability and Society* 16 (3): 393–401.

Jones, Simon R., and Charles Fernyhough. 2007. "A New Look at the Neural Diathesis-Stress Model of Schizophrenia: The Primacy of Social-

Evaluative and Uncontrollable Situations." *Schizophrenia Bulletin* 33 (5): 1171–1177.

Joyce, Kelly A. 2008. *Magnetic Appeal: MRI and the Myth of Transparency.* Ithaca, N.Y.: Cornell University Press.

Jurecic, Ann. 2007. "Neurodiversity." *College English* 69 (5): 421–442.

Kabat-Zinn, Jon, and Richard J. Davidson. 2015. "A Confluence of Streams and a Flowering of Possibilities." In *The Mind's Own Physician: A Scientific Dialogue with the Dalai Lama on the Healing Power of Meditation,* edited by Jon Kabat-Zinn and Richard J. Davidson, 1–19. Oakland, Calif.: Mind and Life Institute / New Harbinger Publications.

Kachka, Boris. 2012. "Proust Wasn't a Neuroscientist. Neither Was Jonah Lehrer." *New York Magazine* (28 October). http://nymag.com/news/features /jonah-lehrer-2012-11/.

Kaitaro, Timo. 2004. "Brain-Mind Identities in Dualism and Materialism: A Historical Perspective." *Studies in History and Philosophy of Biology and Biomedical Sciences* 35: 627–645.

Kapur, Shitij, Anthony G. Phillips, and Thomas Insel. 2012. "Why Has It Taken So Long for Biological Psychiatry to Develop Clinical Tests and What to Do About It?" *Molecular Psychiatry* 17 (12): 1174–1179.

Kaufman, Sharon R., and Lynn M. Morgan. 2005. "The Anthropology of the Beginnings and Ends of Life." *Annual Review of Anthropology* 34: 317–341.

Kawabata, Hideaki, and Semir Zeki. 2004. "Neural Correlates of Beauty." *Journal of Neurophysiology* 91: 1699–1705.

Keedwell, Paul. 2009. "Brain Circuitry." In Ingram 2009, 86–98.

Keim-Malpass, Jessica, Richard H. Steeves, and Christine Kennedy. 2014. "Internet Ethnography: A Review of Methodological Considerations for Studying Online Illness Blogs." *International Journal of Nursing Studies* 51 (12): 1686–1692.

Kelley, William M., C. Neil Macrae, Carrie L. Wyland, Sali Caglar, Souheil Inati, and Todd F. Heatherton. 2002. "Finding the Self? An Event-Related fMRI Study." *Journal of Cognitive Neurosciences* 14: 785–794.

Kellogg, John Harvey. 1887. *First Book in Physiology and Hygiene.* New York: Harper & Brothers.

Kemp, Simon. 1990. *Medieval Psychology.* New York: Greenwood.

Kempton, Matthew J., John R. Geddes, Ulrich Ettinger, Simon C. Williams, and Paul M. Grasby. 2008. "Meta-analysis, Database, and Meta-regression of 98 Structural Imaging Studies in Bipolar Disorder." *Archives of General Psychiatry* 65 (9): 1017–1032.

Kempton, Matthew J., et al. 2011. "Structural Neuroimaging Studies in Major Depressive Disorder: Meta-analysis and Comparison with Bipolar Disorder." *Archives of General Psychiatry* 68 (7): 675–690.

Kenway, Ian M. 2009. "Blessing or Curse? Autism and the Rise of the Internet." *Journal of Religion, Disability, and Health* 13 (2): 94–103.

Keysers, Christian. 2011. *The Empathic Brain: How Mirror Neurons Help You Understand Others.* Amsterdam: Social Brain.

Khamsi, Roxanne. 2013. "Brain Scans Could Become EKGs for Mental Disorders." *Time* (28 June). http://healthland.time.com/2013/06/28/brain -scans-could-become-ekgs-for-mental-disorders/.

Kieseppä, Tuula, et al. 2009. "Major Depressive Disorder and White Matter Abnormalities: A Diffusion Tensor Imaging Study with Tract-Based Spatial Statistics." *Journal of Affective Disorders* 120 (1): 240–244.

Kilner, James M., and Roger N. Lemon. 2013. "What We Know Currently About Mirror Neurons." *Current Biology* 23: R1057–R1062.

Kim, Heejung S., and Joni Y. Sasaki. 2014. "Cultural Neuroscience: Biology of the Mind in Cultural Contexts." *Annual Review of Psychology* 65: 487–514.

Kirby, David. 2003. "Scientists on the Set: Science Consultants and Communication of Science in Visual Fiction." *Public Understanding of Science* 12: 261–278.

Kirmayer, Laurence J. 2002. "Psychopharmacology in a Globalizing World: The Use of Antidepressants in Japan." *Transcultural Psychiatry* 39: 295–322.

Kirmayer, Laurence J., and Daina Crafa. 2014. "What Kind of Science for Psychiatry." *Frontiers in Human Neuroscience* 8, art. 435:1–12.

Kirmayer, Laurence J., and Eugene Raikhel. 2009. "Editorial: From Amrita to Substance D: Psychopharmacology, Political Economy, and Technologies of the Self." *Transcultural Psychiatry* 46: 5–15.

Kirsch, Irving, et al. 2008. "Initial Severity and Antidepressant Benefits: A Meta-analysis of Data Submitted to the Food and Drug Administration." *PLoS Medicine* 5 (2): e45, 0260–0268.

Kirsch, Irving. 2009. *The Emperor's New Drugs: Exploding the Antidepressant Myth.* London: Bodley Head.

Kitanaka, Junko. 2011. *Depression in Japan: Psychiatric Cures for a Society in Distress.* Princeton, N.J.: Princeton University Press.

Kitayama, Shinobu, Sean Duffy, Tadashi Kawamura, and Jeff T. Larsen. 2003. "Perceiving an Object and Its Context in Different Cultures: A Cultural Look at New Look." *Psychological Science* 14: 201–206.

Kitayama, Shinobu, and Jiyoung Park. 2010. "Cultural Neuroscience of the Self: Understanding the Social Grounding of the Brain." *Social Cognitive and Affective Neuroscience* 5 (2/3): 111–129.

Kitayama, Shinobu, and Sarah Huff. 2015. "Cultural Neuroscience: Connecting Culture, Brain, and Genes." In *Emerging Trends in the Social and Behavioral Sciences: An Interdisciplinary, Searchable, and Linkable Resource,* edited by Robert A. Scott and Stephen M. Kosslyn. Wiley. doi:10.1002/9781118900772, 1–16.

Kiverstein, Julian, and Mark Miller. 2015. "The Embodied Brain: Towards a Radical Embodied Cognitive Neuroscience." *Frontiers in Human Neuroscience* 9, art. 237. doi:10.3389/fnhum.2015.00237.

Klein, Colin. 2010. "Images Are Not the Evidence in Neuroimaging." *British Journal for the Philosophy of Science* 61: 265–278.

Kleinman, Arthur. 2009. "Global Mental Health: A Failure of Humanity." *The Lancet* 374: 603–604.

———. 2012. "Medical Anthropology and Mental Health. Five Questions for the Next Fifty Years." In *Medical Anthropology at the Intersections: Histories, Activisms, and Futures,* edited by Marcia C. Inhorn and Emily A. Wentzell. Durham, N.C.: Duke University Press, 116–128.

Klibansky, Raymond, Erwin Panofsky, and Fritz Saxl. 1964. *Saturn and Melancholy: Studies in the History of Natural Philosophy, Religion, and Art.* New York: Basic Books.

Koenigs, Michael, and Daniel Tranel. 2008. "Prefrontal Cortex Damage Abolishes Brand-Cued Changes in Cola Preference." *SCAN* 3:1–6.

Koolschijn, Cédric, et al. 2009. "Brain Volume Abnormalities in Major Depressive Disorder: A Meta-analysis of Magnetic Resonance Imaging Studies." *Human Brain Mapping* 30 (11): 3719–3735.

Kosslyn, Stephen M. 1999. "If Neuroimaging Is the Answer, What Is the Question?" *Philosophical Transactions of the Royal Society of London* 354: 1283–1294.

Kozinets, Robert T. 2010. *Netnography: Doing Ethnographic Research Online.* Los Angeles: Sage.

Krementsov, Nikolai. 2009. "Off with Your Heads: Isolated Organs in Early Soviet Science and Fiction." *Studies in History and Philosophy of Biological and Biomedical Sciences* 40: 87–100.

———. 2014. *Revolutionary Experiments: The Quest for Immortality in Bolshevik Science and Fiction.* New York: Oxford University Press.

Kroeber, Alfred L., and Clyde Kluckhohn. 1952. *Culture: A Critical Review of Concepts and Definitions.* Harvard University Peabody Museum of American Archeology and Ethnology Papers 47. Cambridge, Mass. [Variously reprinted.]

Krow-Lucal, Martha G. 1983. "Balzac, Galdós, and Phrenology." *Anales Galdosianos* 18: 7–14.

Kupferschimdt, Kai. 2013. "Concentrating on Kindness." *Science* 341 (20 September): 1336–1339.

Kwint, Marius, and Richard Wingate. 2012. *Brains: The Mind as Matter.* London: Profile.

Lacasse, Jeffrey R., and Jonathan Leo. 2005. "Serotonin and Depression: A Disconnect Between the Advertisements and the Scientific Literature." *PLoS Medicine* 2 (12): 1211–1216.

Lacey, Simon, et al. 2011. "Art for Reward's Sake: Visual Art Recruits the Ventral Striatum." *Neuroimage* 55 (1): 420–433.

Lage, Andrey. 2006. "Autistas usam remédios para controlar aspectos da doença." *Folha OnLine* (July 27). http://www1.folha.uol.com.br/folha /equilibrio/noticias/ult263u4160.shtml.

Lakoff, Andrew. 2005. *Pharmaceutical Reason: Medication and Psychiatric Knowledge in Argentina*. Cambridge: Cambridge University Press.

———. 2006. "High Contact: Gifts and Surveillance in Argentina." In Petryna, Lakoff, and Kleinman 2006, 111–135.

Lakoff, George. 2008. *The Political Mind: A Cognitive Scientist's Guide to Your Brain and Its Politics*. New York: Penguin.

Lamberton, Robert. 1986. *Homer the Theologian: Neoplatonist Allegorical Reading and the Growth of the Epic Tradition*. Berkeley: University of California Press.

Landsberg, Alison. 2004. *Prosthetic Memory: The Transformation of American Remembrance in the Age of Mass Culture*. New York: Columbia University Press.

Landi, Anne. 2009. "Brain Wave." *ARTnews* (June): 88–93.

Lane, Christopher. 2007. *Shyness: How Normal Behavior Became a Sickness*. New Haven, Conn.: Yale University Press.

Lardreau, Guy. 1988. *Fictions philosophiques et science fiction*. Paris: Actes Sud.

Larkin, Warren, and John Read. 2008. "Childhood Trauma and Psychosis: Evidence, Pathways, and Implications." *Journal of Postgraduate Medicine* 54 (4): 287–293.

Latour, Bruno. 2004. "How to Talk About the Body? The Normative Dimension of Science Studies." *Body and Society* 10 (2/3): 205–229.

Lauer, Gerhard. 2009. "Going Empirical: Why We Need Cognitive Literary Studies." *Journal of Literary Theory* 3: 145–154.

Laureys, Steven, et al. 2010. "Unresponsive Wakefulness Syndrome: A New Name for the Vegetative State or Apallic Syndrome." *BMC Medicine* 8: 68. http://www.biomedcentral.com/1741-7015/8/68.

Lauring, Jon O., ed. 2015. *An Introduction to Neuroaesthetics: The Neuroscientific Approach to Aesthetic Experience, Artistic Creativity, and Arts Appreciation*. Copenahgen: Museum Tusculanum Press.

Lazar, Sara W., Catherine E. Kerr, Rachel H. Wasserman, et al. 2005. "Meditation Experience Is Associated with Increased Cortical Thickness." *Neuroreport* 16 (17): 1893–1897.

Leary, Timothy. 1980. *The Politics of Ecstasy*. Berkeley: Ronin, 1998.

Leary, Timothy, with Robert Anton Wilson and George A. Koopman. 1977. *Neuropolitics: The Sociobiology of Human Metamorphosis*. Los Angeles: Starseed/Peace.

Leder, Helmut. 2001. "Determinants of Preference. When Do We Like What We Know?" *Empirical Studies of the Arts* 19 (2): 201–211.

Lederer, E. Susan, ed. 2002. *Frankenstein: Penetrating the Secrets of Nature.* New Brunswick, N.J.: Rutgers University Press.

Lega, Bradley C. 2006. "An Essay Concerning Human Understanding: How the Cerebri Anatome of Thomas Willis Influenced John Locke." *Neurosurgery* 58: 567–576.

Legrenzi, Paolo, and Carlo Umiltà. 2009. *Neuro-mania: Il cervello non spiega chi siamo.* Bologna: Il Mulino.

Lehrer, Jonah. 2007. *Proust Was a Neuroscientist.* New York: Houghton Mifflin Harcourt.

Lei, Miaomei, Hiroyuki Akama, and Brian Murphy. 2014. "Neural Basis of Language Switching in the Brain: fMRI Evidence from Korean-Chinese Early Bilinguals." *Brain and Language* 138: 12–18.

Leibing, Annette. 2009. "Tense Prescriptions? Alzheimer Medications and the Anthropology of Uncertainty." *Transcultural Psychiatry* 46: 180–206.

Leichsenring, Falk, and Susanne Klein. 2014. "Evidence for Psychodynamic Psychotherapy in Specific Mental Disorders: A Systematic Review." *Psychoanalytic Psychotherapy* 28 (1): 4–32.

Leichsenring, Falk, and Sven Rabung. 2008. "Effectiveness of Long-Term Psychodynamic Psychotherapy." *Journal of the American Medical Association* 300: 1151–1565.

———. 2011. "Long-Term Psychodynamic Psychotherapy in Complex Mental Disorders: Update of a Meta-analysis." *British Journal of Psychiatry* 199 (1): 15–22.

Lende, Daniel H., and Greg Downey, eds. 2012a. *The Encultured Brain: An Introduction to Neuroanthropology.* Cambridge, Mass.: MIT Press.

———. 2012b. "The Encultured Brain—Toward the Future." In Lende and Downey 2012a, 391–420.

Leo, Jonathan, and Jeffrey R. Lacasse. 2008. "The Media and the Chemical Imbalance Theory of Depression." *Society* 45: 35–45.

Lepore, Frederick E. 2001. "Dissecting Genius. Einstein's Brain and the Search for the Neural Basis of Intellect." *Cerebrum* 3 (1). http://www.dana.org/Cerebrum/Default.aspx?id=39337.

LeVay, Simon. 1997. *Albrick's Gold.* London: Headline Book.

Levine, Timothy R., Mary J. Bresnahan, Hee S. Park, et al. 2003. "Self-Construal Scales Lack Validity." *Human Communication Research* 29 (2): 210–252.

Levy, Neil. 2007. *Neuroethics: Challenges for the Twenty-First Century.* New York: Cambridge University Press.

Lichtenstein, Jacqueline, Carole Maigné, and Pierre Arnauld, eds. 2013. *Vers la science de l'art. L'esthétique scientifique en France 1857–1937.* Paris: Presses de l'Université Paris Sorbonne.

Littlefield, Melissa M., 2011. *The Lying Brain: Lie Detection in Science and Science Fiction.* Michigan: University of Michigan Press.

Littlefield, Melissa M., Des Fitzgerald, Kasper Knudsen, James Tonks, and Martin J. Dietz. 2014. "Contextualizing Neuro-Collaborations: Reflections on a Transdisciplinary fMRI Lie Detection Experiment." *Frontiers in Human Neuroscience.* doi:10.3389/fnhum.2014.00149.

Littlefield, Melissa M., and Jenell M. Johnson, eds. 2012. *The Neuroscientific Turn: Transdisciplinarity in the Age of the Brain.* Ann Arbor: University of Michigan Press.

Livingstone, Margaret S. 2000. "Is It Warm? Is It Real? Or Just Low Spatial Frequency?" *Science* 290 (17 November): 1299.

Livingstone, Margaret S., and Bevil R. Conway. 2004. "Was Rembrandt Stereoblind?" *New England Journal of Medicine* 351: 1264–1265.

Lock, Margaret. 2002. *Twice Dead: Organ Transplants and the Reinvention of Death.* Berkeley: University of California Press.

Locke, John. 1690. "Second Treatise of Government." In *Two Treatises of Government,* edited by Peter Laslett. New York: Cambridge University Press.

———. 1988 [1694]. *An Essay Concerning Human Understanding.* 2nd ed. Edited by Peter H. Nidditch. Oxford: Clarendon.

Lodge, David. 1988. *Nice Work.* London: Secker & Warburg.

———. 2001. *Thinks . . .* New York: Viking.

Löfholm, Cecilia Andrée, Lars Brännström, Martin Olsson, and Kjell Hansson. 2013. "Treatment-as-Usual in Effectiveness Studies: What Is It and Does It Matter?" *International Journal of Social Welfare* 22: 25–34.

Loftus, Elizabeth, and Katherine Ketcham. 1994. *The Myth of Repressed Memory: False Memories and Allegations of Sexual Abuse.* New York: St. Martin's.

Logothetis, Nikos K. 2008. "What We Can Do and What We Cannot Do with fMRI." *Nature* 453: 869–878.

Lohmann, Gabriele, Johannes Stelzer, Jane Neumann, Nihat Ay, and Robert Turner. 2013. "'More Is Different' in Functional Magnetic Resonance Imaging: A Review of Recent Data Analysis Techniques." *Brain Connectivity* 3 (3): 223–239.

Lopez-Ibor, J. Juan. 2002. "The WPA and the Fight Against Stigma Because of Mental Illness." *World Psychiatry* 1: 30–31.

Lord, Catherine, and Rebecca Jones. 2012. "Annual Research Review: Re-thinking the Classification of Autism Spectrum Disorders." *Journal of Child Psychology and Psychiatry* 53: 490–509.

Luhrmann, Tanya Marie. 2000. *Of Two Minds: An Anthropologist Looks at American Psychiatry.* New York: Knopf.

———. 2007. "Social Defeat and the Culture of Chronicity; Or, Why Schizophrenia Does So Well Over There and So Badly Here." *Culture, Medicine, and Psychiatry* 31: 135–172.

———. 2012. "Beyond the Brain." *Wilson Quarterly* (Summer): 28–34.

Luria, Alexander Romanovich. 1966. "Vygotski et l'étude des fonctions psychiques supérieures." *Recherches Internationales à la Lumière du Marxisme* 51: 93–103.

———. 1979. *The Making of Mind: A Personal Account of Soviet Psychology.* Edited by Michael Cole and Sheila Cole. Cambridge, Mass.: Harvard University Press.

Lutz, Amy S. F. 2013. "You Do Not Have Asperger's: What Psychiatry's New Diagnostic Manual Means for People on the Autism Spectrum." *Slate* (May 22). http://www.slate.com/articles/health_and_science/medical _examiner/2013/05/autism_spectrum_diagnoses_the_dsm_5_eliminates _asperger_s_and_pdd_nos.html.

Maasen, Sabine, and Barbara Sutter, eds. 2007. *On Willing Selves: Neoliberal Politics and the Challenge of Neuroscience.* Basingstoke: Macmillan.

MacKinnon, Katherine C. 2014. "Contemporary Biological Anthropology in 2013: Integrative, Connected, and Relevant." *American Anthropologist* 116 (2): 352–365.

Macpherson, Crawford Brough. 1962. *The Political Theory of Possessive Individualism: Hobbes to Locke.* Oxford: Oxford University Press.

Maguire, Eleanor A., David G. Gadian, Ingrid S. Johnsrude, et al. 2000. "Navigation-Related Structural Change in the Hippocampi of Taxi Drivers." *PNAS* 97 (8): 4398–4403.

Malabou, Catherine. 2008. *What Should We Do with Our Brain?* Translated by Sebastian Rand. New York: Fordham University Press.

Malane, Rachel Ann. 2005. *Sex in Mind: The Gendered Brain in Nineteenth-Century Literature and Mental Sciences.* New York: Peter Lang.

Marcus, Gary. 2013. "The Problem with the Neuroscience Backlash." *New Yorker* (19 June). http://www.newyorker.com/online/blogs/elements/2013 /06/the-problem-with-the-neuroscience-backlash.html.

Marcus, Joseph A. 1997. "Neuroanthropology." In *The Dictionary of Anthropology,* edited by Thomas Barfield, 340–342. Malden, Mass.: Blackwell.

Mark, Vernon H., and Jeffrey P. Mark. 1991. *Brain Power: A Neurosurgeon's Complete Program to Maintain and Enhance Brain Fitness Throughout Your Life.* Boston: Houghton Mifflin.

Marquardt, Wolfgang. 2015. *Human Brain Project Mediation Report.* Juelich: Mediation of the Human Brain Project c/o Forschungszentrum Juelich GmbH. http://www.fz-juelich.de/SharedDocs/Pressemitteilungen/UK /EN/2015/15-03-09hbp-mediation.html.

Marsen, Sky. 2004. "Against Heritage: Invented Identities in Science Fiction Film." *Semiotica* 152 (1/4): 141–157.

Martín-Aragúz, Antonio, et al., eds. 2010. *Neuroestética.* Madrid: Saned.

Martin, Emily. 2000. "Mind-Body Problems." *American Ethnologist*
27: 569–590.
———. 2007. *Bipolar Expeditions: Mania and Depression in American Culture.*
Princeton, N.J.: Princeton University Press.
———. 2009. "Identity, Identification, and the Brain." Presented at the
workshop "Neurocultures." Max Planck Institute of the History of
Science. Berlin, 20–22 February.
———. 2010. "Self-Making and the Brain." *Subjectivity* 3 (4): 366–381.
Martindale, Colin, Paul Locher, and Vladimir Petrov, eds. 2007. *Evolutionary
and Neurocognitive Approaches to Aesthetics: Creativity and the Arts.* Ami-
tyville, N.Y.: Baywood.
Massaro, Davide, Federica Savazzi, Cinzia Di Dio, et al. 2012. "When Art
Moves the Eyes: A Behavioral and Eye-Tracking Study." *PLoS ONE* 7 (5):
e37285. doi:10.1371/journal.pone.0037285.
Mateo, Marina Martínez, Maurice Cabanis, Nicole Cruz de Echeverría
Loebell, and Sören Krach. 2012. "Concerns About Cultural Neuroscience.
A Critical Analysis." *Neuroscience and Biobehavioral Reviews* 36 (1): 152–161.
Mateo, Marina Martínez, Maurice Cabanis, Julian Stenmanns, and Sören
Krach. 2013. "Essentializing the Binary Self: Individualism and Collectivism
in Cultural Neuroscience." *Frontiers in Human Neuroscience* 7, art. 289: 1–4.
Max, Daniel. T. 2007. "Swann's Hypothesis." *New York Times* (4 November).
http://www.nytimes.com/2007/11/04/books/review/Max-t.html?_r=0.
Mayberg, Helen S. 2007. "Defining the Neural Circuitry of Depression:
Towards a New Nosology with Therapeutic Implications." *Biological
Psychiatry* 61: 729–730.
———. 2014. "Neuroimaging and Psychiatry: The Long Road from Bench to
Bedside." *Hastings Center Report* 44: S31–S36.
Mayberg, Helen, et al. 2005. "Deep Brain Stimulation for Clinical Study of
Treatment-Resistant Depression." *Neuron* 45: 651–660.
McCabe, David P., and Alan D. Castel. 2008. "Seeing Is Believing: The Effect
of Brain Images on Judgments of Scientific Reasoning." *Cognition*
107: 343–352.
McCarthy, Margaret M. 2015. Sex Differences in the Brain," *The Scientist*
(1 October). http://www.the-scientist.com/?articles.view/articleNo/44096
/title/Sex-Differences-in-the-Brain/.
McClure, Samuel M., Jian Li, Damon Tomlin, et al. 2004. "Neural Correlates of
Behavioral Preference for Culturally Familiar Drinks." *Neuron* 44: 379–387.
McEwan, Ian. 2004 [1997]. *Enduring Love.* London: Vintage.
———. 2005. *Saturday.* New York: Doubleday.
McGee, Micki. 2005. *Self-Help, Inc.: Makeover Culture in American Life.* New
York: Oxford University Press.

McGrath, Callie L., et al. 2013. "Toward a Neuroimaging Treatment Selection Biomarker for Major Depressive Disorder." *JAMA Psychiatry* 70 (8): 821–829.

McKinley, Marc. 2011. "Avoiding a Collapse in Thinking: Commentary on Jonathan Shedler's 'The Efficacy of Psychodynamic Psychotherapy.'" http://www.apadivisions.org/division-39/publications/review/2011/01 /psychodynamic-psychotherapy.aspx.

Mellor, Felicity. 2009. "The Politics of Accuracy in Judging Global Warming Films." *Environmental Communication* 3 (2): 134–150.

Meloni, Maurizio. 2011. "The Cerebral Subject at the Junction of Naturalism and Antinaturalism." In Ortega and Vidal 2011, 110–115.

———. 2012. "On the Growing Intellectual Authority of Neuroscience for Political and Moral Theory: Sketch for a Genealogy." In Vander Valk 2012a, 25–49.

———. 2013. "Biology Without Biologism: Social Theory in a Postgenomic Age." *Sociology.* doi:10.1177/0038038513501944.

———. 2014a. "How Biology Became Social, and What It Means for Social Theory." *Sociological Review* 62 (3): 593–614.

———. 2014b. "The Social Brain Meets the Reactive Genome: Neuroscience, Epigenetics, and the New Social Biology." *Frontiers in Human Neuroscience* 8, art. 309.

Men, Weiwei, Dean Falk, Tao Sun, et al. 2013. "The Corpus Callosum of Albert Einstein's Brain: Another Clue to His High Intelligence?" *Brain.* doi:10.1093/brain/awt252.

Menand, Louis. 2002. "What Comes Naturally." *The New Yorker* (22 November). http://www.newyorker.com/archive/2002/11/25/021125crbo_books.

Menninghaus, Winfried. 2008. *Kunst als "Beförderung des Lebens": Perspektiven transzendentaler und evolutionärer Ästhetik.* Munich: Carl Friedrich von Siemens Stiftung.

Merzenich, Michael, Mor Nahum, and Thomas M. van Vleet, eds. 2013. *Changing Brains: Applying Brain Plasticity to Advance and Recover Human Ability.* Amsterdam: Elsevier.

Merzenich, Michael, Thomas M. van Vleet, and Mor Nahum. 2014. "Brain Plasticity-Based Therapeutics." *Frontiers in Human Neuroscience* 8, art. 385.

Metzinger, Thomas. 2009. *The Ego Tunnel: The Science of the Mind and the Myth of the Self.* New York: Perseus.

Meyerding, Jane. 1998. "Thoughts on Finding Myself Differently Brained." http://www.planetautism.com/jane/diff.html.

———. 2003. "The Great 'Why Label?' Debate." http://www.planetautism .com/jane/label.html.

Michael, Emily. 2000. "Renaissance Theories of Body, Soul, and Mind." In *Psyche and Soma: Physicians and Metaphysicians on the Mind-Body Problem from Antiquity to Enlightenment*, edited by John P. Wright and Paul Potter. Oxford: Clarendon, 147–172.

Miller, Gavin. 2014. "Is the Agenda for Global Mental Health a Form of Cultural Imperialism?" *Medical Humanities* 40 (2): 131–134.

Miller, Greg. 2016. "Brain Scans Are Prone to False Positives, Study Says." *Science* 353 (6296): 208–209.

Millet, Aurore. 2009. *Reminiscence* [Catalog]. http://www.auroremillet.com /meta/downloads.html.

Mills, China. 2014. *Decolonizing Global Mental Health: The Psychiatrization of the Majority World*. London: Routledge.

Mitchell, Philip B. 2009. "Winds of Change: Growing Demands for Transparency in the Relationship Between Doctors and the Pharmaceutical Industry." *Medical Journal of Australia* 191: 273–275.

Mlodinow, Leonard. 2012. "Why People Choose Coke Over Pepsi: How Our Brains Create Our Consumer Experience." http://www.psychologytoday .com/blog/subliminal/201205/why-people-choose-coke-over-pepsi.

Molnár, Zoltán. 2004. "Thomas Willis (1621–1675), the Founder of Clinical Neuroscience." *Nature Reviews Neuroscience* 5: 329–335.

Moncrieff, Joanna. 2008. *The Myth of the Chemical Cure: A Critique of Psychiatric Drug Treatment*. Houndmills: Palgrave.

Montanini, Daniel, and Cláudio E. M. Banzato. 2012. "Do estigma da psicose maníaco-depressiva ao incentivo ao tratamento do transtorno bipolar: a evolução da abordagem em dois veículos midiáticos nos últimos 40 anos." *Jornal brasileiro de psiquiatria* 61 (2): 84–88.

Moran, Joseph M., and Jamil Zaki. 2013. "Functional Neuroimaging and Psychology: What Have You Done for Me Lately?" *Journal of Cognitive Neuroscience* 25 (6): 834–842.

Morioka, Masahiro. 2001. "Reconsidering Brain Death: A Lesson from Japan's Fifteen Years of Experience." *Hastings Center Report* 31 (4): 41–46.

Morrison P. Anthony, Paul Hutton, David Shiers, and Douglas Turkington. 2012. "Antipsychotics: Is It Time to Introduce Patient Choice?" *British Journal of Psychiatry* 201: 83–84.

Morrison, P. Anthony, Douglas Turkington, Melissa Pyle, et al. 2014. "Cognitive Therapy for People with Schizophrenia Spectrum Disorder Not Taking Antipsychotic Medication: A Single-Blind Randomised Controlled Trial." *The Lancet* 383 (9926): 1395–1403.

Mowaljarlai, David, Patricia Vinnicombe, Graeme K. Ward, and Christopher Chippindale. 1988. "Repainting of Images in Australia and the Maintenance of Aboriginal Culture." *Antiquity* 62: 690–696.

Mrazek, Alissa. J., Tokiko Harada, and Joan Y. Chiao. 2014. "Cultural Neuroscience of Identity Development." In *The Oxford Handbook of Identity*

Development, edited by Kate C. McLean and Moin Syed. Oxford: Oxford University Press, 423–436.

Munro, Geoffrey D., and Cynthia A. Munro. 2014. "'Soft' Versus 'Hard' Psychological Science: Biased Evaluations of Scientific Evidence That Threatens or Supports a Strongly Held Political Identity." *Basic and Applied Social Psychology* 36 (6): 533–543.

Muzur, Amir, and Iva Rinčić. 2013. "Neurocriticism: A Contribution to the Study of the Etiology, Phenomenology, and Ethics of the Use and Abuse of the Prefix *Neuro-*." *JAHR–European Journal of Bioethics* 4 (7): 545–554.

Nadal, Marcos. 2013. "The Experience of Art: Insights from Neuroimaging." In Finger, Zaidel, Boller, and Bogousslavsky 2013, 153–158.

Nadal, Marcos, and Marcus T. Pearce. 2011. "The Copenhagen Neuroesthetics Conference: Prospects and Pitfalls for an Emerging Field." *Brain and Cognition* 76: 172–183.

Nadesan, Majia H. 2005. *Constructing Autism: Unravelling the "Truth" and Understanding the Social*. London: Routledge.

Nalbantian, Suzanne. 2008. "Neuroesthetics, Neuroscientific Theory, and Illustration from the Arts." *Interdisciplinary Science Reviews* 33 (4): 357–368.

Nan, Yun, Thomas R. Knösche, Stefan Zysset, and Angela D. Friederici. 2008. "Cross-Cultural Music Phrase Processing: An fMRI Study." *Human Brain Mapping* 29: 312–328.

Nantel-Vivier, Amélie, and Robert Pihl. 2008. "Biological Vulnerability of Depression." In *Handbook of Depression in Children and Adolescents*, edited by John R. Z. Abela and Benjamin L. Hankin, 103–123. New York: Guilford.

Nelkin, Dorothy, and M. Susan Lindee. 1995. *The DNA Mystique: The Gene as a Cultural Icon*. New York: Freeman.

Nelson, Amy. 2004. "Declaration from the Autism Community That They Are a Minority Group." November 18. http://www.prweb.com/releases/2004/11/prweb179444.htm.

Netherland, Julie. 2011. "'We Haven't Sliced Open Anyone's Brain Yet': Neuroscience, Embodiment, and the Governance of Addiction." In Pickersgill and Van Keulen 2011, 153–177.

Ng, Brandon W., James P. Morris, and Shigehiro Oishi. 2013. "Cultural Neuroscience: The Current State of Affairs." *Psychological Inquiry* 24: 53–57.

Ng, Sik Hung, Shihui Han, Lihua Mao, and Julian C. Lai. 2010. "Dynamic Bicultural Brains: fMRI Study of Their Flexible Neural Representation of Self and Significant Others in Response to Culture Primes." *Asian Journal of Social Psychology* 13 (2): 83–91.

Nissenbaum, Stephen. 1980. *Sex, Diet, and Debility in Jacksonian America: Sylvester Graham and Health Reform*. Westport, Conn.: Greenwood.

Noë, Alva. 2009. *Out of Our Heads: Why You Are Not Your Brain, and Other Lessons from the Biology of Consciousness*. New York: Hill and Wang.

———. 2015. *Strange Tools: Art and Human Nature*. New York: Hill and Wang.

Northoff, Georg. 2013a. "Gene, Brains, and Environment—Genetic Neuro-imaging of Depression." *Current Opinion in Neurobiology* 23: 133–142.

———. 2013b. "What Is Culture? Culture Is Context-Dependence!" *Culture and Brain* 1 (2/4): 77–99.

Northoff, Georg, Christine Wiebking, Todd Feinberg, and Jaak Panksepp. 2011. "The 'Resting-State Hypothesis' of Major Depressive Disorder—A Translational Subcortical-Cortical Framework for a System Disorder." *Neuroscience and Biobehavioral Reviews* 35 (9): 1929–1945.

Novas, Carlos, and Nikolas Rose. 2000. "Genetic Risk and the Birth of the Somatic Individual." *Economy and Society* 29: 485–513.

Nozick, Robert. 1981. *Philosophical Explanations*. Cambridge, Mass.: Harvard Univeristy Press.

O'Connor, Cliodhna, and Helene Joffe. 2013. "How Has Neuroscience Affected Lay Understandings of Personhood? A Review of the Evidence." *Public Understanding of Science* 22 (3): 254–268.

Ochs, Elinor, and Olga Solomon. 2010. "Autistic Sociality." *Ethos* 38: 69–92.

Oehler-Klein, Sigrid. 1990. *Die Schadellehre Franz Joseph Galls in Literatur und Kritik des 19. Jahrhunderts: Zur Rezeptionsgeschichte einer medizinisch-biologisch begründeten Theorie der Physiognomik und Psychologie*. Stuttgart: Gustav Fischer.

Oliver, Mike. 1990. *The Politics of Disablement*. London: Macmillan.

Olney, Jennifer. 2006. "Exercise May Be Key to Keeping Your Brain Fit." http://www.brainhq.com/media/news/exercise-may-be-key-keeping-your-brain-fit.

Olson, Gary. 2008. "We Empathize, Therefore We Are: Toward a Moral Neuropolitics." *ZNet* (26 July). https://zcomm.org/znetarticle/we-empathize-therefore-we-are-toward-a-moral-neuropolitics-by-gary-olson/.

———. 2013. *Empathy Imperiled: Capitalism, Culture, and the Brain*. New York: Springer.

Onians, John. 2008a. *Neuroarthistory: From Aristotle and Pliny to Baxandall and Zeki*. New Haven, Conn.: Yale University Press.

———. 2008b. "Neuro Ways of Seeing [Interview with Eric Fernie]." *Tate Etc.* 13 (Summer). http://web.archive.org/web/20120203092837/http://www.tate.org.uk/tateetc/issue13/neuroarthistory.htm.

Open Science Collaboration. 2015. "Estimating the Reproducibility of Psychological Science." *Science* 349 (6251): aac4716. doi:10.1126/science.aac4716.

Orsini, Michael. 2009. "Contesting the Autistic Subject: Biological Citizenship and the Autism/Autistic Movement." In *Critical Interventions in the Ethics of Health Care*, edited by Stuart Murray and Dave Holmes. London: Ashgate, 115–130.

————. 2012. "Autism, Neurodiversity, and the Welfare State: The Challenges of Accommodating Neurological Difference." *Canadian Journal of Political Science* 45: 805–882.

Ortega, Francisco. 2011. "Toward a Genealogy of Neuroacesis." In Ortega and Vidal 2011, 27–44.

————. 2014. *Corporeality, Medical Technologies, and Contemporary Culture.* New York: Routledge.

Ortega, Francisco, and Fernando Vidal, eds. 2011. *Neurocultures: Glimpses Into an Expanding Universe.* Berlin: Peter Lang.

Ortega, Francisco, Rafaela Zorzanelli, Lilian Kozslowski Meierhoffer, et al. 2013. "A Construção do Diagnóstico do Autismo em uma Rede Social Virtual Brasileira." *Interface—Comunicação, Saúde, Educação* 17: 119–132.

Osteen, Mark, ed. 2008. *Autism and Representation.* New York: Routledge.

Owen, Adrian M., Adam Hampshire, Jessica A. Grahn, et al. 2010. "Putting Brain Training to the Test." *Nature* 465 (7299): 775–778.

Padden, Carol, and Tom Humphries. 2006. *Inside Deaf Culture.* Cambridge, Mass.: Harvard University Press.

Painter, Nell I. 2010. *The History of White People.* New York: Norton.

Pardo, Michael S., and Dennis Patterson. 2011. "Minds, Brains, and Norms." *Neuroethics* 4: 179–190.

Parlette, Snowdon. 1997. *The Brain Workout Book.* New York: M. Evans and Co.

Patel, Vikram. 2012. "Global Mental Health: From Science to Action." *Harvard Review of Psychiatry* 20 (1): 6–12.

Patel, Vikram, and Mark Winston. 1994. "'Universality of Mental Illness' Revisited: Assumptions, Artefacts, and New Directions." *British Journal of Psychiatry* 165: 437–440.

Patel, Vikram, Helen A. Weiss, Neerja Chowdhary, et al. 2011. "Lay Health Worker–Led Intervention for Depressive and Anxiety Disorders in India: Impact on Clinical and Disability Outcomes Over 12 Months." *British Journal of Psychiatry* 199: 459–466.

Patel, Vikram, Harry Minas, Alex Cohen, and Martin J. Prince, eds. 2014. *Global Mental Health: Principles and Practice.* New York: Oxford University Press.

Paterniti, Michael. 2000. *Driving Mr. Albert: A Trip Across America with Einstein's Brain.* New York: Dial.

Pedersen, David Budtz. 2011. "Revisiting the Neuro-Turn in the Humanities and Natural Sciences." *Pensamiento* 67 (254): 767–786.

Pepperell, Robert. 2011. "Connecting Art and the Brain: An Artist's Perspective on Visual Indeterminacy." *Frontiers in Human Neuroscience* 5, art. 84: 1–12.

Peters, June A., Luba Djurdjinovic, and Diane Baker. 1999. "The Genetic Self: The Human Genome Project, Genetic Counseling, and Family Therapy." *Families, Systems, and Health* 17 (1): 5–25.

Pethes, Nicolas. 2005. "Terminal Men, Biotechnological Experimentation, and the Reshaping of 'the Human' in Medical Thrillers." *New Literary History* 36 (2): 161–185.

Petryna, Adriana, Andrew Lakoff, and Arthur Kleinman, eds. 2006. *Global Pharmaceuticals: Ethics, Markets, Practices*. Durham, N.C.: Duke University Press.

Petryna, Adriana, and Arthur Kleinman. 2006. "The Pharmaceutical Nexus." In Petryna, Lakoff, and Kleinman 2006, 1–32.

Phelan, C. Jo. 2005. "Geneticization of Deviant Behavior and Consequences for Stigma: The Case of Mental Illness." *Journal of Health and Social Behavior* 46 (4): 307–322.

Phillips, Kristopher G., Alan Beretta, and Harry A. Whitaker. 2015. "Mind and Brain: Toward an Understanding of Dualism." In *Brain, Mind, and Consciousness in the History of Neuroscience*, edited by C. U. M. Smith and Harry Whitaker, 355–369. Dordrecht: Springer.

Pickering, Andrew. 2011. *The Cybernetic Brain: Sketches of Another Future*. Chicago: University of Chicago Press.

Pickersgill, Martyn, Sarah Cunningham-Burley, and Paul Martin. 2011. "Constituting Neurologic Subjects: Neuroscience, Subjectivity, and the Mundane Significance of the Brain." *Subjectivity* 4: 346–365.

Pickersgill, Martyn, and Ira Van Keulen, eds. 2012. *Sociological Reflections on the Neurosciences*. Bingley: Emerald.

Pickersgill, Martyn, Paul Martin, and Sarah Cunningham-Burley. 2015. "The Changing Brain: Neuroscience and the Enduring Import of Everyday Experience." *Public Understanding of Science* 24: 878–892.

Pitts-Taylor, Victoria. 2010. "The Plastic Brain: Neoliberalism and the Neuronal Self." *Health* 14 (6): 635–652.

Podgorny, Irina. 2005. "La derrota del genio. Cráneos y cerebros en la filogenia argentina." *Saber y tiempo. Revista de historia de la ciencia* 5 (20): 63–106.

Poldrack, Russell A. 2008. "The Role of fMRI in Cognitive Neuroscience: Where Do We Stand?" *Current Opinion in Neurobiology* 2: 223–227.

Posner, Jonathan, Virginia Rauh, Allison Gruber, et al. 2013. "Dissociable Attentional and Affective Circuits in Medication-Naïve Children with Attention-Deficit/Hyperactivity Disorder." *Psychiatry Research: Neuroimaging* 213: 24–30.

Posner, Jonathan, Christine Park, and Zhishun Wang. 2014. "Connecting the Dots: A Review of Resting Connectivity MRI Studies in Attention-Deficit/Hyperactivity Disorder." *Neuropsychology Review* 24: 3–15.

Powers, Richard. 1996 [1995]. *Galatea 2.2*. New York: Harper Perennial.

———. 2006. *The Echomaker*. New York: Farrar, Straus and Giroux.

———. 2007. "The Brain Is the Ultimate Storytelling Machine, and Consciousness is the Ultimate Story. Interview with Richard Powers." *Believer* (February). http://www.believermag.com/issues/200702/?read=interview_powers.

Presidential Commision 2015. *Gray Matters: Topics at the Intersection of Neuroscience, Ethics, and Society.* Vol. 2. Washington, D.C.: Presidential Commission for the Study of Bioethical Issues.

Press Release. 2011. "Mindfulness Meditation Training Changes Brain Structure in 8 Weeks." http://www.massgeneral.org/about/pressrelease.aspx?id=1329. [Hölzel et al. 2011.]

Prévost, Bertrand. 2003. "Pouvoir ou efficacité symbolique des images." *L'Homme. Revue Française d'Anthropologie* 165: 275–282.

Price, Joseph L., and Wayne C. Drevets. 2010. "Neurocircuitry of Mood Disorders." *Neuropsychopharmacology Reviews* 35: 192–216.

Prince-Hughes, Dawn. 2004. *Songs of the Gorilla Nation: My Journey Through Autism.* New York: Harmony.

Prince, Dawn Eddings. 2010. "An Exceptional Path: An Ethnographic Narrative Reflecting on Autistic Parenthood from Evolutionary, Cultural, and Spiritual Perspectives." *Ethos* 38: 56–68.

Prince, Martin, Vikram Patel, Shekhar Saxena, et al. 2007. "No Health Without Mental Health." *The Lancet* 370: 859–877.

Prince, Martin, Atif Rahman, Rosie Mayston, and Benedict Weobong. 2014. "Mental Health and the Global Health and Development Agendas." In Patel, Minas, Cohen, and Prince 2014, 102–115.

Protevi, John. 2009. *Political Affect: Connecting the Social and the Somatic.* Minneapolis: University of Minnesota Press.

Puccetti, Roland. 1969. "Brain Transplantation and Personal Identity." *Analysis* 29: 65–77.

———. 1973. "Brain Bisection and Personal Identity." *British Journal for the Philosophy of Science* 24: 339–355.

Pugliese, Joseph. 2010. *Biometrics: Bodies, Technologies, Biopolitics.* New York: Routledge.

Putnam, Hilary. 1981. *Reason, Truth, and History.* Cambridge, Mass.: Harvard University Press.

Rachul, Christen, and Amy Zarzeczny. 2012. "The Rise of Neuroskepticism." *International Journal of Law and Psychiatry* 35: 77–81.

Racine, Eric. 2010. *Pragmatic Neuroethics: Improving Treatment and Understanding of the Mind-Brain.* Cambridge, Mass.: MIT Press.

Racine, Eric, Ofek Bar-Ilan, and Judy Illes. 2005. "fMRI in the Public Eye." *Nature Reviews Neuroscience* 6: 159–164.

Radstone, Susannah. 2010. "Cinema and Memory." In *Memory: Histories, Theories, Debates*, edited by Susannah Radstone and Bill Schwartz, 325–342. New York: Fordham University Press.

Rafter, Nicole. 2008. *The Criminal Brain: Understanding Biological Theories of Crime.* New York: New York University Press.

Raichle, Marcus E., Ann M. MacLeod, Abraham Z. Snyder, et al. 2001. "A Default Mode of Brain Function." *PNAS* 98 (2): 676–682.

Raichle, Marcus E., and Abraham Z. Snyder. 2007. "A Default Mode of Brain Function: A Brief History of an Evolving Idea." *NeuroImage* 37: 1083–1090.

Ramachandran, Vilayanur Subramanian, and William Hirstein. 1999. "The Science of Art: A Neurological Theory of Aesthetic Experience." *Journal of Consciousness Studies* 6 (6/7): 15–51.

Randall, Kevin. 2015. "Neuropolitics, Where Campaigns Try to Read Your Mind." *New York Times* (3 November). http://www.nytimes.com/2015/11/04/world/americas/neuropolitics-where-campaigns-try-to-read-your-mind.html?_r=0.

Rapp, Rayna. 2011. "A Child Surrounds This Brain: The Future of Neurological Difference According to Scientists, Parents, and Diagnosed Young Adults." In Pickersgill and Van Keulen 2011, 3–26.

Rawlings, Charlie E., and Eugene Rossitch Jr. 1994. "Franz Josef Gall and His Contribution to Neuroanatomy with Emphasis on the Brain Stem." *Surgical Neurology* 42: 272–275.

Ray, Rebecca D., Amy L. Shelton, Nick G. Hollon, et al. 2010. "Interdependent Self-Construal and Neural Representations of Self and Mother." *Social Cognitive and Affective Neuroscience* 5: 318–323.

Ray, Wayne A., Cecilia P. Chung, Katherine T. Murray, et al. 2009. "Atypical Antipsychotic Drugs and the Risk of Sudden Cardiac Death." *New England Journal of Medicine* 360: 225–235.

Read, John. 2005. "The Bio-bio-bio Model of Madness." *The Psychologists* 18 (10): 596–597.

Read, John, Richard Bentall, and Roar Fosse. 2009. "Time to Abandon the Bio-bio-bio Model of Psychosis: Exploring the Epigenetic and Psychological Mechanisms by Which Adverse Life Events Lead to Psychotic Symptoms." *Epidemiologia e psichiatria sociale* 18 (4): 299–310.

Read, John, and Niki Harré. 2001. "The Role of Biological and Genetic Causal Beliefs in the Stigmatization of 'Mental Patients.'" *Journal of Mental Health* 10 (2): 223–235.

Redies, Christoph. 2007. "A Universal Model of Aesthetic Perception Based on the Sensory Coding of Natural Stimuli." *Spatial Vision* 21 (1/2): 97–117.

Redies, Christoph, Jan Hänisch, Marko Blickhan, and Joachim Denzler. 2007. "Artists Portray Human Faces with the Fourier Statistics of Complex Natural Scenes." *Network: Computation in Neural Systems* 18 (3): 235–248.

Redies, Christoph, Jens Hasenstein, and Joachim Denzler. 2007. "Fractal-like Image Statistics in Visual Art: Similarity to Natural Scenes." *Spatial Vision* 21 (1/2): 137–148.

Redwood, Daniel. 2007. "Meditation, Positive Emotions, and Brain Science: Interview with Richard Davidson Ph.D." http://web.archive.org/web /20100427160604/http://www.healthy.net/scr/interview.aspx?Id=306.

Rees, Tobias. 2010. "Being Neurologically Human Today: Life and Science and Adult Cerebral Plasticity (an Ethical Analysis)." *American Ethnologist* 37 (1): 150–166.

———. 2011. "So Plastic a Brain: On Philosophy, Fieldwork in Philosophy, and the Rise of Adult Cerebral Plasticity." *BioSocieties* 6 (2): 263–267.

Regalado, Antonio. 2015. "Why America's Top Mental Health Researcher Joined Alphabet [Interview with Thomas Insel]." *MIT Technology Review* (21 September). http://www.technologyreview.com/news/541446/why -americas-top-mental-health-researcher-joined-alphabet/.

Reichle, Ingeborg. 2009. *Art in the Age of Technoscience: Genetic Engineering, Robotics, and Artificial Life in Contemporary Art.* Translated by Gloria Custance. New York: Springer.

Reid, Ian C. 2013. "Are Antidepressants Overprescribed? No." *British Medical Journal* 346. doi:10.1136/bmj.f190.

Renard, Maurice. 1921. *Les mains d'Orlac.* Paris: Nilsson.

Rengachary, Setti S., Andrew Xavier, Sunil Manjila, et al. 2008. "The Legendary Contributions of Thomas Willis (1621–1675): The Arterial Circle and Beyond." *Journal of Neurosurgery* 109: 765–775.

Renneville, Marc. 2000. *Le langage des crânes. Une histoire de la phrénologie.* Paris: Les Empêcheurs de tourner en rond.

Richards, Graham. 2002. "The Psychology of Psychology: A Historically Grounded Sketch." *Theory and Psychology* 12: 7–36.

Richardson, Alan. 2004. "Studies in Literature and Cognition: A Field Map." In *The Work of Fiction: Cognition, Culture, and Complexity*, edited by Alan Richardson and Ellen Spolsky, 1–29. Aldershot: Ashgate.

Rios, Clarice, and Barbara C. Andrada. 2015. "The Changing Face of Autism in Brazil." *Culture, Medicine, and Psychiatry* 39 (2): 213–234.

Rizzolatti, Giacomo, and Laila Craighero. 2004. "The Mirror-Neuron System." *Annual Review of Neuroscience* 27: 169–192.

Rizzolatti, Giacomo, and Maddalena Fabbri-Destro. 2010. "Mirror Neurons: From Discovery to Autism." *Experimental Brain Research* 200: 223–237.

Rizzolatti, Giacomo, and Corrado Sinigaglia. 2010. "The Functional Role of the Parietofrontal Mirror Circuit: Interpretations and Misinterpretations." *Nature Reviews Neuroscience* 11 (4): 264–274.

Rocca, Julius. 2003. *Galen on the Brain: Anatomical Knowledge and Physiological Speculation in the Second Century AD.* Leiden: Brill.

Rodriguez, Paul. 2006. "Talking Brains: A Cognitive Semantic Analysis of an Emerging Folk Neuropsychology." *Public Understanding of Science* 15 (3): 301–330.

Roepstorff, Andreas. 2011. "Culture: A Site of Relativist Energy in the Cognitive Sciences." *Common Knowledge* 17: 37–41.

Roepstorff, Andreas, and Chris Frith. 2012. "Neuroanthropology or Simply Anthropology? Going Experimental as Method, as Object of Study, and as Research Aesthetic." *Anthropological Theory* 12 (1): 101–111.

Roepstorff, Andreas, and Kai Vogeley. 2009. "Contextualising Culture and Social Cognition." *Trends in Cognitive Science* 13: 511–516.

Roepstorff, Andreas, Jörg Niewöhner, and Stefan Beck. 2010. "Enculturing Brains Through Patterned Practices." *Neural Networks* 23: 1051–1059.

Rose, Nikolas. 1990. *Governing the Soul: The Shaping of the Private Self.* London: Routledge

———. 1996. *Inventing Our Selves: Psychology, Power, and Personhood.* New York: Cambridge University Press.

———. 2003. "The Neurochemical Self and Its Anomalies." In *Risk and Morality*, edited by Richard Ericson and Aaron Doyle. Toronto: University of Toronto Press, 407–437.

———. 2004. "Becoming Neurochemical Selves." In *Biotechnology: Between Commerce and Civil Society*, edited by Nico Stehr. New Brunswick, N.J.: Transaction, 89–128.

———. 2007. *The Politics of Life Itself: Biomedicine, Power, and Subjectivity in the Twenty-First Century.* Princeton, N.J.: Princeton University Press.

———. 2013a. "The Human Sciences in a Biological Age." *Theory, Culture, and Society* 30 (1): 3–34.

———. 2013b. "What Is Diagnosis For?" Conference paper, "DSM-5 and the Future of Diagnosis." http://nikolasrose.com/index.php/lectures/.

Rose, Nikolas, and Joelle M. Abi-Rached. 2013. *Neuro: The New Brain Sciences and the Management of the Mind.* Princeton, N.J.: Princeton University Press.

———. 2014. "Governing Through the Brain: Neuropolitics, Neuroscience, and Subjectivity." *Cambridge Anthropology* 32 (1): 3–23.

Rosen, Bruce R., and Robert L. Savoy. 2012. "fMRI at 20: Has It Changed the World?" *NeuroImage* 62: 1316–1324.

Rosenbaum, Bent, Susanne Harder, Per Knudsen, et al. 2012. "Supportive Psychodynamic Psychotherapy Versus Treatment as Usual for First-Episode Psychosis: Two-Year Outcome." *Psychiatry: Interpersonal and Biological Processes* 75 (4): 331–341.

Roskies, Adina L. 2002. "Neuroethics for the New Millenium." *Neuron* 35: 21–23.

———. 2007. "Are Neuroimages Like Photographs of the Brain?" *Philosophy of Science* 74: 860–872.

———. 2009. "Brain-Mind and Structure-Function Relationships: A Methodological Response to Coltheart." *Philosophy of Science* 76: 927–939.

———. 2010. "Saving Subtraction: A Reply to Van Orden and Paap." *British Journal of the Philosophy of Science* 61: 635–665.

Roth, Marco. 2009. "The Rise of the Neuronovel." *N+1* 8 (19 October). http://web.archive.org/web/20110714203925/http://www.nplusonemag.com/rise-neuronovel.

Roth, Michael S. 1981. "Foucault's 'History of the Present.'" *History and Theory* 20 (1): 32–46.

Rousseau, George Sebastian. 2007. "'Brainomania': Brain, Mind, and Soul in the Long Eighteenth Century." *British Journal for Eighteenth-Century Studies* 30: 161–191.

Rowland, Margaret. 2015. "Angry and Mad: A Critical Examination of Identity Politics, Neurodiversity, and the Mad Pride Movement." *Journal of Ethics in Mental Health* 1: 1–3.

Rozenblit, Leonid, and Frank Keil. 2002. "The Misunderstood Limits of Folk Science: An Illusion of Explanatory Depth." *Cognitive Science* 26: 521–562.

Rubin, Sue. 2005. "Acceptance Versus Cure." http://www.cnn.com/CNN/Programs/presents/shows/autism.world/notebooks/sue/notebook.html.

Rugg, Michael D., and Sharon L. Thompson-Schill. 2013. "Moving Forward with fMRI Data." *Perspectives on Psychological Science* 8 (1): 84–87.

Rusconi, Elena, and Timothy Mitchener-Nissen. 2014. "The Role of Expectations, Hype, and Ethics in Neuroimaging and Neuromodulation Futures." *Frontiers in Systems Neuroscience* 8, art. 214.

Sacher, Julia, Jane Neumann, Tillmann Fünfstück, et al. 2012. "Mapping the Depressed Brain: A Meta-analysis of Structural and Functional Alterations in Major Depressive Disorder." *Journal of Affective Disorders* 140: 142–148.

Sacks, Oliver. 1985. *The Man Who Mistook His Wife for a Hat and Other Clinical Tales*. New York: Touchstone.

———. 1995. *An Anthropologist on Mars*. New York: Vintage.

———. 2013. "Speak, Memory." *The New York Review of Books* (21 February). http://www.nybooks.com/articles/archives/2013/feb/21/speak-memory/.

Sahlins, Marshall. 2000. "Sentimental Pessimism and Ethnographic Experience; Or, Why Culture Is Not a Disappearing 'Object.'" In *Biographies of Scientific Objects*, edited by Lorraine Daston. Chicago: University of Chicago Press, 152–202.

Sass, Hans-Martin. 1989. "Brain Life and Brain Death: A Proposal for a Normative Agreement." *Journal of Medicine and Philosophy* 14: 45–59.

Sbriscia-Fioretti, Beatrice, Cristina Berchio, David Freedberg, et al. 2013. "ERP Modulation During Observation of Abstract Paintings by Franz Kline." *PLoS ONE* 8 (10): e75241. doi:10.1371/journal.pone.0075241.

Schaeffer, Jean-Marie. 1997. "La relation esthétique comme fait anthropologique." *Critique* 53: 691–708.

———. 2009. *Adieu à l'esthétique*. Paris: PUF.

————. 2010. *Théorie des signaux coûteux, esthétique et art.* Trois-Rivières: Tangence.

Scharinger, Christian, Ulrich Rabl, Lukas Pezawas, and Siegfried Kasper. 2011. "The Genetic Blueprint of Major Depressive Disorder: Contributions of Imaging Genetics Studies." *World Journal of Biological Psychiatry* 12: 474–488.

Scheper-Hugues, Nancy. 1984. "The Margaret Mead Controversy: Culture, Biology, and Anthropological Inquiry." *Human Organization* 43 (1): 85–93.

Schick, Ari. 2005. "Neuro Exceptionalism?" *American Journal of Bioethics* 5 (2): 36–38.

Schlaepfer, Thomas E., Bettina H. Bewernick, Sarah Kayser, et al. 2014. "Deep Brain Stimulation of the Human Reward System for Major Depression—Rationale, Outcomes, and Outlook." *Neuropsychopharmacology* 39: 1303–1314.

Schleim, Stephan, and Jonathan P. Roiser. 2009. "fMRI in Translation: The Challenges Facing Real-World Applications." *Frontiers in Human Neuroscience* 3, art. 63: 1–7.

Schnittker, Jason. 2008. "An Uncertain Revolution: Why the Rise of a Genetic Model of Mental Illness Has Not Increased Tolerance." *Social Science and Medicine* 67 (9): 1370–1381.

Schreyach, Michael. 2007. "'I Am Nature': Science and Jackson Pollock." *Apollo* 7: 35–43.

Schwartz, Jeffrey M., and Sharon Begley. 2002. *The Mind and the Brain: Neuroplasticity and the Power of Mental Force.* New York: HarperCollins.

Senn, Bryan, and John Johnson. 1992. *Fantastic Cinema Subject Guide: A Topical Index to 2,500 Horror, Science Fiction, and Fantasy Films.* Jefferson, N.C.: McFarland.

Shakespeare, Tom. 2006. *Disability Rights and Wrongs.* Abingdon: Routledge.

Shapin, Steven. 2008. *The Scientific Life: A Moral History of a Late Modern Vocation.* Chicago: University of Chicago Press.

Shapiro, Joseph P. 1993. *No Pity: People with Disabilities Forging a New Civil Rights Movement.* New York: Random House.

————. 2006. "Autism Movement Seeks Acceptance, Not Cures." http://www.npr.org/templates/story/story.php?storyId=5488463.

Shedler, Jonathan. 2010. "The Efficacy of Psychodynamic Psychotherapy." *American Psychologist* 65 (2): 98–109.

Sherwood, Katherine. 2009. *Golgi's Door* [exhibition catalog]. Washington, D.C.: National Academy of Sciences.

Shoemaker, Sidney. 1963. *Self-Knowledge and Self-Identity.* Ithaca, N.Y.: Cornell University Press.

Shorter, Edward. 2013. *How Everyone Became Depressed: The Rise and Fall of the Nervous Breakdown.* New York: Oxford University Press.

Shuttleworth, Sally. 1996. *Charlotte Brontë and Victorian Psychology.* New York: Cambridge University Press.

Shweder, Richard A. 1991. *Thinking Through Cultures: Expeditions in Cultural Psychology.* Cambridge, Mass.: Harvard University Press.

———. 2001. "Culture: Contemporary Views." In *International Encyclopedia of the Social and Behavioral Sciences,* edited by Neil J. Smelser and Paul B. Baltes. Oxford: Elsevier, 3151–3158.

Siegle, Greg J., Wesley K. Thompson, Amanda Collier, et al. 2012. "Toward Clinically Useful Neuroimaging in Depression Treatment." *Archives of General Psychiatry* 69 (9): 913–924.

Silberman, Steve. 2015. *NeuroTribes: The Legacy of Autism and the Future of Neurodiversity.* New York: Avery.

Silverman, Chloe. 2008a. "Brains, Pedigrees, and Promises: Lessons from the Politics of Autism Genetics." In Gibbon and Novas 2008b, 38–55.

———. 2008b. "Fieldwork on Another Planet: Social Science Perspectives on the Autism Spectrum." *BioSocieties* 3 (3): 325–341.

———. 2012. *Understanding Autism: Parents, Doctors, and the History of a Disorder.* Princeton, N.J.: Princeton University Press.

Simon, Herbert. 1994. "Literary Criticism: A Cognitive Approach." *Stanford Humanities Review* 4 (1).

Simpson, Donald. 2005. "Phrenology and the Neurosciences: Contributions of F. J. Gall and J. G. Spurzheim." *ANZ Journal of Surgery* 75: 475–482.

Sinclair, Jim. 1993. "Don't Mourn for Us." *Voice* 1 (3). http://www.autreat.com/dont_mourn.html.

———. 1999. "Why I Dislike "Person First" Language." http://autismmythbusters.com/general-public/autistic-vs-people-with-autism/jim-sinclair-why-i-dislike-person-first-language/.

———. 2005. "Autism Network International: The Development of a Community and Its Culture." http://www.autreat.com/History_of_ANI.html.

Singel, Ryan. 2003. "He Thinks, Therefore He Sells." http://archive.wired.com/culture/lifestyle/news/2003/10/60757?currentPage=all.

Singer, Judy. 1999. "Why Can't You Be Normal for Once in Your Life? From a 'Problem with No Name' to the Emergence of a New Category of Difference." In Corker and French 1999, 59–67.

———. 2007. "Light and Dark. Correcting the Balance." https://archive.is/pu1O6.

Singh, Ilina. 2013. "Brain Talk: Power and Negotiation in Children's Discourse About Self, Brain, and Behavior." *Sociology of Health and Illness* 35 (6): 813–827.

Singh, Ilina, and Nikolas Rose. 2006. "Neuro-Forum: An Introduction." *BioSocieties* 1: 97–102.

———. 2009. "Biomarkers in Psychiatry: Promises and Perils in the Real World." *Nature* 460 (7252): 202–207.

Singh, Krish D. 2012. "Which 'Neural Activity' Do You Mean? fMRI, MEG, Oscillations, and Neurotransmitters." *NeuroImage* 62: 1121–1130.

Siodmak, Curt. 1992 [1942]. *Donovan's Brain*. New York: Leisure Books.

Skolnick Weisberg, Deena, Frank C. Keil, Joshua Goodstein, et al. 2008. "The Seductive Allure of Neuroscience Explanations." *Journal of Cognitive Neuroscience* 20: 470–477.

Skov, Martin. 2006. "A Short Bibliographic Guide to the Emerging Field of Bioaesthetics." http://brainethics.wordpress.com/2006/09/27/a-short -bibliographic-guide-to-the-emerging-field-of-bioaesthetics/.

Skov, Martin, and Oshin Vartanian, eds. 2009a. *Neuroaesthetics*. Amityville, N.Y.: Baywood.

———. 2009b. "Introduction: What Is Neuroaesthetics?" In Skov and Vartanian 2009a, 1–7.

Slaby, Jan, Philipp Haueis, and Suparna Choudhury. 2012. "Neuroscience as Applied Hermeneutics. Towards a Critical Neuroscience of Political Theory." In Vander Valk 2012a, 50–73.

Smith, Gwenn S., ed. 2015. *Handbook of Depression in Alzheimer's Disease*. Amsterdam: IOS.

Smith, Jennifer. 2009. "Building a Better Brain." *Isthmus* (July 27). http://www .isthmus.com/isthmus/article.php?article=25405.

Smith, Martin. 2012. "Brain Death: Time for an International Consensus." *British Journal of Anaesthesia* 108 (S1): i6–i9.

Smith, Roger. 1997. *The Fontana History of the Human Sciences*. London: Fontana.

———. 2007. *Being Human: Historical Knowledge and the Creation of Human Nature*. Manchester: Manchester University Press.

Snodgrass, Jeffrey G. 2014. "Ethnography of Online Cultures." In *Handbook of Methods in Cultural Anthropology*, edited by H. Russell Bernard and Clarence C. Gravlee. Lanham, Md.: Rowman & Littlefield, 465–496.

Sokolow, Jayme A. 1983. *Eros and Modernization: Sylvester Graham, Health Reform, and the Origins of Victorian Sexuality in America*. London: Associated Universities Press.

Solomon, Andrew. 2008. "The Autism Rights Movement." *New York Magazine* (May 25). http://nymag.com/news/features/47225.

Solso, Robert L. 2000. "The Cognitive Neuroscience of Art." *Journal of Consciousness Studies* 7–8/9: 75–81.

———. 2001. "Brain Activities in a Skilled Versus a Novice Artist: An fMRI Study." *Leonardo* 34 (1): 31–34.

Solymosi, Tibor, and John R. Shook, eds. 2014. *Neuroscience, Neurophilosophy, and Pragmatism: Brains at Work with the World*. New York: Palgrave Macmillan.

Spence, Des. 2013. "Are Antidepressants Overprescribed? Yes." *British Medical Journal* 346 (7907): 16.

Spence, Donald P. 1984. *Narrative Truth and Historical Truth: Meaning and Interpretation in Psychoanalysis*. New York: Norton.

Spiers, Hugo J., and Daniel Bendor. 2014. "Enhance, Delete, Incept: Manipulating Hippocampus-Dependent Memories." *Brain Research Bulletin* 105: 2–7.

Spolsky, Ellen. 2002. "Darwin and Derrida: Cognitive Literary Theory as a Species of Post-Structuralism." *Poetics Today* 23: 43–62.

Spotts, Dane, and Nancy Atkins. 1999. *Super Brain Power. 28 Minutes to a Supercharged Brain*. Seattle: LifeQuest.

Starr, Gabrielle. 2012. "Evolved Reading and the Science(s) of Literary Study: A Response to Jonathan Kramnick." *Critical Inquiry* 38: 418–425.

Stein, Dan J., Yanling He, Anthony Phillips, et al. 2015. "Global Mental Health and Neuroscience: Potential Synergies." *Lancet Psychiatry* 2: 178–185.

Steinberg, Laurence. 2008. "A Social Neuroscience Perspective on Adolescent Risk-Taking." *Developmental Review* 28: 78–106.

Sterling, Bruce, ed. 1990 [1986]. *Mirrorshades: The Cyberpunk Anthology*. Glasgow: Paladin GraftonBooks.

Stern, Madeleine Bettina. 1971. *Heads and Headlines: The Phrenological Fowlers*. Norman: University of Oklahoma Press.

Stiles, Anne. 2006a. "Robert Louis Stevenson's *Jekyll and Hyde* and the Double Brain." *Studies in English Literature, 1500–1900* 46 (4): 879–900.

———. 2006b. "Cerebral Automatism, the Brain, and the Soul in Bram Stoker's *Dracula*." *Journal of the History of the Neurosciences* 15 (2): 131–152.

———, ed. 2007. *Neurology and Literature, 1860–1920*. New York: Palgrave Macmillan.

Stollfuß, Sven. 2014. "The Rise of the Posthuman Brain: Computational Neuroscience, Digital Networks, and the 'In Silico Cerebral Subject.'" *Trans-Humanities* 7 (3): 79–102.

Strasser, Peter. 2014. *Diktatur des Gehirns. Für eine Philosophie des Geistes*. Paderborn: Fink.

Sumeet, Jain, and Sushrut Jadhav. 2009. "Pills That Swallow Policy: Clinical Ethnography of a Community Mental Health Program in Northern India." *Transcultural Psychiatry* 46: 60–85.

Summerfield, Derek. 2008. "How Scientifically Valid Is the Knowledge Base of Global Mental Health?" *British Medical Journal* 336 (7651): 992–994.

———. 2012. "Afterword: Against 'Global Mental Health.'" *Transcultural Psychiatry* 49 (3/4): 519.

———. 2014. "A Short Conversation with Arthur Kleinman About His Support for the Global Mental Health Movement." *Disability and the Global South* 1 (2): 406–411.

Swain, John, and Colin Cameron. 1999. "Unless Otherwise Stated: Discourses of Labeling and Identity in Coming Out." In Corker and French 1999, 68–78.

Tabbi, Joseph. 2008. "Afterthoughts on *The Echo Maker.*" In Burn and Dempsey 2008, 219–229.

Tadd, James Liberty. 1900. *New Methods in Education.* London: Sampson Low, Marston & Co.

Tallis, Raymond. 2004. *Why the Mind Is Not a Computer: A Pocket Lexicon of Neuromythology.* Exeter: Imprint Academic.

———. 2007. "Not All in the Brain." *Brain* 130 (11): 3050–3054.

———. 2008a. "The Neuroscience Delusion." *Times Literary Supplement* (9 April). http://tomraworth.com/talls.pdf.

———. 2008b. "The Limitations of a Neurological Approach to Art [Review of Onians 2007]." *The Lancet* 372 (5 July): 19–20.

———. 2009. "Neurotrash." *New Humanist* 124 (6). http://newhumanist.org.uk/2172/neurotrash.

Tan, Li-Hai, Angela R. Laird, Karl Li, and Peter T. Fox. 2005. "Neuroanatomical Correlates of Phonological Processing of Chinese Characters and Alphabetic Words: A Meta-analysis." *Human Brain Mapping* 25: 83–91.

Tang, Yi-Yuan, Britta K. Hölzel, and Michael I. Posner. 2015. "The Neuroscience of Mindfulness Meditation." *Nature Reviews Neuroscience* 16: 213–225.

Tang, Yi-Yuan, and Michael I. Posner. 2013. "Editorial: Special Issue on Mindfulness Neuroscience." *SCAN* 8: 1–3.

Tang, Yi-yuan, Wutian Zhang, Kewei Chen, et al. 2006. "Arithmetic Processing in the Brain Shaped by Cultures." *PNAS* 103 (28): 10775–10780.

Tannen, Susan. n.d. "Mental fitness—Exercises for the Brain." http://web.archive.org/web/20070221201506/http://www.bellydoc.com/articles/article7.htm.

Tauber, Alfred. 2012. "The Biological Notion of Self and Non-self." *Stanford Encyclopedia of Philosophy.* http://plato.stanford.edu/entries/biology-self/.

Taylor, Charles. 1989. *Sources of the Self: The Making of the Modern Identity.* Cambridge: Mass.: Harvard University Press.

Taylor, Richard P. 2002. "Order in Pollock's Chaos." *Scientific American* (December): 116–121.

Taylor, Richard P., Adam P. Micolich, and David Jonas. 1999. "Fractal Analysis of Pollock's Drip Paintings." *Nature* 399 (3 June): 422.

Teahan, John F. 1979. "Warren Felt Evans and Mental Healing: Romantic Idealism and Practical Mysticism in Nineteenth-Century America." *Church History* 48 (1): 63–80.

Temkin, Owsei. 1973. *Galenism: Rise and Decline of a Medical Philosophy.* Ithaca, N.Y.: Cornell University Press.

Theil, Stefan. 2015. "Why the Human Brain Project Went Wrong—and How to Fix It." *Scientific American*, 1 October. http://www.scientificamerican.com/article/why-the-human-brain-project-went-wrong-and-how-to-fix-it/.

Thiel, Udo. 2011. *The Early Modern Subject: Self-Consciousness and Personal Identity from Descartes to Hume*. New York: Oxford University Press.

Thoma, Nathan C., Dean McKay, Andrew J. Gerber, et al. 2012. "A Quality-Based Review of Randomized Controlled Trials of Cognitive-Behavioural Therapy for Depression: An Assessment and Metaregression." *American Journal of Psychiatry* 169: 22–30.

Thomas, Julia Adeney. 2015. "Who Is the 'We' Endangered by Climate Change?" In *Endangerment, Biodiversity, and Culture*, edited by Fernando Vidal and Nélia Dias. New York: Routledge, 241–260.

Thompson, Paul M., Jason L. Stein, Sarah E. Medland, et al. 2014. "The ENIGMA Consortium: Large-Scale Collaborative Analyses of Neuroimaging and Genetic Data." *Brain Imaging and Behavior* 8 (2): 153–182.

Thrailkill, Jane F. 2011. "Ian McEwan's Neurological Novel." *Poetics Today* 32 (1): 171–201.

Tinio, Pablo P. L., and Jeffrey K. Smith, eds. 2014. *The Cambridge Handbook of the Psychology of Aesthetics and the Arts*. New York: Cambridge University Press.

Tofts, Darren, Annemarie Jonson, and Alessio Cavallaro, eds. 2004. *Prefiguring Cyberculture: An Intellectual History*. Sydney: MIT Press.

Tougaw, Jason. 2012. "Brain Memoirs, Neuroscience, and the Self: A Review Article." *Literature and Medicine* 30 (1): 171–192.

———. 2016. "Amnesia and Identity in Contemporary Literature." In *Memory in the Twenty-First Century: New Critical Perspectives from the Arts, Humanities, and Sciences*, edited by Sebastian Groes, 280–285. New York: Palgrave Macmillan.

Toyokawa, Satoshi, Monica Uddin, Karestan C. Koenen, and Sandro Galea. 2012. "How Does the Social Environment 'Get Into the Mind'? Epigenetics at the Intersection of Social and Psychiatric Epidemiology." *Social Science and Medicine* 74: 67–74.

Tracy, Harry M. 2016. "The Neuro Funding Rollecoaster." *Cerebrum* (June). http://www.dana.org/Cerebrum/2016/The_Neuro_Funding_Rollercoaster/.

Tsur, Reuven. 1992. *Toward a Theory of Cognitive Poetics*. Amsterdam: North-Holland.

Turner, D. Trevor, Mark van der Gaag, Eirini Karyotaki, and Pim Cuijpers. 2014. "Psychological Interventions for Psychosis: A Meta-analysis of Comparative Outcome Studies." *American Journal of Psychiatry* 171: 523–538.

Turner, Erick H., Annette M. Matthews, Eftihia Linardatos, et al. 2008. "Selective Publication of Antidepressant Trials and Its Influence on Apparent Efficacy." *New England Journal of Medicine* 358: 252–260.

Tylor, Edward B. 1871. *Primitive Culture: Researches Into the Development of Mythology, Philosophy, Religion, Art, and Custom*. London: John Murray.

Umiltà, Maria Alessandra, Cristina Berchio, Mariateresa Sestito, et al. 2012. "Abstract Art and Cortical Motor Activation: An EEG Study." *Frontiers in Human Neuroscience* 6, art. 311: 1–9.

Uttal, William R. 2003. *The New Phrenology: The Limits of Localizing Cognitive Processes in the Brain.* Cambridge, Mass.: MIT Press.

———. 2015. *Macroneural Theories in Cognitive Neuroscience.* New York: Psychology Press.

Valenstein, Elliot. S. 1998. *Blaming the Brain: The Truth About Drugs and Mental Health.* New York: The Free Press.

Valentine, Gill, Tracey Skelton, and Ruth Butler. 2003. "Coming Out and Outcomes: Negotiating Lesbian and Gay Identities with, and in, the Family." *Environment and Planning D: Society and Space* 21 (4): 479–499.

Van Orden, Guy C., and Kenneth R. Paap. 1997. "Functional Neuroimages Fail to Discover Pieces of the Mind in the Parts of the Brain." *Philosophy of Science* 64 (Proceedings): S85–S94.

van Praag, Herman M. 2000. "Nosologomania: A Disorder of Psychiatry." *World Journal of Biological Psychiatry* 1: 151–158.

———. 2005. "Can Stress Cause Depression?" *World Journal of Biological Psychiatry* 6 (Suppl 2): 5–22.

———. 2008. "Kraepelin, Biological Psychiatry, and Beyond." *European Archives of Psychiatry and Clinical Neuroscience* 258 (Suppl 2): 29–32.

———. 2010. "Biological Psychiatry: Still Marching Forward in a Dead End." *World Psychiatry* 9 (3): 164–165.

van Praag, Herman M., Rene S. Kahn, Gregory M. Asnis, et al. 1987. "Denosologization of Biological Psychiatry or the Specificity of 5-HT Disturbances in Psychiatric Disorders." *Journal of Affective Disorders* 13: 1–8.

Van Wyhe, John. 2002. "The Authority of Human Nature: The *Schädellehre* of Franz Joseph Gall." *British Journal for the History of Science* 35: 17–42.

———. 2004. *Phrenology and the Origins of Victorian Scientific Naturalism.* Aldershot: Ashgate.

Vander Valk, Frank, ed. 2012a. *Neuroscience and Political Theory: Thinking the Body Politic.* Routledge: London

———. 2012b. "Introduction." In Vander Valk 2012a, 1–22.

Vartanian, Oshin. 2014. "Empirical Aesthetics: Hindsight and Foresight." In Tinio and Smith 2014, 6–34.

Veer, Ilya M., Christian F. Beckmann, Marie-José van Tol, et al. 2010. "Whole Brain Resting-State Analysis Reveals Decreased Functional Connectivity in Major Depression." *Frontiers in Systems Neuroscience* 4, art. 41. doi:10.3389/fnsys.2010.00041.

Vidal, Fernando. 2009a. "Brainhood, Anthropological Figure of Modernity." *History of the Human Sciences* 22 (1): 5–36.

———. 2009b. "Ectobrains in the Movies." In *The Fragment: An Incomplete History*, edited by William Tronzo, 193–211. Los Angeles: Getty Research Institute.

———. 2011. *Sciences of the Soul: The Early Modern Origins of Psychology.* Translated by Saskia Brown. Chicago: University of Chicago Press.

———. 2016. "Frankenstein's Brain: 'The Final Touch.'" *SubStance* 45 (2): 88–117.

Vidal, Fernando, and Francisco Ortega. 2011. "Approaching the Neurocultural Spectrum: An Introduction." In Ortega and Vidal 2011, 7–27.

Vrecko, Scott. 2006. "Folk Neurology and the Remaking of Identity." *Molecular Interventions* 6: 300–303.

Vuilleumier, Patrik, Jorge L. Armony, Jon Driver, and Raymond J. Dolan. 2003. "Distinct Spatial Frequency Sensitivities for Processing Faces and Emotional Expressions." *Nature Neuroscience* 6 (6): 624–631.

Wajman, José Roberto, Paulo H. Bertolucci, Leticia Mansur, and Serge Gauthier. 2015. "Culture as a Variable in Neuroscience and Clinical Neuropsychology: A Comprehensive Review." *Dementia and Neuropsychologia* 9 (3): 203–218.

Waldman, Paul. 2013. "David Brooks and the Anti-Neuroscience Backlash." *The American Prospect* (18 June). http://prospect.org/article/david-brooks -and-anti-neuroscience-backlash.

Walsh, Pat, Mayada Elsabbagh, Patrick Bolton, and Ilina Singh. 2011. "In Search of Biomarkers for Autism: Scientific, Social, and Ethical Challenges." *Nature Reviews Neuroscience* 12: 603–612.

Walton, Alice. 2015. "7 Ways Meditation Can Actually Change the Brain." http://www.forbes.com/sites/alicegwalton/2015/02/09/7-ways-meditation -can-actually-change-the-brain/.

Waltz, M. 2005. "Reading Case Studies of People with Autistic Spectrum Disorders: A Cultural Studies Approach to Issues of Disability Representation." *Disability and Society* 20 (5): 421–435.

Wang, Lin, Daniel F. Hermens, Ian B. Hickie, and Jim Lagopoulos. 2012. "A Systematic Review of Resting-State Functional-MRI Studies in Major Depression." *Journal of Affective Disorders* 142: 6–12.

Warnick, Jason E., and Dan Landis, eds. 2015. *Neuroscience in Intercultural Contexts.* New York: Springer.

Watters, Ethan. 2010. *Crazy Like Us: The Globalization of the American Psyche.* New York: Simon and Schuster.

Wazana, Ashley. 2000. "Physicians and the Pharmaceutical Industry: Is a Gift Ever Just a Gift?" *Journal of the American Medical Association* 283 (3): 373–380.

Weber, Matthew J., and Sharon L. Thompson-Schill. 2010. "Functional Neuroimaging Can Support Causal Claims About Brain Function." *Journal of Cognitive Neuroscience* 22 (11): 2415–2416.

Weinmann, Stefan, John Read, and Volkmar Aderhold. 2009. "Influence of Antipsychotics on Mortality in Schizophrenia: Systematic Review." *Schizophrenia Research* 113: 1–11.

Weintraub, Kit. 2005. "A Mother's Perspective." http://web.archive.org/web /20051216174906/http://asatonline.org/resources/library/moms _perspective.html.

Wells, Carol G. 1989. *Right Brain Sex: Using Creative Visualization to Enhance Sexual Pleasure.* New York: Simon & Schuster.

Wendell, Susan. 1996. *The Rejected Body: Feminist Philosophical Reflections on Disability.* New York: Routledge.

Western, Drew. 2008. *The Political Brain: The Role of Emotion in Deciding the Fate of the Nation.* Philadelphia: Public Affairs.

Whelan, Robert, and Hugh Garavan. 2014. "When Optimism Hurts: Inflated Predictions in Psychiatric Neuroimaging." *Biological Psychiatry* 75: 746–748.

Whitaker, Robert. 2010. *Anatomy of an Epidemic: Magic Bullets, Psychiatric Drugs, and the Astonishing Rise of Mental Illness in America.* New York: Crown.

White, Ross. 2013. "The Globalisation of Mental Illness." *The Psychologist* 26 (3): 182–185.

WHO. 2001. *The World Health Report 2001—Mental Health: New Understanding, New Hope.* Geneva: World Health Organization.

———. 2008. *Mental Health Gap Action Programme (mhGAP): Scaling Up Care for Mental, Neurological, and Substance Abuse Disorders.* Geneva: WHO.

———. 2013. "Mental Health Action Plan 2013–2020." Sixty-Sixth World Health Assembly. Resolution WHA66/8.

Whorton, James C. 1982. *Crusaders for Fitness: The History of American Health Reformers.* Princeton, N.J.: Princeton University Press.

Wickelgren, Ingrid. 2005. "Autistic Brains out of Synch?" *Science* 308: 1856–1858.

Wigan, Arthur L. 1985 [1844]. *A New View of Insanity: The Duality of the Mind Proved by the Structure, Functions, and Diseases of the Brain and by the Phenomena of Mental Derangement, and Shewn to Be Essential to Moral Responsibility.* Malibu: Joseph Simon.

Wijdicks, Eelco F. M. 2012. "The Transatlantic Divide Over Brain Death Determination and the Debate." *Brain* 135 (4): 1321–1331.

Wilfond, Benjamin S., and Vardit Ravitsky. 2005. "On the Proliferation of Bioethics Subdisciplines: Do We Really Need 'Genethics' and 'Neuroethics'?" *American Journal of Bioethics* 5 (2): 20–21.

Wilkes, Kathleen. 1988. *Real People: Personal Identity Without Thought Experiments.* Oxford: Clarendon.

Williams, J. Simon, Stephen Katz, and Paul Martin. 2011. "The Neuro-Complex: Some Comments and Convergences." *MediaTropes* 3 (1): 135–146.

Williams, Raymond. 1985. *Keywords: A Vocabulary of Culture and Society.* Rev. ed. London: Fontana.

Willis, Thomas. 1681 [1664]. *The Anatomy of the Brain and Nerves.* Translated by Samuel Pordage. Birmingham, Ala.: The Classics of Neurology & Neurosurgery Library, 1983.

———. 1683 [1971]. *Two Discourses Concerning the Soul of Brutes . . . [De anima brutorum,* 1672]. Translated by Samuel Pordage. Gainsville, Fla.: Scholars' Facsimiles & Reprints.

Wing, Lorna. 1997. "The History of Ideas on Autism: Legends, Myths, and Reality." *Autism* 1: 13–23.

Winter, Arthur, and Ruth Winter. 1987. *Build Your Brain Power: The Latest Techniques to Preserve, Restore, and Improve Your Brain's Potential.* New York: St. Martin's Press.

Witelson, Sandra F., Debra L. Kigar, and Thomas Harvey. 1999. "The Exceptional Brain of Albert Einstein." *The Lancet* 353: 2149–2153.

Wojciehowski, Hannah Chapelle, and Vittorio Gallese. 2011. "How Stories Make Us Feel: Toward an Embodied Narratology." *California Italian Studies* 2 (1). http://escholarship.org/uc/item/3jg726c2.

Wolbring, Gregor. 2007. "Neurodiversity, Neuroenhancement, Neurodisease, and Neurobusiness." *Innovation Watch* (15 May). http://www.innovationwatch.com/choiceisyours/choiceisyours-2007-04-30.htm.

Wolpe, Paul Root. 2002. "The Neuroscience Revolution." *The Hastings Center Report* 32 (4): 8.

Wright, T. R. 1982. "From Bumps to Morals: The Phrenological Background to George Eliot's Moral Framework." *Review of English Studies* 33 (129): 24–46.

Xie, Changchun, Zheng Wu, Weizhong Li, et al. 2008. "Neural Correlates of Depression in Subjects with Amnestic Mild Cognitive Impairment." *Alzheimer's and Dementia* 4 (4), Supplement 1: T259–T260.

Young, Kay, and Jeffrey L. Saver. 2001. "The Neurology of Narrative." *SubStance: A Review of Theory and Literary Criticism* 30 (1/2): 72–84.

Young, Robert Maxwell. 1990. *Mind, Brain, and Adaptation in the Nineteenth Century: Cerebral Localization and Its Biological Context from Gall to Ferrier.* New York: Oxford University Press.

Yuste, Rafael. 2015. "'Cuando entendamos el cerebro, la humanidad se entenderá a sí misma' [Interview with Núria Jar Benabarre]." *El País* (25 May). http://elpais.com/elpais/2015/05/21/ciencia/1432224592_729117.html.

Zaidel, Dahlia W. 2013. "Art and Brain: The Relationship of Biology and Evolution to Art." In Finger, Zaidel, Boller, and Bogousslavsky 2013, 217–233.

Zalewski, Daniel. 2009. "Ian McEwan's Art of Unease." *The New Yorker* (23 February). http://www.newyorker.com/reporting/2009/02/23/090223fa_fact_zalewski.

Zawidzki, Tadeusz, and William P. Bechtel. 2005. "Gall's Legacy Revisited: Decomposition and Localization in Cognitive Neurosciences." In *The Mind as a Scientific Object: Between Brain and Culture*, edited by Christina E. Erneling and David Martel Johnson. Oxford: Oxford University Press, 293–318.

Zeki, Semir. 1998. "Art and the Brain." *Daedalus* 127 (2): 71–103.

———. 2000. "L'artiste à sa manière est un neurologue." *La Recherche* hors-série 4 (November): 98–100.

———. 2001. "Artistic Creativity and the Brain." *Science* 263 (6 July): 51–52.

———. 2002. "Neural Concept Formation and Art. Dante, Michelangelo, Wagner." *Journal of Consciousness Studies* 9: 53–76.

———. n.d. "Statement on Neuroaesthetics." http://www.neuroaesthetics.org /statement-on-neuroaesthetics.php.

Zeki, Semir, and Tomohiro Ishizu. 2013. "The 'Visual Shock' of Francis Bacon: An Essay in Neuroesthetics." *Frontiers in Human Neuroscience* 7:850. doi:10.3389/ fnhum.2013.00850.

Zhou, Haotian, and John Cacioppo. 2010. "Culture and the Brain: Opportunities and Obstacles." *Asian Journal of Social Psychology* 13:59–71.

Zhu, Ying, Li Zhang, Jin Fan, and Shihu Han. 2007. "Neural Basis of Cultural Influence on Self-Representation." *NeuroImage* 34:1310–1316.

Zhu, Ying, and Shihu Han. 2008. "Cultural Differences in the Self: From Philosophy to Psychology and Neuroscience." *Social and Personality Psychology Compass* 2 (5): 1799–1811.

Zunshine, Lisa, ed. 2010. *Introduction to Cognitive Cultural Studies*. Baltimore, Md.: Johns Hopkins University Press.

Zwijnenberg, Robert. 2011. "Brains, Art, and the Humanities." In Ortega and Vidal 2011, 293–309.

forms of living

Stefanos Geroulanos and Todd Meyers, *series editors*

Printed and bound by CPI Group (UK) Ltd, Croydon, CR0 4YY

27/10/2024

145803327-0004